D1061150

Social Neuroscience

Foundations in Social Neuroscience
edited by John T. Cacioppo, Gary G. Berntson, Ralph Adolphs, C. Sue Carter, Richard J. Davidson, Martha K. McClintock, Bruce S. McEwen, Michael J. Meaney, Daniel L. Schacter, Esther M. Sternberg, Steve S. Suomi, and Shelley E. Taylor

Essays in Social Neuroscience
edited by John T. Cacioppo and Gary G. Berntson

Social Neuroscience: People Thinking about Thinking People
edited by John T. Cacioppo, Penny S. Visser, and Cynthia L. Pickett

Social Neuroscience

People Thinking about Thinking People

edited by John T. Cacioppo, Penny S. Visser, and Cynthia L. Pickett

A Bradford Book
The MIT Press
Cambridge, Massachusetts
London, England

MIT Press books may be purchased at special quantity discounts for business or sales promotional use. For information, please email special_sales@mitpress.mit.edu or write to Special Sales Department, The MIT Press, 55 Hayward Street, Cambridge, MA 02142.

This book was set in Stone Sans and Stone Serif by SNP Best-set Typesetter Ltd., Hong Kong, and was printed and bound in the United States of America.

Library of Congress Cataloging-in-Publication Data
Social neuroscience : people thinking about thinking people / edited by John T. Cacioppo, Penny S. Visser, and Cynthia L. Pickett.
p. cm.—(Social neuroscience series)
"A Bradford book."
Includes bibliographical references and index.
ISBN-13 978-0-262-03335-0 (hbk. : alk. paper)
ISBN 0-262-03335-6 (hbk. : alk. paper)
1. Cognitive neuroscience. 2. Neuropsychology. 3. Social psychology. I. Cacioppo, John T. II. Visser, Penny S. III. Pickett, Cynthia L. IV. Series.

QP360.5.S636 2006
153—dc22

2005041691

10 9 8 7 6 5 4 3 2

Contents

Foreword: Science's Newest Brain Child, Social Neuroscience

One can do little about the timing of one's birth. So it is a matter of unearned good fortune to be present when the evolution of one's science happens to be in a moment of some note in its development. I found myself contemplating this matter at a conference organized by John Cacioppo, the papers from which form the basis of this volume. A few years too early and I may have dispassionately watched, not just from the sidelines but a long way away, the rumblings and grumblings of the shaping of social neuroscience. A few years too late, and I may have become deeply involved, but without the thrill of being with the first generation while also acutely aware of its distinct intellectual traditions, soon to meld invisibly into the new science. It is with examined delight that I introduce a volume that above all indicates the many paths and travelers that make up social neuroscience, the excellence across generations of elder and younger, and the uncharted territory ahead.

If many paths lead to social neuroscience, it is in no small measure because somebody was there breaking ground, clearing brush, pointing ahead to the next spot suitable for a rest. John T. Cacioppo is such a person, and he predates the first generation of social neuroscientists by a few decades. His own intellectual development was remarkable in that he was self-taught and drew from every strand that linked behavior, brain, and social world. John is able to stack level-of-analysis upon level-of-analysis, shunning none in favor of another. He is both a superb reductionist and a committed integrationist. To him nothing is more satisfactory than to see it all line up. John's passion for social neuroscience, his prescience regarding its inevitability, and his pulling it all toward the center so that nothing topples off the path of discovery are among the reasons that many of us are able to be cotravelers.

Solo path breakers remain exactly that until other travelers take note and decide to crowd the path. A crowd has indeed gathered over the past decade, conversations about social neuroscience are frequent, and the decibel level is high enough sometimes to draw in others and sometimes to be complained about by passersby. We now have a critical mass, an intellectual core to speak of; many of the chapter in this volume could not have been imagined, let alone produced, even as recently as the start of this century. Younger entrants to the field are Stakanovites who are also outspoken in their vision of the new science. They come from different places (their advisors did not attend the same meetings or even recognized each others' names) but they are bound by a desire to understand social animals and to do so by observing the activity of a three-pound organ tucked between the ears. As far as I can tell, there is no stopping.

One of the most frequently occurring human acts involves a person thinking and feeling, consciously or not, about himself or herself, others, or larger social groups. Each of us performs countless numbers of such mental acts every day, and social neuroscience is one place to examine them, to fashion individual jigsaw pieces, one at a time, readying them for a future when integration will be possible at a higher level. In the past decade, social neuroscience shed light on aspects of social life as diverse as social regulation, social rejection, impression formation, the specialty of self-knowledge and social cognition broadly, self-awareness, emotion regulation, and attitudes, beliefs, and memory involving social groups. In this volume, Cacioppo, Penny S. Visser, and Cynthia L. Pickett have paid special attention to gathering diverse methods and subject populations while keeping the focus on social thinking and feeling systems.

In the early days (e.g., 2000!), I recall a cognitive neuroscientist being legitimately surprised at being included in a social neuroscience workshop because the stimuli in her research happened to be human faces. Now, already, stronger credentials would be necessary. The fundamental questions that draw social neuroscientists together have little or nothing to do with the type of stimulus chosen; one could study face perception and have no interest whatsoever in social cognition. Instead, social neuroscience concerns all the ways in which human beings influence and are influenced by the presence, actual or imagined, of other humans. It is the act of making sense of oneself and others, and events surrounding that act, that draw people to study social neuroscience. Just as Thurstone and Likert made instruments to allow early behavioral measures of attitudes and beliefs, technologies of today allow the same topics to be studied by measuring mental and brain function. If a brain imaging study, for instance, looks somewhat infantile compared with traditional behavioral ones, there's a reason. New technologies have to be honed and it is interesting that the steps cannot easily be skipped—one must go through the same slow building layer by layer and hope for breakthroughs that will allow faster advances than are currently visible. Everybody agrees that nothing truly novel is here yet; a large group, myself included, believe they will come.

Amidst the hullabaloo, a few matters deserve to be mentioned, even if they are no-brainers to social neuroscientists themselves. First, in doing social neuroscience, areas of work that become relevant have no boundary; many fields become of interest because many ideas and combinations of them have the potential to contribute and transform. Just look at discussions about naming—anything involving neuroscience now encounters the same ridiculous problem confronted by feminist parents deciding on the surname of their child. Social neuroscience? Social cognitive neuroscience? Social-personality neuroscience (as more individual difference work becomes possible)? And what about social cognitive developmental neuroscience? This problem will sort itself out as it has in other fields, but to me it points out just how important it is going to be to be a "lumper." This field will be kind to those who are naturals, or ready to jump into the lump.

Second, the importance of going deeply into a particular area is going to be no different than anywhere else in science, and early "splitters" will gain a great deal of leverage and even create new domains of study. However, unlike the past, going deep cannot be a solitary journey. Collaborations may look more like baseball teams, and may even bring a sense of unease, because every single piece of knowledge is not going to be equally accessible to every expert on the team. Psychologists, neuroscientists, and physicists will have to develop appropriate levels of faith in the other's expertise as well as posing challenges to another's familiar assumptions.

Third, developing respect for disciplines not one's own should not be underestimated. For example, take the group of social psychologists who for a hundred years have tried to understand mental constructs such as belief, attitude, self, and all matters that concern the understanding of oneself, individual others, and social groups. They did this when it was not fashionable to study mental constructs and they do it now when it apparently is. It is a matter of some sadness when the hammer of new technology crudely fixes a problem when a far more delicate and superior job could be done with what is already available. But such excesses are likely in any new field, and to minimize it, those who have the benefit of experience should speak up. This is not yet a field in which only those who "do it" should have opinions. Onlookers who are sympathetic to the broad enterprise of social neuroscience, but with deep knowledge of the phenomenon, must be commentators.

The brain, as an object of study, belongs to anybody who has one. It would be silly to assume that it is only for those who study what is innate, what is biological, what is genetic. The past few decades have given us as much evidence about the plasticity of the brain and much else that was considered to be rigid and unchangeable, as it has the reverse: showing the determining power of seemingly ephemeral entities such as social situations. Among the more important potential contributions of social neuroscience is the likelihood that misguided debates about nature versus nurture that still fill popular books (and journal articles to a lesser extent) will die away.

Many years ago, I commented that those who were least likely to fall prey to a closet dualism (a more common syndrome than expressed attitudes toward dualism might suggest) were psychobiologists, because they did their work at the close intersection of biology and experience. Social neuroscience, practiced not by a few individuals from a single orientation, but by larger units of collaborators, has the great potential of wiping off the cobwebs of twentieth-century simplicities about human nature and human nurture.

Mahzarin R. Banaji
Cambridge, Massachusetts

Preface

The dominant metaphor for the scientific study of the human mind during the latter half of the twentieth century was the computer, a solitary device with massive information-processing capacities. At the dawn of the twenty-first century, this metaphor seems dated. Computers today are massively interconnected devices with capacities that extend far beyond the resident hardware and software of a solitary computer. It is suddenly apparent that the telereceptors of the human brain have provided wireless broadband interconnectivity to humans for millennia. Just as computers have capacities and processes that are transduced through but extend far beyond the hardware of a single computer, the human brain has come to be recognized as having evolved to promote social and cultural capacities and processes that are transduced through, but that extend far beyond, a solitary brain.

The notion that humans are inherently social creatures is no longer contestable, either. Human infants will not survive to contribute to the gene pool unless they receive care and nurturance over an extended period of time. As adults, humans are still not isolative by nature, nor are they particularly strong, fast, or stealthy relative to other species. Humans are an adaptable and formidable species because of their ability to think, to develop and use tools, and to work together. It may be that the genetic constitution of species characterized by a negligible period of dependency is reducible to the reproductive success of individual members of the species. The genetic constitution of *Homo sapiens*, however, derives not simply from an individual's reproductive success but more critically from the success of one's children to reproduce. Hunter-gatherers who, in times of famine or danger, chose not to return to share their food or protection may have survived to hunt and reproduce again, but their genes were also less likely to be propagated by their offspring. In contrast, those who yearned to return or assist others despite personal deprivation, threats, or hardship, and individuals who protected and nurtured those close to them, were more likely to have offspring who survived to contribute to the human gene pool. In short, humans have evolved a brain and biology whose functions include formation and maintenance of social recognition, attachments, alliances, and collectives; and development of communication, deception, and reasoning about the mental states of others.

The field of social neuroscience has been stimulated by these recognitions and by the development of new methods that permit more thorough plumbing of the brain and mind. Special issues of journals on the topic of social neuroscience have appeared or are in the works, in *Neuropsychologia* (2003), *Journal of Personality and Social Psychology* (2003), *Political Psychology* (2003), *Biological Psychiatry* (2002), *Journal of Cognitive Neuroscience* (2004), and *Neuroimage* (in press), and numerous reviews of the field have appeared since the early 1990s in outlets ranging from the *American Psychologist* and the *Annual Review of Neuroscience* to *Current Opinion in Neurobiology* and *Trends in Cognitive Sciences*.

The current volume builds on this burgeoning literature in three ways. First, rather than addressing the vast array of questions that fall under the rubric of social neuroscience, it focuses specifically on the neurobiological underpinnings of social information processing; more specifically, on mechanisms underlying people thinking about thinking people. Recent work in evolutionary biology and in social neuroscience suggests there may be something special about social cognition. Three distinctions can be drawn: (1) cognitive operations that represent general information processes acting on social stimuli, (2) cognitive operations that evolved from the adaptive value they conferred to social information processing but that have been exapted for general information processes, and (3) cognitive operations that are specific to social stimuli. Contributions to this book bear on these distinctions, as well.

Second, too often the treatments of social neuroscience have been limited to work within a given disciplinary perspective, when among the strengths of social neuroscience is its multidisciplinary approach. This book therefore draws heavily on the work of psychologists, neurobiologists, psychiatrists, radiologists, *and* neurologists, revealing how the future of the field lies in the confluence of these perspectives.

Third, a hallmark of social neuroscience is the use of multiple methods that bridge disciplines and levels of analysis. Accordingly, the volume draws on research using many methods, including functional brain imaging techniques, patients with brain lesions, comparative analyses, and developmental data. As in any book this size, gaps in coverage exist. Although animal and patient data are represented, we barely scratch this literature. The goal, however, was to provide an illustrative rather than exhaustive treatment of the topic of people thinking about thinking people. We hope this volume contributes to further integrations across what have historically been viewed as disparate literatures.

Work on this book was made possible by funding from the National Science Foundation (NSF) BCS-0086314. The NSF has supported efforts to bridge the abyss across the social, cognitive, and neurosciences for more than a quarter-century, and we are indebted for its support. We also thank Tom Stone, the cognitive science, linguistics, philosophy, and psychology editor at MIT Press, for his efforts and backing; and to Judy Runge and Marsha Greaves for their logistical and administrative assistance. Finally, we thank Gary Berntson, Marc Raichle, Gün Semin, Wendi Gardner, and John Brehm for their advice and counsel while we were formulating this book.

Social Neuroscience

1 Reasoning about Brains

Gary G. Berntson

The chapters in this book raise interesting and important questions concerning the brain circuits and mechanisms underlying social reasoning, and the extent to which these circuits and mechanisms are specialized and/or specific for social reasoning. Social reasoning is a broad and diverse class of affective and cognitive activities that include but are not limited to person perception, social categorization and reasoning, and socioemotional regulation.

It is clear that social cognition is especially salient for humans, as well as other animals, and may differ qualitatively as well as quantitatively from nonsocial cognition. The special (brain) mechanisms debate has a long and enduring history, including the possibility of specialized perceptual and cognitive mechanisms for social communication and language. For example, neonatal and infant chimpanzees show a defensive-like pattern of cardioacceleration to conspecific threat vocalizations, even if isolated from the mother at birth and thus deprived of learning opportunities (Berntson & Boysen, 1989). This defensive response was highly specific to the communicative signal value of threat vocalization, as it was not seen to white noise or other chimpanzee vocalizations that were matched for intensity, duration, and frequency composition. Moreover, infant orangutans did not display this response to the chimpanzee threat vocalization, indicating that cardioacceleration is not an obligatory reaction to some simple physical property of the stimulus per se. These findings are in keeping with the view that special processing mechanisms may be constitutionally endowed through evolutionary history, and may play an important role in social perception and evaluative judgment. The example seems fitting in the current context, as Premack and Woodruff (1978) first introduced the construct "theory of mind" in reference to chimpanzees.

Another example of the special nature of social perception comes from studies of stress and immune function. Social stressors are particularly salient as reactivators of latent herpes simplex virus type 1 infections in humans (HSV1), increasing the probability of "cold sore" eruptions around stressful periods or events. The availability of an animal model of such stress reactivation would be important to allow studies of the underlying mechanisms of this stress-disease link. Unfortunately, standard laboratory stressors such as restraint and shock are ineffective in triggering HSV1 reactivation in mice. Research in our laboratory, however, discovered that reactivation could be achieved by the *social* stress associated with social reorganization (Padgett et al., 1998). This social stressor was not merely more intense, as it yielded a corticosteroid response comparable with that of restraint stress. Rather, it appeared to have something distinct about it. Subsequent work showed that social stress results in a specific and selective pattern of glucocorticoid resistance, which could have substantial implications for health (Quan et al., 2001).

This discussion offers two examples of an extensive literature documenting powerful and selective effects of social factors on psychological and physiological functioning. The question remains, however, as to whether such selective effects imply special mechanisms. This is a topic to which we will return shortly.

Functional Brain Localization

Before considering the issue of localization of function, it may be worth while to consider briefly precisely what we mean by functional localization. In fact, functions cannot really be localized at all, as they are not space-occupying entities. Moreover, they are not properties of neuronal circuits or even the outputs of these circuits. Functions are at least one step removed from outputs, as they represent the operations, utilities, or consequences of outputs within a given context. When one speaks of localization of function, it is important to recognize that this is a proxy for the fact that functions can arise from, or be causally related to, operations of specific neuronal circuits. It is in this sense that we use the construct of functional localization.

The history of brain localization is a colorful one, framed in the extreme by the classic contest of the nineteenth century between the ultra-localizationist and phrenologist Francis Gall and his antagonist Pierre Fluorens, who advanced a more holistic and integrative view of brain function (Phillips et al., 1984). This issue was also featured in a florid debate at the seventeenth annual meeting of the International Medical Congress in London in 1881, with David Ferrier advancing the localizationist view, and Friedrich Goltz stridently challenging that position (see Phillips et al., 1984; Young, 1968). The controversy extended well into the twentieth century, with

Karl Lashley formulating principles of mass action and equipotentiality of cortical function, while highly specific representations were increasingly being identified in sensorimotor and even behavioral systems (Hothersall, 2004; Berntson & Beattie, 1975).

There remains a natural tension between localizationistic and holistic perspectives in the contemporary literature, but developments and insights during the latter half of the twentieth century led to partial resolution of this issue that, as is often the case, lies somewhere between the extremes. This resolution can be credited in part to Donald O. Hebb (1950), who articulated how highly specific, but distributed, interconnected circuits (*cell assemblies* and *phase sequences*) could underlie memory and cognition. It is now clear that rather precise and focal topographical maps, with a high degree of functional specificity, exist within sensory and motor systems. On the other hand, more complex cognitive processes increasingly appear to entail multiple neural operations mediated by distributed, interacting cortical-subcortical circuits. This is not to say that these circuits are nonspecific or equipotential, just that they do not have a punctuate cortical representation (Sarter, Berntson, & Cacioppo, 1996; Uttal, 2003). Moreover, these distributed circuits may anatomically *overlap* with others engaged in different processes, and elements of distributed systems may in fact participate in many circuits.

An additional complication in functional localization is what has been termed the *category error*, which is the mismatch that can occur when mapping facts, constructs, and theories from the psychological domain (e.g., concerning social attribution processes) onto those from the biological domain (cells, synapses, structures, circuits, patterns of activity etc.). Because psychological constructs, for the most part, were developed in the absence of understanding of their neural mechanisms, and vice versa, it is unlikely that the terms and concepts of the two domains would show a simple isomorphism or one-to-one correspondence (Sarter et al., 1996). The absence of such isomorphism limits logical interpretation (Cacioppo & Tassinary, 1990).

An example may illustrate these complexities. Classic studies employing electrical stimulation of the brain of free-moving animals show that activation of local brain regions can evoke highly organized, integrated, and motivated species-characteristic behaviors such as eating, drinking, grooming, defensive responses, and predatory-like behavior. Distribution fields for eliciting such diverse behaviors often overlapped, however, and some electrode loci could evoke several behaviors depending on the environmental context. Although this was interepreted by some to reflect lack of neuroanatomical specificity in motivational neurosubstrates (Valenstein, 1975), a

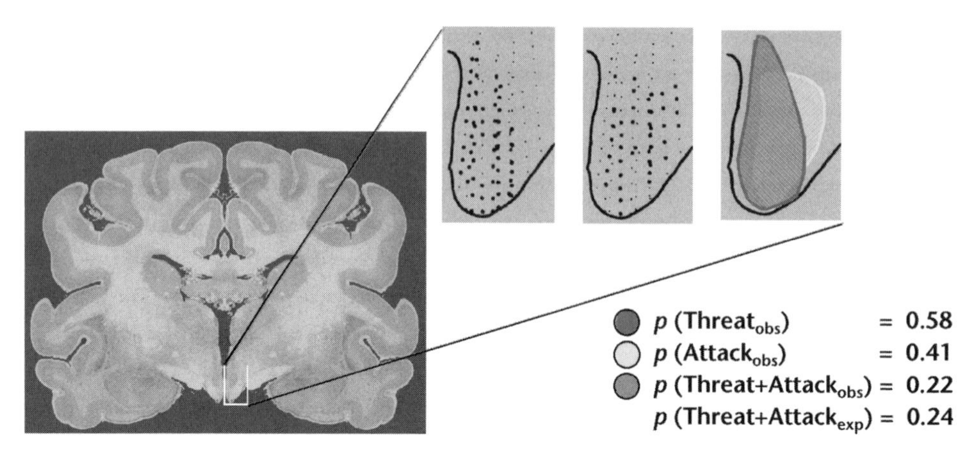

Figure 1.1

Stimulation-induced behaviors. Distribution of effective (large dots) and ineffective (small dots) electrode points yielding defensive threat and predatory attacklike responses in a hypothalamic-stimulation mapping study in animals. Coronal section of the brain on the left illustrates the relevant area which is expanded to the right. Observed probabilities of threat [p(Threat$_{obs)}$] and attack [p(Attack$_{obs)}$] are illustrated, together with the probability of both threat and attack responses elicited by the same electrode [p(Threat+Attack$_{obs)}$]. The latter compares closely with the expected probability [p(Threat+Attack$_{exp)}$] based on the joint occurrence of independent events. Results suggest an overlapping distribution of specific differentiated behavioral systems. Results further show that stimulation of a single locus can induce more than one response and that stimulation of divergent points within a distribution field can induce the same behavior. See plate 1 for color version.

quite different explanation became apparent (Berntson & Beattie, 1975). Figure 1.1 (see plate 1) illustrates the results of a mapping study of stimulation of the hypothalamus (expanded insert) in which some stimulation loci could evoke defensive threat-like responses, predatory-like stalking attack, or both. These responses are behaviorally distinct, have different adaptive significance, and project to separate mechanisms at the level of the brainstem (Berntson, 1972, 1973). Note that effective points for the evocation of threat overlap extensively with those that yielded attack, and many loci could evoke both responses. This is the sort of finding that led to arguments about nonspecificity in motivational substrates.

Despite this overlap, however, the distributions were doubly dissociated in the perimeters of the field (figure 1.1). This raises the possibility that distinct but interdigitated circuits may be represented within the central area of overlap. A stochastic approach offered an opportunity to test these alternate hypotheses. If a common behavioral substrate were responsible for the two elicited behaviors, one would expect to see a statistical dependency between responses, whereas if separate but overlapping circuits were involved there should be a statistical independence among the behaviors. In the latter case, the probability of a single locus evoking one behavior (threat), given the other (attack), should be equal to the independent probability of that response; that is, there should not be statistical dependence. In contrast, if an identity or a functional linkage existed between substrates, the probability of evoking one response, given the other, should be greater than the overall probability of that response.

As illustrated in figure 1.1, results support statistical independence of these circuits. The expected probability of the joint occurrence of independent events is the product of the independent probabilities of the two events (two coin flips producing two heads $= 0.5 * 0.5 = 0.25$). Within the overlap area, the observed probability of loci yielding both responses was equivalent to the expected probability based on independent systems. No functional coupling appears to be present between these systems, despite their overlapping distribution.

Functional Localization and Brain Imaging

The findings and considerations outlined above suggest caveats in drawing inferences from imaging studies (for more complete discussions see Cacioppo, Tassinary, & Berntson, 2000; Sarter et al., 1996). First, neurobehavioral systems, especially for complex processes, may be represented in *distributed circuits*, and activation of different anatomical loci within this system may yield comparable results. In the study outlined above, activation of either the dorsal or ventral region of the hypothalamic field for threat evoked comparable threat responses, despite minimal direct spread of the stimulating current. That is, *different neural activations can yield the same behavioral outcome.*

An identical pattern of brain activation can also yield more than one behavioral outcome, as when stimulation of a single locus evokes both threat and attack concurrently. To state the converse, to the extent to which functional systems overlap, *the same pattern of brain activation across contexts and occasions (within the limits of resolution) does not necessarily imply the same behavioral process.* Brain imaging in cognitive and social neuroscience often begins with an observation of the following sort: psychological state or process x (ψ_x) results in activation of area i (Φ_i). If that relation can be shown to be reliable and cross-situational, one could offer the following proposition: if (ψ_x), then (Φ_i). But the inferences we often want to draw are of the form: if (Φ_i), then (ψ_x). That is, to infer a psychological state from the pattern of brain activation is an example of the logical flaw of affirming the consequent. The only condition under which that inference would hold is if there was a one-to-one relation or isomorphism between ψ_x and Φ_i (Cacioppo et al., 2000). We return to this special case below.

Finally, another inherent complication arises in the analysis of brain imaging results. Given the complexity of social processes and the distributed nature of neural substrates, broad segments of the cerebrum would be expected to be active during social psychological events, and indeed this is the case. In large part, that is why imaging approaches entail a comparison condition and a subtractive analytical model, to isolate a more limited aspect of brain circuits involved in the social event. The very aim of this subtractive approach is to preclude from consideration those brain areas that contribute both to the target social event and the comparison condition, despite the fact that those areas might play an important role in the social process. In fact, occluded areas may play a more fundamental role in target social process than those that show selective activation during that condition.

To some extent the substractive approach is necessitated by the difficulty of dealing with complex patterns of results and the constraints of statistical approaches. Identifying a region of interest does facilitate hypothesis testing and statistical analyses, but it also imposes a restricted vantage on neural mechanisms.

Distributed Brain Mechanisms for Social Cognition

The complex and distributed nature of neural substrates of social processes is clearly documented in this book. Perusal of this volume in fact reveals that most major telencephalic structures outside classic sensory and motor systems are implicated in social processes. Moreover, there are subcortical systems that regulate all areas of the cortical mantle and thus affect social processes indirectly. An example is the basal forebrain cortical cholinergic system, which supplies cholinergic innervation to all cortical areas and regulates cortical excitability and cortically dependent cognitive and affective operations (Berntson, Sarter, & Cacioppo, 2003; Sarter, Bruno, & Givens, 2003). This system is an especially important component of circuits underlying social processes, as revealed by the severe cognitive and social decline of people with Alzheimer's disease that is attributable in part to degeneration of the basal forebrain cortical cholinergic system.

So How Do We Proceed?

The distributed nature of social processing poses a challenge to understanding the neurology of social cognition and social behavior. Contributing to this challenge is that social processes are complex and intricate, comprising many integrated substrates for attention, sensory and perceptual processing, motivation, memory access and associative linkage, expectations, decision making, and response processes. Constructs such as theory of mind and social regulation have complex mappings onto multiple neurobehavioral substrates. Moreover, category errors can arise when social constructs are developed independent of understanding of neural processes, and where understandings of neural systems and processes have developed with minimal regard to the complexity of social functions.

Social psychological concepts and theories do not necessarily correspond to neurological processes and codes. The ultimate development of social neuroscience will await a convergence of constructs across disciplines and levels of organization and analysis; from the social and behavioral to the anatomical to the cellular and genetic (Cacioppo & Berntson, 1992; Cacioppo et al., 2000). Relations between psychological and biological processes cannot be comprehended fully by investigations at a single level of organization. An integrative, interdisciplinary approach across levels of organization and analysis can add specific perspectives as different levels of analysis can reveal distinct patterns of order in data, and because different levels of organization are known to interact. Multilevel research will be necessary to form bridges among disciplines, calibrate and refine constructs to minimize category errors, and ultimately, achieve a truly interdisciplinary social neuroscience. A need remains for research with a more restricted focus, of course, and it is not necessary that all researchers pursue multilevel analyses, but the ultimate goal should be toward multilevel understandings.

Basic principles of multilevel analysis, which frame issues and organize research efforts, have been articulated (Cacioppo & Berntson, 1992; see also Cacioppo et al., 2000). These include:

1. The principle of *multiple determinism* stipulates that an event or process at one level of organization, especially at more molar levels, will have numerous antecedents within and across levels of analysis. Because mappings across more divergent levels of analysis become increasingly complex, the *corollary of proximity* suggests that multilevel analysis may be more straightforward for more proximal levels or organization. Bridges that are built among adjacent levels will ultimately facilitate broader mappings across more disparate levels.

2. The principle of *reciprocal determinism* asserts that mutual, reciprocal influences may exist among levels of organization; that is, the direction of causation is not one way. Consequently, although the multilevel perspective entails the ability to relate social constructs to neural and cellular events, it does not represent a unidirectional reductionism, as an equally important set of social determinants affect neural systems.

3. Finally, the principle of *nonadditive determinism* specifies that properties of the whole cannot always be predicted by knowledge of properties of the parts. Although single-level analyses continue to have an important place, no single level of analysis can yield a comprehensive understanding of behavioral processes.

An illustration of the utility of the multilevel approach comes from literature on arousal. Classic studies of Morruzzi, Magoun, and Lindsley on the ascending reticular activating system offered a potential neural substrate for the construct of arousal, which appeared as a powerful determinant of sleep and waking states as well as behavior (see Magoun, 1963). The construct of generalized arousal promised a parsimonius account of a broad range of behavioral and performance variables, and quickly became entrenched in the psychological literature. It became apparent quite early, however, that arousal may be differentiated, and in any event did not have a simple, monotonic relation to behavior. Duffy (1962) wrote, "There appears to be some degree of 'generality' and some degree of 'specificity' in activation, the extent of each remaining an unsolved problem" (p. 322). Nevertheless, the construct continues to be employed in the psychological literature, partly because of its utility, despite overwhelming evidence against either the generalized nature of arousal or the existence of a uniform neural activating substrate.

Although a clear differentiation of arousal dimensions has not emerged from the psychological literature, continuing work in psychobiology offers important guidance in this domain. It is now clear that there is not a single reticular activating system, but rather, that many ascending anatomical systems subserve arousal- or activation-like functions (Sarter, Bruno, & Berntson, 2003). These parallel systems are anatomically as well as neurochemically distinct and have different behavioral effects. The basal forebrain cortical cholinergic projection, for example, is especially implicated in cortical processing of behaviorally relevant signals, whereas ascending noradrenergic projections from the locus coeruleus are more involved in classic arousal effects and target selection (Sarter, Bruno, & Berntson, 2003). In contrast, dopaminergic projections are involved in motivation and response initiation, whereas ascending serotonergic, glutamatergic, and histaminergic projections have yet additional effects on distinct aspects of behavior. These findings offer an excellent opportunity for multilevel research to clarify the nature and dimensions of arousal processes in a neurologically meaningful fashion, and to characterize more fully the psychological significance of the many ascending activating systems.

Overview

The chapters in this book represent important developments in social neuroscience, in which multilevel research efforts provide reciprocal benefits to both biological and psychological perspectives. Brain imaging can be a powerful tool for studying social processes, but its ultimate value will lie in the integration and convergence of its results with other interdisciplinary approaches. These include studies of the neuropsychological consequences of brain damage, emerging methodology of transcranial magnetic brain stimulation, and even animal research in which more invasive brain measures and manipulations can be performed.

To return to the earlier question of whether social processes reflect special brain mechanisms, it is simply too early to tell. We do not know enough about either social psychological processes or brain mechanisms to answer this question definitively at the present time. Perhaps the more important questions at this point are how this issue might best be approached, and what useful neurosocial constructs and theories should look like. Brain localization can inform neuropsychological theories, but meaningful neurosocial theories will not be theories about places, nor will their critical elements and conceptual relations be couched in the language of space. Rather, they will have to incorporate fundamental underlying processes that subserve social psychological phenomena. For this, multilevel approaches that can calibrate constructs and theories across levels of organization will be indispensable.

Acknowledgment

Preparation of this paper was supported in part by grant HL 54428 from the National Heart, Lung, and Blood Institute.

References

Berntson, G. G. (1972). Blockade and release of hypothalamically and naturally elicited aggressive behaviors in cats following midbrain lesions. *Journal of Comparative and Physiological Psychology, 81*, 541–554.

Berntson, G. G. (1973). Attack, grooming and threat elicited by stimulation of the pontine tegmentum in cats. *Physiology and Behavior, 11*, 81–87.

Berntson, G. G. & Beattie, M. S. (1975). Functional differentiation within hypothalamic behavioral systems in the cat. *Physiological Psychology, 3*, 183–188.

Berntson, G. G. & Boysen, S. T. (1989). Specificity of the cardiac response to conspecific vocalizations in chimpanzees (*Pan troglodytes*). *Behavioral Neuroscience, 103*, 235–245.

Berntson, G. G., Sarter, M., & Cacioppo, J. T. (2003). Ascending visceral regulation of cortical affective information processing. *European Journal of Neuroscience, 18*, 2103–2109.

Cacioppo, J. T. & Berntson, G. G. (1992). Social psychological contributions to the decade of the brain: Doctrine of multilevel analysis. *American Psychologist, 47*, 1019–1028.

Cacioppo, J. T., Berntson, G. G., Sheridan, J. F., & McClintock, M. K. (2000). Multi-level integrative analyses of human behavior: The complementing nature of social and biological approaches. *Psychological Bulletin, 126*, 829–843.

Cacioppo, J. T. & Tassinary, L. G. (1990). Inferring psychological significance from physiological signals. *American Psychologist, 45*, 16–28.

Cacioppo, J. T., Tassinary, L. G., & Berntson, G. G. (2000). Psychophysiological science. In J. T. Cacioppo, L. G. Tassinary, & G. G. Berntson (Eds.), *Handbook of Psychophysiology* (pp. 3–23). Cambridge: Cambridge University Press.

Duffy, E. (1962). *Activation and Behavior.* New York: Wiley.

Hebb, D. O. (1949). *The Organization of Behavior: A Neuropsychological Theory.* New York: Wiley.

Hothersall, D. (2004). *History of Psychology* (4th Ed.). Boston: McGraw Hill.

Magoun, H. W. (1963). *The Waking Brain.* Springfield, IL: Charles C Thomas.

Padgett, D. A., Sheridan, J. F., Dorne, J., Berntson, G. G., Candelora, J., & Glaser, R. (1998). Social stress and the reactivation of latent herpes simplex virus-type 1. *Proceedings of the National Academy of Sciences of the United States of America, 95*, 7231–7235.

Phillips, C. G., Zeki, S., & Barlow, H. B. (1984). Localization of function in the cerebral cortex. Past, present and future. *Brain, 107*, 327–361.

Premack, D. & Woodruff, G. (1978). Does the chimpanzee have a theory of mind? *Behavioral Brain Sciences, 1*, 515–526.

Quan, N., Avitsur, R., Stark, J. L., He, L., Shah, M., Caligiuri, M., et al. (2001). Social stress increases the susceptibility to endotoxic shock. *Journal of Neuroimmunology, 115*, 36–45.

Sarter, M., Berntson, G. G., & Cacioppo, J. T. (1996). Brain imaging and cognitive neuroscience: Towards strong inference in attributing function to structure. *American Psychologist*, 51, 13–21.

Sarter, M., Bruno, J. P., & Berntson, G. G. (2003). Reticular activating system. In L. Nadel (Ed.), *Encyclopedia of Cognitive Science,* Vol. 3 (pp. 963–967). London: Nature Publishing Group.

Sarter, M., Bruno, J. P., & Givens, B. (2003). Attentional functions of cortical cholinergic inputs: What does it mean for learning and memory? *Neurobiology of Learning and Memory, 80,* 245–256.

Uttal, W. R. (2003). *The New Phrenology: The Limits of Localizing Cognitive Processes in the Brain.* Cambridge: MIT Press.

Valenstein, E. S. (1975). Brain stimulation and behavior control. *Nebraska Symposium on Motivation, 22,* 251–292.

Young, R. M. (1968). The functions of the brain: Gall to Ferrier (1808–1886), *Isis, 59,* 251–268.

2 Neurological Substrates of Emotional and Social Intelligence: Evidence from Patients with Focal Brain Lesions

Antoine Bechara and Reuven Bar-On

Based on the Bar-On model, the construct of emotional intelligence is defined as a multifactorial array of interrelated emotional, personal, and social competencies that influence our ability to cope actively and effectively with daily demands (Bar-On, 1997b, 2000). Most conceptualizations of this construct address one or more of the following basic components: (1) ability to be aware of and express emotions; (2) ability to be aware of others' feelings and to establish interpersonal relationships; (3) ability to manage and regulate emotions; (4) ability to cope realistically and flexibly with the immediate situation and solve problems of a personal and interpersonal nature; (5) and ability to generate positive affect in order to be sufficiently self-motivated to achieve personal goals.

This construct is closely related to the construct of social intelligence. Such conceptual proximity is evident in the way social intelligence was first defined by Thorndike (1920) as the ability to perceive one's own and others' internal states, motives, and behaviors, and to act toward them optimally on the basis of that information. Because of the similarity between the concepts, some psychologists suggested that they may relate to different aspects of the same construct and could actually be referred to as "emotional and social intelligence" (Bar-On, 2000, 2001; Bar-On et al., 2003).

Empirical evidence for these notions is based on studies of patients with focal brain lesions, which suggest that (1) emotional and social intelligence is different from cognitive intelligence, in that these two major components of general intelligence are supported by separate neural substrates; (2) neural systems that support emotional and social intelligence overlap with neural systems subserving the processing of emotions and feelings, but not neural systems associated with "cold" cognition; and (3) damage to neural structures that subserve emotions and feelings, but not those that subserve cold cognition, are associated with changes in emotional experience and social functioning. Furthermore, these lesions are associated with low scores on measures of emotional and social intelligence, whereas scores on measures of cognitive intelligence fall within the normal range (Bar-On et al., 2003).

The Neuroanatomy of Emotions and Feelings

According to Damasio, an important distinction exists between emotions and feelings (Damasio, 1994, 1999, 2003). The specific object or event that predictably causes an emotion is designated as an emotionally competent stimulus. Responses toward the body proper enacted in a body state involve physiological modifications. These modifications range from changes in internal milieu and viscera that may not be perceptible to an external observer (endocrine release, heart rate, smooth muscle contraction) to changes in the musculoskeletal system that may be obvious to an external observer (posture, facial expression, specific behaviors such as freezing, flight and fight, etc.). The ensemble of these enacted responses in the body proper and in the brain constitutes an emotion.

Responses aimed at the brain lead to (1) central nervous system release of certain neurotransmitters (dopamine, serotonin, acetylcholine, noreadrenaline), (2) an active modification of the state of somatosensory maps such as those of the insular cortex, and (3) modification of the transmission of signals from the body to somatosensory regions. The ensemble of signals as mapped in somatosensory regions of the brain itself provide the essential ingredients for what is ultimately perceived as a feeling, a phenomenon perceptible to the individual in whom the signals are enacted. Thus emotions are what an outside observer can see or, at least, can measure. Feelings are what the individual senses or experiences subjectively.

After the initial debate of the James–Lang versus Cannon–Bard theories of emotion, neuroanatomist James Papez demonstrated that emotion is not a function of a specific brain center. Rather, it is a function of a neural circuitry that involves several brain structures interconnected through several neural pathways. Subsequent work by Paul McLean provided further elaboration on the original Papez circuit by adding the prefrontal cortex, parahippocampal gyrus, amygdala, and several other subcortical and brainstem structures. The culmination of all this work resulted in what has become known as the limbic system.

Damasio's description of neural systems that subserve emotions and feelings (1999, 2003) is consistent and overlaps considerably with the neuroanatomy of the limbic system as described originally. The functional anatomy of this system includes the following.

An emotion begins with appraisal of an emotionally competent object. An emotionally competent object is basically the object of one's emotion. It can be present or recalled from memory. In neural terms, images related to the emotional object are represented in one or more of the brain's sensory processing systems. Regardless of how short this presentation is, signals related to the presence of that object are made available to a number of emotion-triggering sites elsewhere in the brain. Two of these sites are the amygdala and orbitofrontal cortex (figure 2.1). Evidence suggests that differences may be present in the way these two structures process emotional information: the amygdala is more engaged in triggering emotions when the emotional object is present in the environment; the orbitofrontal cortex is more important when the emotional object is recalled from memory. However, the general function is the same; that is, to trigger emotional responses from emotionally competent stimuli.

To create an emotional state, the activity in triggering sites must be propagated to execution sites by means of neural connections. Emotion execution sites are visceral motor structures that include the hypothalamus, basal forebrain, and some nuclei in the brainstem. Feelings, on the other hand, result from neural patterns that represent changes in the body's response to an emotional object. Signals from body states are relayed back to the brain, and representations of these body states are formed at the level of visceral sensory nuclei in the brainstem. Representations of these body signals also form at the level of the insular cortex and lateral somatosensory cortices (SII and SI areas) (figure 2.1). It is most likely that reception of body signals at the level of the brainstem does not give rise to conscious feelings as we know them, but reception of these signals at the level of the cortex does so. The anterior insular cortex, especially on the right side, plays a special role in mapping visceral states and in bringing interoceptive signals to conscious perception (Damasio, 1999; Craig, 2002). There is some debate as to whether the right anterior insular cortex is sufficient for translating visceral states into subjective feeling and self-awareness (Craig, 2002), or whether this translation requires additional regions such as the anterior cingulate cortex (Damasio, 1999).

Many emotional disturbances are linked to focal lesions to structures outline earlier as subserving the mechanisms of emotions and conscious feelings. The following is an outline of the disturbances and alterations of emotional experience that ensue with each type of lesion.

Lesions of the Amygdala

Patients with amygdala damage, especially bilateral damage (figure 2.2), express one form of emotional lopsidedness: negative emotions such as anger and fear are less frequent and less intense than positive emotions (Damasio, 1999). Many laboratory experiments established problems in these patients with processing emotional

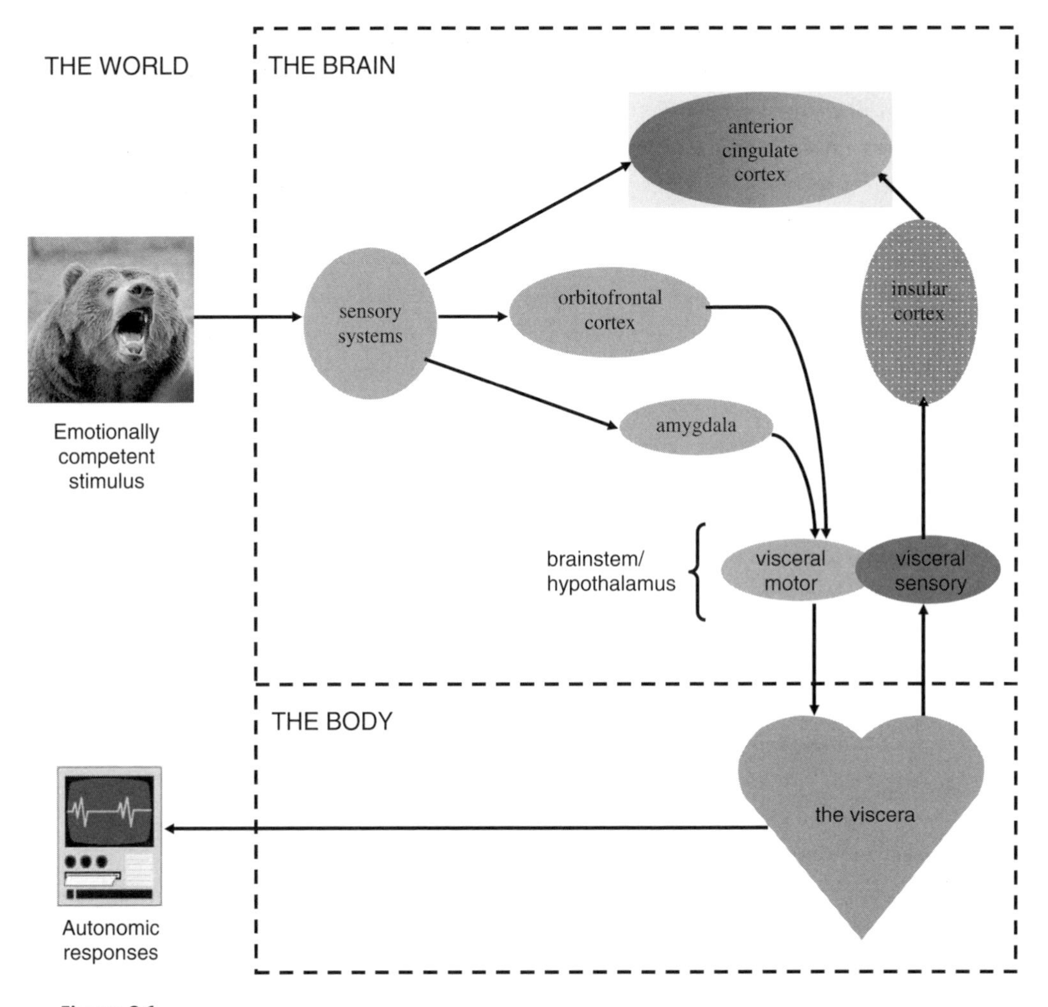

Figure 2.1
Information related to the emotionally competent object is represented in one or more of the brain's sensory processing systems. This information, which can be derived from the environment or recalled from memory, is made available to the amygdala and orbitofrontal cortex, which are trigger sites for emotion. Emotion execution sites include the hypothalamus, basal forebrain, and nuclei in the brainstem tegmentum. Only the visceral response is represented, although emotion comprises endocrine and somatomotor responses as well. Visceral sensations reach the anterior insular cortex by passing through the brainstem. Feelings result from the re-representation of changes in viscera in relation to the object or event that incited them. The anterior cingulate cortex is a site where this second-order map is realized.

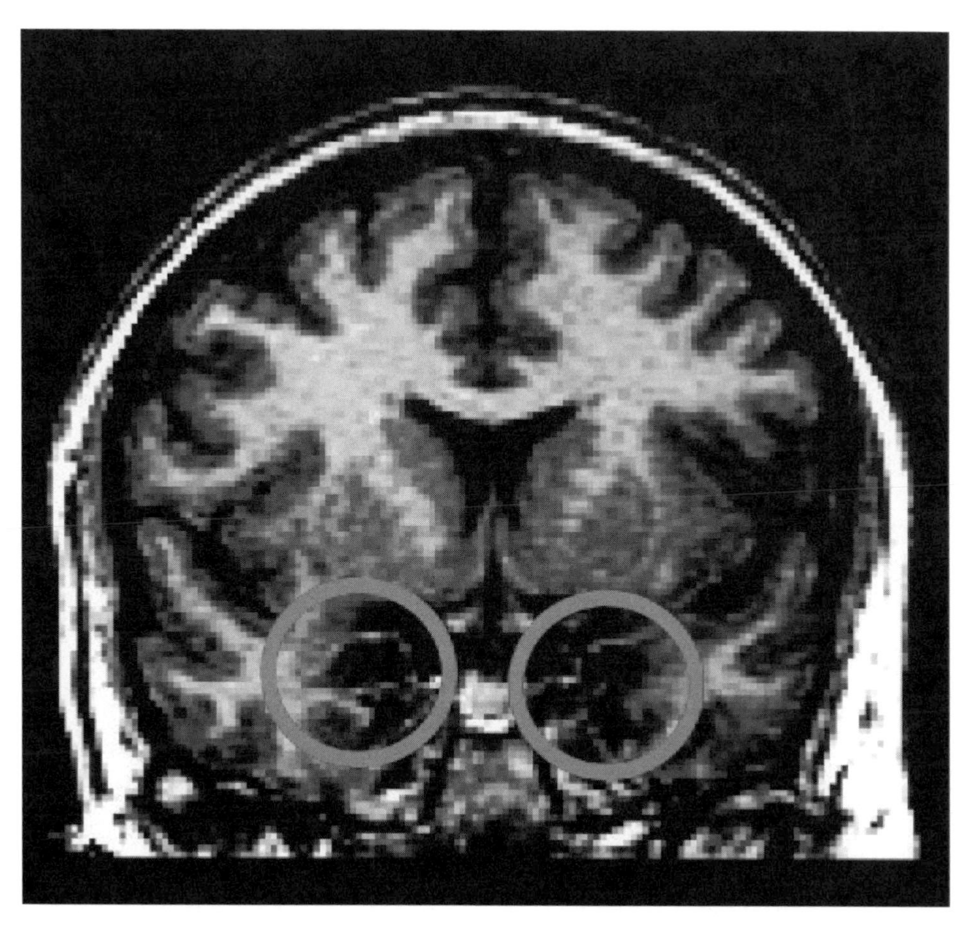

Figure 2.2
A coronal section through the amygdala taken from the three-dimensional reconstruction of a brain of a patient with bilateral amygdala damage. The reconstruction was based on magnetic resonance data and obtained with Brainvox. Circles highlight the region showing bilateral amygdala damage.

information, especially in relation to fear (Adolphs et al., 1995, 1998; LaBar et al., 1995; Phelps et al., 1998). It is important to note that when damage occurs early in life, these patients grow up to have many abnormal social behaviors and functions (Tranel and Hyman, 1990; Adolphs et al., 1995).

More specifically, laboratory experiments reveal that the amygdala plays a critical role in a specific induction mechanism of emotional states. Emotion may be induced in two different ways: (1) in an impulsive or automatic way, which we referred to as primary induction, and (2) in a reflective or thoughtful way that we referred to as secondary induction (Bechara et al., 2003).

Primary inducers are stimuli or entities that are innate or learned to be pleasurable or aversive. Once they are present in the immediate environment, they automatically, quickly, and obligatorily elicit an emotional response. An example of primary inducers is an encounter with a feared object such as a snake or a stimulus predictive of a snake. Semantic information such as winning or losing a large sum of money, which instantly, automatically, and obligatorily elicits an emotional response, is another example.

Secondary inducers are generated by recall of a personal or hypothetical emotional event (thoughts and memories about the primary inducer) that, when they are brought to working memory, slowly and gradually begin to elicit an emotional response. Examples of secondary inducers include the emotional response elicited by the memory of encountering or being bitten by a snake, the memory or imagination of gaining or losing a large sum of money, and recall or imagination of the death of a loved one.

Evidence suggests that the amygdala is a critical substrate in the neural system necessary for triggering emotional states from primary inducers. It couples the features of primary inducers, which can be processed subliminally by way of the thalamus (LeDoux, 1996; Morris et al., 1999) or explicitly by early sensory and high-order-association cortices, with representations (conscious and nonconscious) of the emotional state (feeling) associated with the inducer. This emotional state is evoked by effector structures such as the hypothalamus and autonomic brainstem nuclei that produce changes in internal milieu and visceral structures together with other effector structures such as the ventral striatum, periacquenductal gray, and other brainstem nuclei. These last produce changes in facial expressions and specific approach or withdrawal behaviors.

Several studies support this notion. Monkeys with mesial temporal lesions that include the amygdala have an increased tendency to approach "emotionally competent" stimuli such as snakes (Kluver & Bucy, 1939; Zola-Morgan et al., 1991; Aggleton,

1992), suggesting that the object of fear can no longer evoke a state of fear. Animal studies also showed that conditioning is highly dependent on the integrity of the amygdala system (Davis, 1992a, b; Gaffan, 1992; LeDoux, 1993a, b, 1996; Malkova et al., 1997; Amorapanth et al., 2000). Human patients with bilateral amygdala lesions have reduced, but not completely blocked, autonomic reactivity to aversive loud sounds (Bechara et al., 1999). These patients also do not acquire conditioned autonomic responses to the same aversive loud sounds, even when the damage is unilateral (Bechara et al., 1995; LaBar et al., 1995). Amygdala lesions in humans also reduce autonomic reactivity to a variety of stressful stimuli (Lee et al., 1988, 1998).

Not only innate and unconditioned emotional stimuli, bilateral amygdala damage in humans also interferes with the emotional response to cognitive information that through learning has acquired properties that automatically and obligatorily elicit emotional responses. Examples of this cognitive information are learned concepts such as winning and losing. The announcement that you have won a Nobel prize, an Oscar, or a lottery ticket can instantly, automatically, involuntarily, and obligatorily elicit an emotional response. Emotional reactions to gains and losses of money, for example, are learned responses because we were not born with them; however, through development and learning, they become automatic. We do not know how this transfer occurs. However, we have presented evidence that patients with bilateral amygdala lesions failed to trigger emotional responses in reaction to the winning or losing of various sums of money (Bechara et al., 1999).

Results of functional neuroimaging studies corroborate those from lesion studies. For instance, activation of the amygdala was shown in classical conditioning experiments (LaBar et al., 1998). Other functional neuroimaging studies revealed amygdala activation in reaction to winning and losing money (Zalla et al., 2000). It is also interesting that humans tend automatically, involuntarily, and obligatorily to show a pleasure response when they solve a puzzle or uncover a solution to a logical problem. In functional neuroimaging experiments in which human subjects were asked to find solutions to series of logical problems, amygdala activations were associated with the "aha" in reaction to finding the solution to a given logical problem (Parsons & Oshercon, 2001).

In essence, the amygdala links the features of a stimulus with the expressed emotional-affective value (representations of emotional states) of that stimulus (Malkova et al., 1997). However, the amygdala appears to respond only when stimuli are actually present in the environment (Whalen, 1998). All these findins are consistent with the Kluver–Bucy syndrome described in monkeys (Kluver & Bucy, 1939; Zola-Morgan et al., 1991; Aggleton, 1992), although it is important to note that the condition in monkeys was not exactly the same as that in humans; the impact of the damage in humans is much milder.

Lesions of the Insular-Somatosensory Cortex

The typical clinical condition of patients with parietal damage (involving insula, somatosensory, and adjacent cortices), especially on the right side, which demonstrates alterations in emotional experience, is anosognosia (figure 2.3). Anosognosia, or denial of illness or failure to recognize an illness, is characterized by apathy and placidity. It is most commonly associated with right hemisphere lesions (as opposed to left), particularly inferior parietal cortices that include primary and secondary

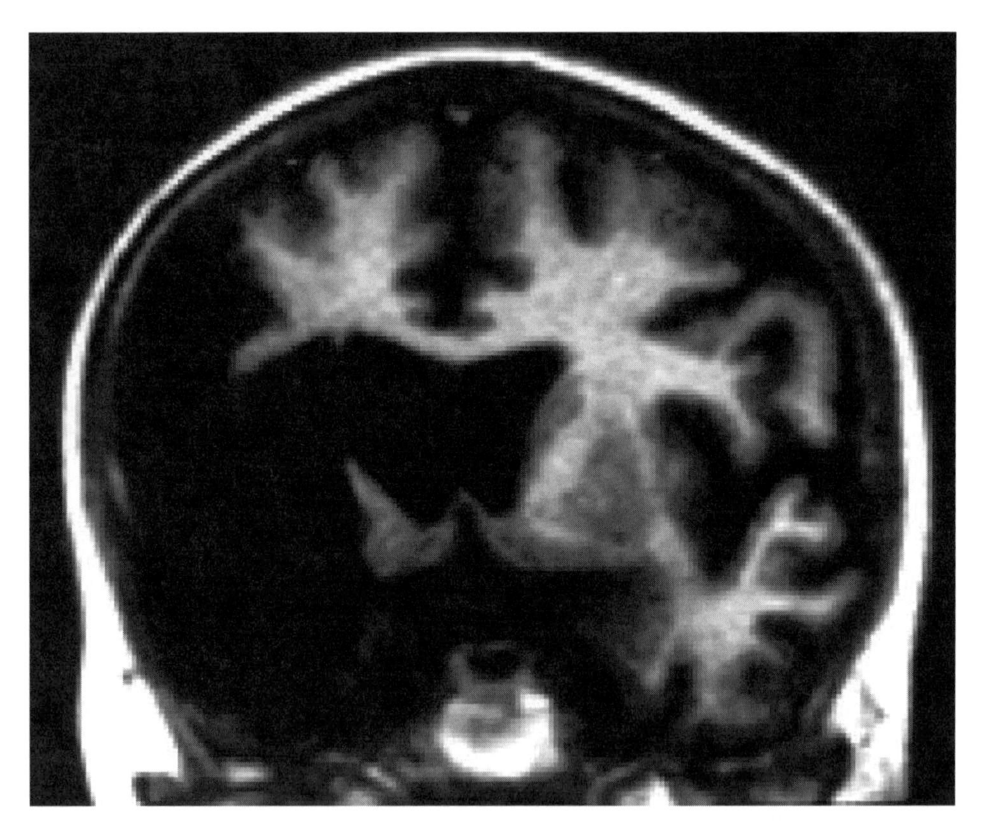

Figure 2.3
A coronal section through the brain of a patient with anosognosia. It shows extensive damage in the right parietal region that includes the insula and somatosensory cortices (SII, SI). The left parietal region is intact.

somatosensory cortices and insula. The most frequent causes of this condition are stroke, glial tumors, and head injury.

The classic example of anosognosia is paralysis on the left side of the body, with the patient unable to move hand, arm, and leg, and unable to stand or walk. When asked how they feel, patients report that they feel fine and seem oblivious to the problem. In patients with stroke, unawareness is typically most profound during the first few days after onset. Within a few days or a week, these patients will begin to acknowledge that they have suffered a stroke and that they are weak or numb, but they minimize the implications of the impairment. They might say, "I am right handed, I don't care if my left hand moves or not." In the chronic epoch (3 months or more after onset), they may provide a more accurate account of their physical disabilities; however, defects in appreciation of acquired cognitive limitations may persist for months or years. Patients with similar damage on the left side of the brain are usually cognizant of the deficit and often feel depressed.

Many laboratory experiments identified problems processing emotional information in these patients, such as empathy and recognition of emotions in facial expressions (Adolphs et al., 1996). Not only are these patients oblivious to their own condition and how they feel, but they are oblivious to the emotions of others and how they feel. Furthermore, although the paralysis and neurological handicap limit social interactions and mask potential abnormal social behaviors, when these patients were allowed extensive social interactions they exhibited severe impairments in judgment and failure to observe social convention. One example is Supreme Court Justice William O. Douglas, who is described in Damasio's *Descartes' Error* (1994).

More specifically, once emotional states from primary inducers are induced, signals from these states are relayed to the brain. The signals then lead to development of state patterns in brainstem nuclei, such as the parabrachial nucleus (PBN), and in somatosensory cortices (insular-SII, SI cortices, cingulated cortices). After an emotional state has been triggered by a primary inducer and experienced at least once, a pattern for it is formed. Presentation of a stimulus that evokes thoughts and memories about a specific primary inducer will then operate as a secondary inducer. Secondary inducers are presumed to reactivate the pattern of the emotional state belonging to a specific primary inducer and generate a weaker activation of the emotional state than if it were triggered by an actual primary inducer. For example, imagining the loss of a large sum of money (secondary inducer) reactivates the pattern of the emotional state belonging to an actual experience of money loss (primary inducer). However, the emotional state generated by the imagined loss is usually weaker than one triggered by an actual loss.

Reception of body signals at the level of visceral sensory nuclei in the brainstem does not give rise to conscious feelings, as we know them, but reception of these signals at the level of the insular-somatosensory cortices and posterior cingulated cortices is perceived as a feeling. Evidence from functional neuroimaging studies suggests that the posterior cingulated and retrosplenial cortex are consistently activated in feeling states (Maddock, 1999; Damasio et al., 2000), which suggests that this region plays a role in the generation of feelings from autobiographical memory. Support for the idea that the insula and SII and SI cortices are necessary for feeling to occur is supported by clinical observations in subjects with focal brain lesions (Berthier et al., 1988; Damasion, 1994, 1999). We also have preliminary evidence showing that when a primary inducer, such as an aversive loud sound, induces a somatic response in normal subjects, measured as changes in skin conductance response and heart rate, subjects provide a high subjective rating of the noise as "to loud." Of interest, subjects with right hemisphere damage trigger an emotional response to the loud sound similar to that of normal controls, but do not report feeling the sound as to loud (unpublished observations).

Lesions of the Orbitofrontal-Anterior Cingulate Cortex

Many investigators confuse the anatomical terms orbitofrontal and ventromedial prefrontal cortex. The orbitofrontal region of the prefrontal cortex includes the rectus gyrus and orbital gyri, which constitute the inferior surface of the frontal lobes lying immediately above the orbital plates. Lesions of this region are not usually restricted to the orbitofrontal cortex, but extend into neighboring cortex and involve different-size sectors of the ventromedial prefrontal (VM) region. The VM region includes the medial and various sectors of the lateral orbitofrontal cortex, and the anterior cingulated, thus encompassing Brodmann's areas (BA) 25; lower 24 and 32; medial aspects of 11, 12, and 10; and white matter subjacent to all of these areas. Therefore, in most studies of patients with lesions in this region, we refer to the damage as involving the VM region, and not strictly the orbitofrontal region (figure 2.4).

Patients with orbitofrontal-anterior cingulated damage have various degrees of disturbances in emotional experience, depending on the location and extent of damage. If damage is localized, especially in the more anterior sector of the VM region, patients have many manifestations including alterations of emotional experience and social functioning. Previously well-adapted individuals become unable to observe social conventions and decide advantageously on personal matters, and their ability to express emotion and to experience feelings in appropriate social situations becomes compro-

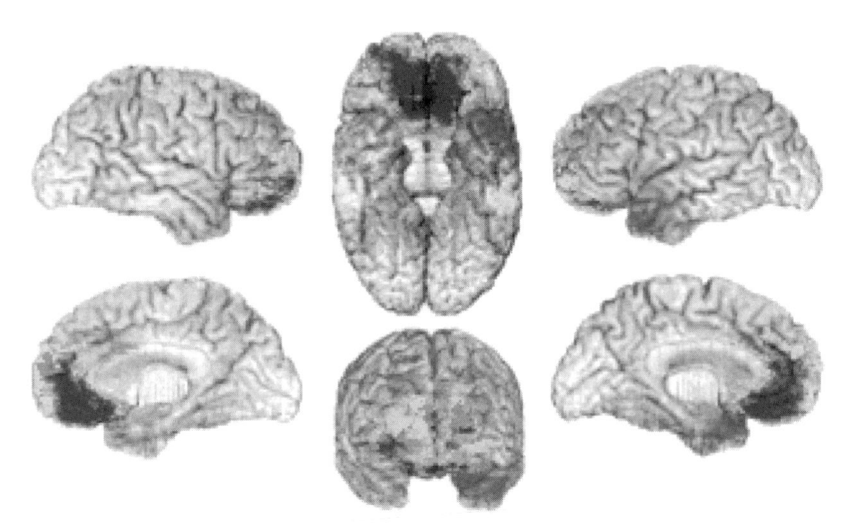

Figure 2.4
Overlap of lesions in a group of patients with bilateral damage to the ventromedial region of the prefrontal cortex, which includes orbitofrontal and anterior cingulate areas discussed in the text.

mised (Bechara et al., 2000, 2002). If the damage is more extensive, especially when it involves parts of the anterior cingulated, patients experience additional problems in impulse control, disinhibition, and antisocial behavior. For instance, they may utter obscene words, make improper sexual advances, or say the first thing that comes to mind, without considering the social correctness of what they say or do. For example, they may urinate in a completely inappropriate social setting when the urge arises, without regard to rules of decency.

With more extensive damage, the patient may suffer a condition known as akinetic mutism, especially when damage involuves the anterior cingulate gyrus and supplementary motor area. The condition is a combiation of mutism and akinesia. Lesions may result from strokes related to impairment of blood supply in the anterior cerebral artery territories, and in some cases, from rupture of aneurysms of the anterior communication artery or anterior cerebral artery. It may also result from parasagittal tumors (meningiomas of the falx cerebri). Lesions can be unilateral or bilateral. There is no difference between those on the left and those on the right side. Unilateral and bilateral lesions appears to differ only in relation to course of recovery: with unilateral lesions, the condition persists for one to two weeks; with bilateral lesions, it may persist for many months. The patient with akinetic mutism makes no effort to communicate verbally or by gesture. Movements are limited to the eyes (tracking moving targets) and to body or arm movements connected with daily necessities (eating, pulling bed sheets, getting up to go to the bathroom). Speech is exceptionally sparse, with only rare isolated utterances, but it is linguistically correct and well articulated (although generally hypophonic). With extensive prompting, the patient may repeat words and short sentences.

Provided that the amygdala and insular-somatosensory cortices were normal during development, emotional states associated with secondary inducers develop normally. Generating emotional states from secondary inducers depends on cortical circuitry in which the VM cortex plays a central role. Evidence suggests that the VM region is a critical substrate in the neural system necessary for triggering emotional states from secondary inducers. It serves as a convergence-divergence zone, in which neuron ensembles can couple (1) a certain category of event based on memory records in high-order-association cortices to (2) effector structures that execute the emotional response and to (3) neural patterns related to the nonconscious (in the PBN) or conscious (in the insular-SII, SI, posterior cingulated cortices) feeling of the emotional state. In other words, the VM cortex couples knowledge of secondary inducer events to emotional state patterns related to "what it feels like" to be in a given situation.

Several studies support this notion. We conducted investigations using paradigms for evoking emotional states, namely, from the internal generation of images related to emotional situations (emotional imagery). We predicted that patients with VM lesions would have reduced ability to experience emotions when they recalled specific emotional situations from their personal life. Subjects were asked to think about a situation in their lives in which they felt each of the following emotions: happiness, sadness, fear, and anger. After a brief description of each memory was obtained, the subject was given a physiological test and asked to image and reexperience each emotional experience while their skin conductance response (SCR) and heart rate were monitored. As a control condition, subjects were asked to recall and image a non-emotional set of events. At the conclusion of the task, emotional as well as neutral, each subject was asked to rate how much emotion was felt (on a scale of 0 to 4). Using this procedure, patients with bilateral VM lesions were clearly able to retrieve previous experiences associated with all four emotions; that is, they recalled emotion-laden events that occurred before their brain lesion such as weddings, funerals, car accidents, and family disputes. However, they had difficulty reexperiencing the actual emotions of these situations as reflected by low physiological activity and low subjective rating of feeling the emotions. This was especially marked in the case of sadness (Tranel et al., 1998). This suggests that damage to the VM cortex weakens the ability to reexperience an emotion by recalling an emotional event.

Findings from functional neuroimaging studies are consistent with these results. They showed activations or deactivations in the VM region during recall and imagery of personal emotional events (Lane et al., 1997; Mayberg et al., 1999; Damasio et al., 2000; Lane & McRae, 2004).

Having said this, it is important to note that patients with VM damage are not emotionless. They can evoke stronger responses to anger from recall of personal experiences, consistent with observations made in monkeys with orbital lesions (Butter et al., 1963, 1968; Butter & Snyder, 1972). The lesions produced a clear reduction in aggressive behavior of these monkeys, but the reduction was situational. In other words, the animals could still demonstrate aggression when brought back to the colony where they had been dominant, suggesting that this capacity had not been eliminated, and that the lesions did not produce a consistent state of bluntness of affect in the monkeys.

Furthermore, in fear-conditioning paradigms, we showed that patients with VM lesions could acquire SCR conditioning with an aversive loud noise (Bechara et al., 1999). Not all patients were able to acquire conditioned SCRs, however; it seemed to depend on whether the anterior cingulate (especially area 25) and/or basal forebrain was involved in the lesion. This suggestion is preliminary and we are still investigating the issue. The sparing of emotional conditioning by VM lesions is consistent with conditioning studies in animals showing that the VM cortex is not necessary for acquiring fear conditioning (Morgan & LeDoux, 1995). This is in contrast to the amygdala, which appears essential for coupling a stimulus with an emotional state induced by a primary aversive unconditioned stimulus such as electric shock (Davis, 1992a, b; Kim et al., 1993; Kim & Davis, 1993; LeDoux, 1993a, b, 1996). This also applies to learning the association between stimuli and the value of a particular reward (Malkova et al., 1997). Human studies show that lesions of the amygdala impair emotional conditioning with an aversive loud sound (Bechara et al., 1995; LaBar et al., 1995), and functional neuroimaging studies reveal that the amygdala is activated during such conditioning tasks (LaBar et al., 1998).

Lesions within the Brainstem

Patients with damage to brainstem structures involving effector structures in the neural circuitry of emotions and feelings (figure 2.1) also experience disturbances in emotions and emotion regulation. One example is pathologic laughing and crying. This condition is also referred to as forced, or spasmodic, laughing and crying, or the syndrome of pseudobulbar palsy. It may occur after bilateral interruption of corticobulbar fibers (direct and indirect) from cortex to motor nuclei of the cranial nerves (motor V, VII, ambiguus, hypoglossal). Thus it appears that disinhibition or loss of control of effector structures occurs in the brainstem from higher cortical structures.

The disorder is observed after strokes, amyotrophic lateral sclerosis, or multiple sclerosis. The prominent neurological sign is weakness (of the upper motor neuron type) in muscles supplied by corresponding cranial nerve nuclei. Most important is that these patients have inappropriate outbursts of laughter and crying: with the slightest provocation or sometimes for no apparent reason, a patient is thrown into hilarious laughter that may last for many minutes to the point of exhaustion. Far more often, the opposite happens: the mere mention of the patient's family or the sight of the doctor provokes a spasm of uncontrollable crying. Of interest, most often patients report that they do not have the feeling accompanying the emotion they are expressing: they cry, but they do not feel sad; they laugh, but they do not feel happy.

A more recent case study challenged this traditional view and suggested that the condition is linked to the cerebellum (Parvizi et al., 2001). Specifically, the condition occurs when the lesion disrupts the cerebro-ponto-cerebellar pathways. As a consequence, the cerebellar structures, which automatically adjust the execution of laughter or crying to the cognitive and situational context of a potential stimulus, become deprived of the input of important cortical information about that context, thus resulting in inadequate expression of the proper emotion (Parvizi et al., 2001). In any case, the corticobulbar and cerebellar views are not contradictory, except that they differ in terms of the anatomical site of the lesion that produces the condition. Both views agree that the disorder represents a disturbance in emotional expression rather than a primary disturbance of feelings.

These patients are extremely rare, and difficult access to them has prevented conduction of controlled laboratory studies designed to examine the effects of such brainstem lesions on emotion, emotion regulation, and social behavior.

Emotional and Social Intelligence versus Cognitive Intelligence

In essence, all the patients previously described had problems feeling and/or expressing emotions and observing social convention. In striking contrast, those with extensive damage in areas that spare the neural circuitry that is critical for emotions and feelings, may express a variety of cognitive problems related to memory, language, perception, and so on. However, very seldom do they express improper social behaviors. On that basis, we conducted a study in which we tested the hypotheses that (1) emotional and social intelligence is different from cognitive intelligence, in that the two are supported by separate neural substrates, and (2) damage to neural structures that subserve emotions and feelings, but not neural structures that subserve cold cognition, is associated with low scores on measures of emotional and social intelligence (Bar-On et al., 2003).

The experimental group consisted of twelve patients with focal brain lesions in one of the target neural areas known to be critical for processing emotions and feelings. Six patients had bilateral damage to the orbitofrontal cortex that also extended posteriorly and included some parts of the anterior cingulate. Three had damage in the right parietal cortex that involved the insular and surrounding somatosensory cortices, especially the inferior parietal lobule. Three remaining patients had bilateral damage to the amygdala. Bilateral selective amygdala damage is extremely rare, and two of these patients had additional damage in the hippocampus and surrounding medial temporal lobe cortex.

The control group consisted of eleven patients with lesions similar in size to those of the experimental group, but within areas that are outside those known to be critical for emotions and feelings. Four had damage within the right dorsolateral prefrontal cortex, and two within the left dorsolateral prefrontal cortex. Five others had damage in the parieto-occipital cortex on the left side. We note that these patients had damage on the left in areas that correspond to those associated with anosognosia on the right side.

All patients were drawn from the Division of Cognitive Neuroscience's Patient Registry at the University of Iowa. Those in the experimental group had stable focal lesions. Damage in the orbitoforntal-anterior cingulate subgroup was due to stroke or surgical removal of a meningioma. In the right insular-somatosensory region, damage was due to stroke. Amygdala damage was due to herpes simplex encephalitis (2 patients) or Urbach–Weithe disease (1 patient). In the control group, the lesions were also stable and focal, and damage was due to stroke. The experimental and control groups were matched with respect to gender, age, level of education and handedness (table 2.1).

We used three measures in examining these patients: the Bar-On EQ-i for assessing emotional intelligence; semistructured interviews for assessing social functioning; and standard neuropsychological tests for assessing cognitive intelligence, perception, memory, and executive functioning.

Emotional Intelligence: The Bar-On EQ-i

The Bar-On Emotional Quotient inventory (EQ-i) is a self-report measure of emotionally and socially intelligent behavior that provides an estimate of one's underlying

Table 2.1
Demographic Data on Participating Patients

Demographic data	Control group	Experimental group	Z	p level (2-tailed)
Number	11	12	—	—
Men	4	7	—	—
Women	7	5	—	—
Age (yrs)	47.1	43.5	0.46	0.644
Age range (yrs)	24–74	21–63	—	—
Education (yrs)	14.6	13.7	0.85	0.398
Handedness				
Right	10	12	—	—
Left	1	0	—	—

The Mann-Whitney U test was applied to compare average age and years of education between groups for significant differences.

emotional and social intelligence (Bar-On, 1997a, b). A more detailed discussion of the psychometric properties of this instrument and how it was developed is found elsewhere (Bar-On, 1997a, 2000, 2004). In brief, the EQ-i comprises 133 items and employs a 5-point response format ranging from very seldom or not true of me (1) to very often true of me or true of me (5). The subject's responses render a total EQ score and the following five composite scale scores comprising fifteen subscale scores in all: (1) intrapersonal EQ (self-regard, emotional self-awareness, assertiveness, Independence, and self-actualization); (2) interpersonal EQ (empathy, social responsibility, and interpersonal relationship); (3) stress management EQ (stress tolerance and impulse control); (4) adaptability EQ (reality testing, flexibility, and problem solving); and (5) general mood EQ (optimism and happiness). Brief definitions of the competencies measured by the fifteen subscales appear in table 2.2.

The EQ-i has a built-in correction factor that automatically adjusts scale scores, based on scores obtained from its two validity indexes: positive impression and negative impression scales, which determine if the individual is attempting to respond in either an exaggerated positive or negative manner, respectively. This is an important psychometric factor for self-report measures in that it reduces the distorting effects of response bias, thereby increasing the accuracy of the results. Also, this correction factor is of particular importance in the current application of the EQ-i, because some patients' self-awareness of their acquired deficits is limited (anosognosia). Raw scores are computer-tabulated and automatically converted into standard scores based on a mean of 100 and standard deviations of 15.

Table 2.2
Brief Definitions of the Emotional Intelligence (EI) Competencies That Are Measured by the EQ-i Subscales

EQ-i subscales	EI competencies measured
Intrapersonal	
Self-regard	To perceive, understand, and accept oneself accurately.
Emotional self-awareness	To be aware of and understand one's emotions.
Assertiveness	To express one's emotion and oneself effectively and constructively.
Independence	To be self-reliant and free of emotional dependency on others.
Self-actualization	To strive to achieve personal goals and actualize one's potential.
Interpersonal	
Empathy	To be aware of and understand how others feel.
Social responsibility	To identify with one's social group and cooperate with others.
Interpersonal relationship	To establish mutually satisfying relationships and relate well to others.
Stress management	
Stress tolerance	To manage emotions effectively and constructively.
Impulse control	To control emotions effectively and constructively.
Adaptability	
Reality testing	To validate one's feeling and thinking objectively with reality.
Flexibility	To adapt and adjust one's feelings and thinking to new situations.
Problem solving	To solve problems of a personal and interpersonal nature effectively.
General mood	
Optimism	To be positive and look at the brighter side of life.
Happiness	To feel content with oneself, others, and life in general.

Social Functioning

Postmorbid employment status, social functioning, interpersonal relationships, and social status were evaluated with a series of semistructured interviews and rating scales (Tranel et al., 2002). Briefly, this evaluation entailed a comprehensive assessment of those realms in each patient by a clinical neuropsychologist. The neuropsychologist was unaware of the objectives and design of the study. For each social domain evaluated, the extent of social change or impairment for each patient was rated on a three-point scale on which 1 corresponded to no change (no impairment), 2 to moderate change (moderate impairment), and 3 to severe change (severe impairment). For each patient, a total social change score was calculated by summing the scores from each of the four domains. Higher overall scores are indicative of greater levels of change (social impairment).

Cognitive Intelligence, Perception, Memory, and Executive Functioning

Cognitive intelligence (IQ) was measured with the Wechsler Adult Intelligence Scale (WAIS-III). Perception was measured with the Benton Facial Discrimination Test and the Benton Judgment of Line Orientation Test. Memory was measured with the Rey Auditory Verbal Learning Test (RAVLT), the Benton Visual Retention Test–revised (BVRT), and the Complex Figure Test. Executive functioning was measured with the Wisconsin Card-Sorting Test (WCST), the Trail-Making Test (TMT), and Controlled Oral Word Association Test (COWA). Possible existence of psychopathology was assessed with the Minnesota Multiphasic Personality Inventory (MMPI). More detailed description and references to these tests are found in the description of the original study (Bar-On et al., 2003).

The results of these measures were intriguing. Tests of cognitive intelligence, perception, memory, and executive functioning revealed that patients in the experimental and control groups were not significantly different from each other (table 2.3). Also, no significant difference was seen between groups with respect to MMPI scores (table 2.4), indicating absence of psychopathology in and between both groups. In addition to lack of significant difference between groups regarding level of cognitive intelligence, it is also important to point out that no significant correlation was found between cognitive intelligence and emotional intelligence for the clinical sample.

In striking contrast to cognitive intelligence, postmorbid social functioning was significantly worse for the experimental group compared with the control group (table 2.5). Most important, table 2.6 shows that those patients had significantly lower emotional intelligence than those in the control group. A review of these results suggest that the key emotional intelligence competencies involved appear to be the ability to be aware of oneself (self-regard), to express oneself (assertiveness), to manage and control emotions (stress tolerance, impulse control), to adapt to change (flexibility), and to solve problems of a personal nature as they arise (problem solving), as well as the ability to motivate oneself and mobilize positive affect (self-actualization, optimism, happiness).

The very small samples in both groups do not allow closer examination of differences between them and especially between the three subgroups forming the experimental group (patients with damage to the orbitofrontal cortex-anterior cingulate, insular-somatosensory cortices, and amygdala). However, in light of the proposed function of each neural structure in processing emotions and feelings, as described earlier, one can speculate on how each subgroup might differ in terms of scores on specific competencies of each subscale of the EQ-i if the samples were larger.

Table 2.3
Neuropsychological Test Scores for the Control Group (n = 11) and Experimental Group (n = 12)

Neuropsychological tests	Control group	Experimental group	Z	p level (2-tailed)
Cognitive intelligence				
WAIS-III				
Full IQ	97.7	105.3	1.17	0.241
Verbal IQ	99.2	107.9	1.32	0.186
Performance IQ	95.7	102.8	1.42	0.155
Perception				
Benton faces	44.4	43.9	0.50	0.620
Benton lines	25.4	24.6	0.81	0.420
Memory				
WAIS-III				
Digit span	11.0	10.8	0.16	0.869
BVRT				
Correct	7.4	7.4	0.12	0.908
Errors	3.8	4.0	0.58	0.565
RAVLT				
Trials 1 to 5	10.1	12.0	1.53	0.127
30-minute recall	8.1	9.3	0.46	0.648
Recognition	28.9	28.8	0.91	0.361
Complex figure (Rey-O)				
Copy	32.2	30.9	0.55	0.585
Delay	20.7	19.5	0.22	0.827
Executive functioning				
Trail-making test A	34.8	39.6	0.06	0.957
Trail-making test B	86.8	79.9	0.33	0.745
WCST				
Perseverative errors	10.2	17.7	0.44	0.662
Categories	5.8	5.1	0.71	0.476
COWA	37.0	40.4	0.53	0.596

See text for abbreviations of these tests.

The Mann–Whitney U test was applied to compare scores for significant differences between groups.

Table 2.4
MMPI Test Scores for the Control Group (n-11) and Experimental Group (n = 12)

Personality tests	Control group	Experimental group	Z	p level (2-tailed)
MMPI Scale				
1 (Hs)	57.3	57.6	0.34	0.732
2 (D)	53.0	63.0	1.26	0.209
3 (Hy)	64.3	57.6	0.69	0.493
4 (Pd)	58.7	61.9	1.03	0.304
5 (Mf)	52.7	55.0	0.57	0.568
6 (Pa)	50.3	57.1	0.92	0.359
7 (Pt)	49.3	61.3	1.38	0.168
8 (Sc)	52.0	63.3	1.48	0.139
9 (Ma)	54.3	53.9	0.23	0.818
0 (Si)	49.7	52.9	0.34	0.731

The Mann–Whitney U test was applied to compare scores for significant differences between groups.
Hs, hypochondriasis; D, depression; Hy, conversion hysteria; Pd, psychopathic deviate; Mf, mesculanity-femininity; Pa, paranoia; Pt, psychasthenia; Sc, schizophrenia; Ma, hypomania; Si, social introversion.

Table 2.5
Differences in Postmorbid Social Functioning between Control (n = 11) and Experimental Groups (n = 12)

Postmorbid changes in	Control group	Experimental group	Z	p level (2-tailed)
Employment	01.17	02.70	–3.11	0.002
Social functioning	01.00	02.30	–2.84	0.005
Interpersonal relationships	01.00	02.40	–2.87	0.004
Social status	01.00	01.90	–3.40	0.001
Total social change	04.17	09.30	–3.00	0.003

The Mann–Whitney U test was applied to compare scores for significant differences.

Table 2.6
Differences in Emotional Intelligence between Control (n = 11) and Experimental Groups (n = 12)

EQ-i scales (emotional intelligence)	Control group	Experimental group	Z	p level (2-tailed)
Total EQ	101.1	082.3	3.33	0.001
Intrapersonal EQ	100.0	081.8	3.23	0.001
Self-regard	099.1	083.8	2.40	0.016
Emotional self-awareness	100.9	090.1	1.48	0.139
Assertiveness	103.6	082.6	3.21	0.001
Independence	097.7	087.3	1.58	0.115
Self-actualization	099.4	086.8	2.25	0.024
Interpersonal EQ	099.6	091.6	1.36	0.175
Empathy	098.6	089.8	1.24	0.216
Social responsibility	101.5	095.3	1.14	0.254
Interpersonal relationship	098.8	092.8	0.83	0.406
Stress management EQ	104.8	089.1	2.62	0.009
Stress tolerance	100.1	083.2	2.56	0.011
Impulse control	108.3	096.9	2.13	0.033
Adaptability EQ	100.0	086.3	2.28	0.023
Reality testing	099.8	091.0	1.08	0.280
Flexibility	100.3	086.8	2.38	0.017
Problem solving	100.6	088.3	2.16	0.031
General mood EQ	099.9	083.3	3.27	0.001
Optimism	099.0	083.5	3.02	0.003
Happiness	100.9	085.8	2.71	0.007

Emotional intelligence was assessed by the EQ-i, and the Mann–Whitney U test was applied to compare scores for significant differences.

Specifically, given that the self-regard and emotional self-awareness competencies are mostly concerned with perception and awareness of one's emotions, we would anticipate that patients with right insular-somatosensory cortex damage (those with anosognosia) would be most incapacitated in this domain. Since assertiveness, independence, and self-actualization competencies involve expression of one's emotions and triggering of certain drives and motivations to achieve certain goals, we would anticipate that amygdala-damaged patients would be most affected in these domains. Finally, all the competencies measured by the interpersonal, stress-management, adaptability, and general mood subscales involve emotional and social competencies that we propose to be dependent on the ventromedial prefrontal cortex region of the brain. Therefore, patients with damage to the orbitofrontal-anterior cingulate region would be expected to be the most affected in these domains.

It is important to note that the specific type of brain injury sustained by subjects in to the experimental group usually produce a certain degree of anosognosia (lack of self-awareness specifically related to acquired impairments). Given reliance of the EQ-i on self-report, the question arises as to whether this instrument can provide a valid measure of emotional and social intelligence when used with this particular group of clinical subjects. The issue of self-awareness of acquired impairments is critical in situations in which self-report measures typically fail to detect acquired impairments in tested individuals. This often occurs in patients with VM lesions, as previously noted, in which case collateral information is usually required to document changes in personality and social behavior (Barrash et al., 2000).

However, the EQ-i proved to be successful in detecting abnormal levels of emotional and social intelligence in target subjects. This suggests that the instrument has adequate construct validity when used with individuals who otherwise may be unaware of the limitations of their own emotional and social abilities. Indeed, low scores, particularly on scales of self-regard (accurate self-awareness) and assertiveness (self-expression), most likely mean that these subjects possess low self-awareness consistent with their symptomatology; and their scores on these two scales would have been even lower without the EQ-i's correction factor that automatically adjusts scale scores to compensate for various types of inaccuracies in responding that occur for one reason or another.

Conclusion

Only patients with injury to the neural circuitry subserving emotions and feelings (experimental group) seem to have low emotional intelligence as well as disturbances in social functioning, in spite of normal levels of cognitive intelligence and in the absence of psychopathology. The experimental group had many disturbances in social functioning, and they scored significantly lower on the EQ-i, indicating impaired emotional and social intelligence. Yet, no differences were observed between experimental and control groups with respect to IQ, executive functioning, perception, memory, or signs of psychopathology. Similarly, the groups did not differ with respect to demographic factors. Thus, differences with respect to emotional and social intelligence (EQ) exist in spite of the fact that the two groups were well matched on demographic and cognitive grounds. In addition, no significant correlation was noted between IQ and EQ in the clinical sample. Together, these results suggest that complex cognitive processes that subserve social competence, which appears to constitute a distinct form of intelligence, do not draw on neural processes specialized for social information.

Rather, these processes depend on known brain mechanisms related to expression of emotions and experience of feelings. Furthermore, these studies suggest that emotional and social intelligence has neural roots that may be associated with these known brain mechanisms underlying the expression of emotions and the experience of feelings. Impairment of these mechanisms may be manifest in low levels of emotional intelligence that consist of a wide array of emotional and social competencies, which can have an ill effect on one's ability to cope effectively with daily demands.

Our findings suggest only minimal overlap between emotional intelligence and cognitive intelligence. This was also confirmed by David Van Rooy (personal communication, April 2003), who suggests that no more than 5 percent of the variance of emotional intelligence can be explained by cognitive intelligence based on a meta-analysis of ten studies (n > 5000). Thus emotional intelligence and cognitive intelligence may not be strongly related, reinforcing the notion that they may be separate constructs. Not only is this assumption statistically supported by studies presented in this chapter and those reviewed by Van Rooy and colleagues (Van Rooy & Viswesvaran, in press; Van Rooy et al., submitted), but neurological evidence also suggests that neural systems governing emotional and social intelligence and those governing cognitive intelligence are most likely localized in different areas of the brain. More succinctly, the ventromedial prefrontal cortex appears to be governing basic aspects of emotionally and socially intelligent behavior (Bar-On et al., 2003; Lane & McRae, 2004), whereas the dorsolateral prefrontal cortex is thought to govern basic aspects of cognitive functioning (Duncan, 2001). These findings are based on converging results from patients with damage to the ventromedial prefrontal region who received significantly low scores on the EQ-i (Bar-On et al., 2003), as well as from functional magnetic resonance imaging and positron emission tomography scanning of nonclinical subjects that revealed activity in the same region while they were engaged in activity related to emotional awareness (Lane & McRae, 2004). On the other hand, functional neuroimaging revealed concentrated activity in the dorsalateral region of the prefrontal cortex of nonclinical subjects who were engaged in various cognitive exercises (Duncan, 2001).

In other studies, emotional intelligence was significantly related to the ability to exercise personal judgment in decision making (Bar-On et al., 2003). Specifically, on the Iowa gambling task, the experimental group made significantly more disadvantageous decisions than the control group, and their personal judgment grew worse rather than better as time went on; subjects failed to make advantageous choices and to learn from experience. This apparent link between poor personal judgment in decision making and deficiencies in emotional intelligence, in spite of average or above average levels of cognitive intelligence, may help explain why the concept of emotional and social intelligence is highly connected with human performance (Bar-On, 2000). To perform well and be successful in one's personal and professional life apparently requires the ability to make emotionally and socially intelligent decisions more than simply a high IQ.

Acknowledgment

Studies described in this chapter were supported by NIDA grants DA11779-02, DA12487-03, and DA16708, and by NINDS grant NS19632-23.

References

Adolphs, R., Damasio, H., Tranel, D., & Damasio, A. R. (1996). Cortical systems for the recognition of emotion in facial expressions. *Journal of Neuroscience, 16*, 7678–7687.

Adolphs, R., Tranel, D., Damasio, H., & Damasio, A. R. (1995). Fear and the human amygdala. *Journal of Neuroscience, 15*, 5879–5892.

Adolphs, R., Tranel, D., & Damasio, A. R. (1998). The human amygdala in social judgment. *Nature, 393*, 470–474.

Aggleton, J. P. (1992). The functional effects of amygdala lesions in humans: A comparison with findings from monkeys. In J. P. Aggleton (Ed.), *The Amygdala: Neurobiological Aspects of Emotion, Memory, and Mental Dysfunction* (pp. 485–504), New York: Wiley–Liss.

Amorapanth, P., LeDoux, J. E., & Nader, K. (2000). Different lateral amygdala outputs mediate reactions and actions elicited by a fear-arousing stimulus. *Nature Neuroscience, 3*, 74–79.

Bar-On, R. (1997a). *The Bar-On Emotional Quotient Inventory (EQ-i): A Test of Emotional Intelligence*. Toronto, Canada: Multi-Health Systems.

Bar-On, R. (1997b). *The Bar-On Emotional Quotient Inventory (EQ-i): Technical Manual*. Toronto, Canada: Multi-Health Systems.

Bar-On, R. (2000). Emotional and social intelligence: Insights from the emotional quotient inventory (EQ-i). In R. Bar-On, J. D. A. Parker (Eds.), *Handbook of Emotional Intelligence* (pp. 363–388). San Francisco: Jossey–Bass.

Bar-On, R. (2001). Emotional intelligence and self-actualization. In J. Ciarrochi, J. Forgas, & J. Mayer (Eds.), *Emotional Intelligence in Everyday Life: A Scientific Inquiry*. New York: Psychology Press.

Bar-On, R. (2004). The Bar-On emotional quotient inventory (EQ-i): Rationale, description, and summary of psychometric properties. In G. Geher (Ed.), *Measurment of Emotional Intelligence: Common Ground and Controversy* (pp. 111–142). Hauppauge, NY: Nova Science Publishers.

Bar-On, R., Tranel, D., Denburg, N., & Bechara, A. (2003). Exploring the neurological substrate of emotional and social intelligence. *Brain, 126*, 1790–1800.

Barrash, J., Tranel, D., & Anderson, S. W. (2000). Acquired personality disturbances associated with bilateral damage to the ventromedial prefrontal region. *Developmental Neuropsychology*, 18, 355–381.

Bechara, A., Damasio, H., & Damasio, A. R. (2000). Emotion, decision-making, and the orbitofrontal cortex. *Cerebral Cortex, 10*, 295–307.

Bechara, A., Damasio, H., & Damasio, A. (2003). The role of the amygdala in decision-making. In P. Shinnick-Gallagher, A. Pitkanen, A. Shekhar, & L. Cahill (Eds.), *The Amygdala in Brain Function: Basic and Clinical Approaches* (pp. 356–369). New York: Annals of the New York Academy of Science.

Bechara, A., Damasio, H., Damasio, A. R., & Lee, G. P. (1999). Different contributions of the human amygdala and ventromedial prefrontal cortex to decision-making. *Journal of Neuroscience, 19*, 5473–5481.

Bechara, A., Tranel, D., & Damasio, A. R. (2002). *The Somatic Marker Hypothesis and Decision-Making*. In F. Boller & J. Grafman (Eds.), *Handbook of Neuropsychology: Frontal Lobes*, 2nd edition (pp. 117–143). Amsterdam: Elsevier.

Bechara, A., Tranel, D., Damasio, H., Adolphs, R., Rockland, C., & Damasio, A. R. (1995). Double dissociation of conditioning and declarative knowledge relative to the amygdala and hippocampus in humans. *Science, 269*, 1115–1118.

Berthier, M., Starkstein, S., & Leiguarda, R. (1988). Asymbolia for pain—A sensory-limbic disconnection syndrome. *Annals of Neurology, 24*, 41–49.

Butter, C. M. & Synder, D. R. (1972). Alternations in aversive and aggressive behaviors following orbital frontal lesions in rhesus monkeys. *Acta of Neurobiological Experiments, 32*, 525–565.

Butter, C. M., Mishkin, M., & Mirsky, A. F. (1968). Emotional responses toward humans in monkeys with selective frontal lesions. *Physiology and Behavior, 3*, 213–215.

Butter, C. M., Mishkin, M., & Rosvold, H. E. (1963). Conditioning and extinction of a food-rewarded response after selective ablations of frontal cortex in rhesus monkeys. *Experimental Neurology, 7*, 65–75.

Craig, A. D. (2002). How do you feel? Interoception: The sense of the physiological condition of the body. *Nature Reviews Neuroscience, 3*, 655–666.

Damasio, A. R. (1994). *Descartes' Error: Emotion, Reason, and the Human Brain*. New York: Grosset/Putnam.

Damasio, A. R. (1999). *The Feeling of what Happens: Body and Emotion in the Making of Consciousness*. New York: Harcourt Brace.

Damasio, A. R. (2003). *Looking for Spinoza: Joy, Sorrow, and the Feeling Brain*. New York: Harcourt.

Damasio, A. R., Grabowski, T. G., Bechara, A., Damasio, H., Ponto, L. L. B., Parvizi, J., et al. (2000). Subcortical and cortical brain activity during the feeling of self-generated emotions. *Nature Neuroscience, 3*, 1049–1056.

Davis, M. (1992a). The role of the amygdala in fear and anxiety. *Annual Review of Neuroscience, 15*, 353–375.

Davis, M. (1992b). The role of the amygdala in conditioned fear. In J. P. Aggleton (Ed.), *The Amygdala: Neurobiological Aspects of Emotion, Memory, and Mental Dysfunction*. New York: Wiley–Liss.

Duncan, J. (2001). An adaptive coding model of neural function in prefrontal cortex. *Nature Reviews Neuroscience, 2*, 820–829.

Gaffan, D. (1992). Amygdala and the memory of reward. In J. P. Aggleton (Ed.), *The Amygdala: Neurobiological Aspects of Emotion, Memory, and Mental Dysfunction*. New York: Wiley–Liss.

Kim, J. J., Rison, R. A., & Fanselow, M. S. (1993). Effects of amygdala, hippocampus, and periaqueductal gray lesions on short and long term contextual fear. *Behavioral Neuroscience, 107*, 1093–1098.

Kim, M. & Davis, M. (1993). Lack of a temporal gradient of retrograde amnesia in rats with amygdala lesions assessed with the fear-potentiated startle paradigm. *Behavioral Neuroscience, 107*, 1088–1092.

Kluver, H. & Bucy, P. C. (1939). Preliminary analysis of functions of the temporal lobes in monkeys. *Archives of Neurological Psychiatry, 42*, 979–997.

LaBar, K. S., Gatenby, J. C., Gore, J. C., LeDoux, J. E., & Phelps, E. A. (1998). Human amygdala activation during conditioned fear acquisition and extinction: A mixed-trial fMRI study. *Neuron, 20*, 937–945.

LaBar, K. S., LeDoux, J. E., Spencer, D. D., & Phelps, E. A. (1995). Impaired fear conditioning following unilateral temporal lobectomy in humans. *Journal of Neuroscience, 15*, 6846–6855.

Lane, R. & McRae, K. (2004). Neural substrates of conscious emotional experience: A cognitive-neuroscientific perspective. In B. Amsterdam & J. Benjamins (Eds.), *Consciousness, Emotional Self-regulation and the Brain* (pp. 87–122).

Lane, R., Reiman, E., Ahern, G., Schwartz, G., & Davidson, R. (1997). Neuroanatomical correlates of happiness, sadness, and disgust. *American Journal of Psychiatry, 154*, 926–933.

LeDoux, J. (1996). *The Emotional Brain: The Mysterious Underpinnings of Emotional Life*. New York: Simon & Schuster.

LeDoux, J. E. (1993a). Emotional memory systems in the brain. *Behavioral Brain Research, 58*, 69–79.

LeDoux, J. E. (1993b). Emotional memory: In search of systems and synapses. *New York Academy of Sciences, 702*, 149–157.

Lee, G. P., Arena, J. G., Meador, K. J., Smith, J. R., Loring, D. W., & Flanigin, H. F. (1988). Changes in autonomic responsiveness following bilateral amygdalotomy in humans. *Neuropsychiatry, Neuropsychology, and Behavioral Neurology, 1,* 119–129.

Lee, G. P., Bechara, A., Adolphs, R., Arena, J., Meador, K. J., Loring, D. W., et al. (1998). Clinical and physiological effects of stereotaxic bilateral amygdalotomy for intractable aggression. *Journal of Neuropsychiatry and Clinical Neurosciences, 10,* 413–420.

Maddock, R. J. (1999). The retrosplenial cortex and emotion: New insights from functional neuroimaging of the human brain. *Trends in Neurosciences, 22,* 310–320.

Malkova, L., Gaffan, D., & Murray, E. A. (1997). Excitotoxic lesions of the amygdala fail to produce impairment in visual learning for auditory secondary reinforcement but interfere with reinforcer devaluation effects in rhesus monkeys. *Journal of Neuroscience, 17,* 6011–6020.

Mayberg, H. S., Liotti, M., Brannan, S. K., McGinnis, S., Mahurin, R. K., Jerabek, P. A., et al. (1999). Reciprocal limbic-cortical function and negative mood: Converging PET findings in depression and normal sadness. *American Journal of Psychiatry, 156,* 675–682.

Morgan, M. A. & LeDoux, J. E. (1995). Differential contribution of dorsal and ventral medial prefrontal cortex to the acquisition and extinction of conditioned fear in rats. *Behavioral Neuroscience, 109,* 681–688.

Morris, J. S., Ohman, A., & Dolan, R. J. (1999). A subcortical pathway to the right amygdala mediating "unseen" fear. *Proceedings of the National Academy of Sciences of the United States of America, 96,* 1680–1685.

Parsons, L. & Oshercon, D. (2001). New evidence for distinct right and left brain systems for deductive versus probabilistic reasoning. *Cerebral Cortex, 11,* 954–965.

Parvizi, J., Anderson, S. W., Coleman, O. M., Damasio, H., & Damasio, A. R. (2001). Pathological laughter and crying: A link to the cerebellum. *Brain, 124,* 1708–1719.

Phelps, E. A., LaBar, K. S., Anderson, A. K., O'Connor, K. J., Fulbright, R. K., & Spencer, D. D. (1998). Specifying the contributions of the human amygdala to emotional memory: A case study. *Neurocase, 4,* 527–540.

Thorndike, E. L. (1920). Intelligence and its uses. *Herper's Magazine, 140,* 227–235.

Tranel, D., Bechara, A., Damasio, H., & Damasio, A. R. (1998). Neural correlates of emotional imagery. *International Journal of Psychophysiology, 30,* 107.

Tranel, D., Bechara, A., & Denburg, N. L. (2002). Asymmetric functional roles of right and left ventromedial prefrontal cortices in social conduct, decision-making, and emotional processing. *Cortex, 38,* 589–612.

Tranel, D. & Hyman, B. T. (1990). Neuropsychological correlates of bilateral amygdala damage. *Archives of Neurology, 47,* 349–355.

Van Rooy, D. L. & Viswesvaran, C. (2004). Emotional intelligence: A meta-analytic investigation of predictive validity and nomological net. *Journal of Vocational Behavior, 65*, 71–95.

Van Rooy, D. L., Viswesvaran, C., & Pluta, P. (in press). An evaluation of construct validity: What is this thing called emotional intelligence? *Human Performance.*

Whalen, P. J. (1998). Fear, vigilance, and ambiguity: Initial neuroimaging studies of the human amygdala. *Current Directions in Psychological Science, 7*, 177–188.

Zalla, T., Koechlin, E., Pietrini, P., Basso, G., Aquino, P., Sirigu, A., et al. (2000). Differential amygdala responses to winning and losing: A functional magnetic resonance imaging study in humans. *European Journal of Neuroscience, 12*, 1764–1770.

Zola-Morgan, S., Squire, L. R., Alverez-Royo, P., & Clower, R. P. (1991). Independence of memory functions and emotional behavior: Separate contributions of the hippocampal formation and the amygdala. *Hippocampus, 1*, 207–220.

3 Neural Substrates of Self-Awareness

Debra A. Gusnard

The capacity for self-awareness, or reflexivity, as it was dubbed by William James (1890), may be conceptualized as referring to the apparently unique ability of the self to take itself as the object of its own view. Thus, it inheres aspects of both process (as the knower or active subject engaged in experience) and content (that which is known or is the object of experience). Such reflexivity involves being able to represent one's own states (motoric, mental, emotional, perceptual, visceral), *as one's own states*. This capacity for self-awareness is a fundamental characteristic of human beings whose workings have long fascinated and challenged investigators from a variety of disciplines. Yet, until recently, very little could be said about its neural substrates.

Cognitive neuroscience has emerged as a field with potential for providing useful information in this regard. Historically, however, it has tended to deal with the subpersonal domain, addressing phenomena such as sensory and motor systems, and functional domains and processes such as language, memory, and attention, using various information-processing models. Until recently, little emphasis was placed on or consideration given to the first-person perspective or motivation and interests of individuals. The subpersonal agenda has been successfully furthered by functional imaging methodology, which has traditionally been grounded in the transformation of individual subjects' data into a common atlas space and then averaged. Relatively recently, however, certain technical advances in imaging methodology (stronger magnets and other means for improving signal-to-noise from more limited amounts of imaging data) have begun to promote another investigative movement, which involves consideration of individual differences in the context of various imaging paradigms. At the same time, growing numbers of investigators have begun to use functional imaging techniques to explore the neural correlates of various aspects of social cognition, in which not only first-person but also second-person dynamics clearly play an important role. Such trends are providing impetus to a relatively new questioning of what functional imaging and other neuroscientific methodologies might contribute to our understanding of issues in these personal and interpersonal domains.

Some ethologists are proponents of the possibility that other species, particularly some nonhuman primates, may share some capacities for aspects of self-awareness (Povinelli & Cant, 1995; Premack & Premack, 1983), but the degree of complexity of human self-awareness is unarguably species specific. Such cross-species comparisons, however, not only illustrate evolutionary considerations for the development of an increasingly multidimensional capacity for self-awareness, they help suggest what the functionality of such a capacity is likely to be. It has been proposed that the capacity for consciousness more generally, for example, facilitates the planning and rehearsal of intended behavior (Edelman & Tononi, 2000), with salient operations thus being able to be performed within a "global workspace" (Dehaene & Naccache, 2001). Such planning and rehearsal depend on the ability to make perceptual discriminations on which additional operations can be performed.

Of significance, investigators from diverse research traditions independently converged on the notion that self-awareness, too, fundamentally involves processes of discrimination and evaluation in the service of guiding and optimizing certain behaviors. For example, self-awareness at the level of consciousness of a body image has been attributed to arboreal higher primates, and is proposed to have arisen to enhance the planning and reduce the risk of clambering through the trees of such animals, who have a significantly larger body size than smaller monkeys and nonprimates (Povinelli & Cant, 1995) At the level of humans, the capacity for self-awareness has clearly extended well-beyond a role for helping to ensure bodily integrity in maneuvering through a challenging physical environment. Various modern philosophers grappling with the subject of human self-awareness (e.g., (Roessler & Eilan, 2003)) have emphasized relationships between self-awareness and the subject's status as a discriminating decision-making agent, having an embodied and thus specific first-person perspective and capacity for control and (in the normal awake state) an experience of ownership of his intentional action. Work in human developmental psychology also underscores the fact that increasingly complex capacities for self-awareness are achieved at later developmental stages that parallel increasing competences in perceptual discrimination, categorization, and self-regulation (Harris, 1998).

In the social psychological literature, in particular (Carver, 2003), some debate has addressed specific effects that self-awareness, or the processing of self-representational content, has on subsequent behavior and experience. It is generally viewed, however, that self-awareness, or self-focused attention as it is sometimes called, is typically associated with an evaluation of the agent's behavior, whether that be relative to a personal standard or one that the agent perceives as socially derived.

Obviously, such self-focused attention and scrutiny do not always dominate the flow of subjective experience, as they clearly do not have to. Much of our behavior, particularly automatic or well-learned behavior, is in fact more effectively executed without our direct conscious supervision and evaluation of our ongoing performance. However, given our nature as highly developed social creatures with a significant amount of our behavior involving interpersonal communication and depending on social learning, circumstances in which we are induced to become more self-aware are clearly ubiquitous. Notably, these circumstances include not only novel ones in which we must effect behaviors relatively naively, but also those in which we perceive ourselves as an object of social attention. It is in such circumstances, as well as in the anticipation of such circumstances, that self-awareness is likely to come to the fore and self-scrutiny may be particularly acute.

Discrimination and evaluation are thus important aspects of self-awareness processes and may be related to dynamic interactions in the thalamocortical system involving reentrant connection of new perceptual categorizations with previously established categorizations in memory systems. Stabilization in an act of awareness has been proposed to require hundreds of milliseconds to seconds (Edelman, 1989), a process occurring along with the continual updating of these memory systems. This processing lag, however slight, which involves a bootstrapping of memory and continuing perception, thus inheres an element of assessment. In the case of self-awareness, such assessment or evaluation may be fleeting or it may be elaborated and associated with a pronounced affective response as the agent regards some aspect of himself or his behavior as being a particularly bad or good fit relative to some standard or goal of his own or one that he views as socially derived. In other words, he may or may not be provoked to become particularly self-conscious.

Recent functional imaging data are relevant to consideration of these notions of our highly evolved and socially refined capacity for self-awareness. Notably, this work illustrates not only potential brain regions in which relevant representational content may typically undergo computation, but it also emphasizes dynamic considerations of self-awareness. In particular, it provides data suggesting neural mechanisms that potentially account for the fact that self-awareness does not continually dominate the flow of subjective experience but rather is a property of the foreground of our experience at some moments to a greater degree than others. It also emphasizes some of the complexities of the coordination of brain network processes that appear likely to be involved in such phenomena.

Functional Imaging and Self-Awareness

Functional Imaging Signals

Before describing the imaging findings specifically relevant to considerations of self-awareness, it will be useful to review some basic features of the functional imaging signal. Understanding such features lays the groundwork for understanding functional imaging activations and deactivations as well as ongoing baseline activity that are, in turn, directly relevant to imaging considerations of self-awareness.

The imaging signal fundamentally involves a change in local blood flow within the brain (Raichle, 1998). This is accompanied by an increase in the metabolism of glucose that parallels in magnitude and spatial extent the change in blood flow (see Hand & Greenberg in Woolsey et al., 1996; Ueki, Linn, & Hossmann, 1988). The change in the local consumption of oxygen is much smaller in magnitude than the increase in either blood flow or glucose utilization (Fox & Raichle, 1986; Fox et al., 1988), which results in an increase in the local blood oxygen content that forms the basis for the blood oxygen level dependent (BOLD) signal of functional magnetic resonance imaging (fMRI) (Kwong et al., 1992; Ogawa et al., 1990). In addition, a spatially localized increase in the fMRI signal directly and monotonically reflects an increase in neural activity (Logothetis et al., 2001). The time course of this signal, which is based on the vascular response to the neural activity, is a temporally delayed and dispersed reflection of underlying neural activity, beginning about two seconds after the latter's onset and reaching a plateau in about seven seconds (see also Boynton et al., 1996).

Historically, initial interest in and interpretation of functional brain imaging studies focused on correlates of imaging signal increases (activations). However, researchers are now increasingly aware of the fact that signal decreases (deactivations) also occur. Although focal decreases in imaging signal intensity have been noted for some time, their interpretation has been challenging and has resulted in different opinions as to their origin and significance (Tootell et al., 1998; Smith, Singh, & Greenlee, 2000; Shmuel et al., 2002). Of interest, some of these focal decreases appear to occur exclusively between or within perceptual systems, such that decreases in activity may appear in auditory or somatosensory cortices when engaging in a task involving visual perception, for example (Haxby et al., 1994; Kawashima, O'Sullivan, & Roland, 1995; Ghatan et al., 1998), or within portions of the somatosensory or visual systems while other portions of these systems exhibit increases in activity during tasks that engage them (Drevets et al., 1995; Shmuel et al. 2002, 2001; Tootell, Tsao, & Vanduffel, 2003). Such decreases have been posited to reflect the suppression or gating of information processing in areas not engaged in task performance (Drevets et al., 1995), facilitating

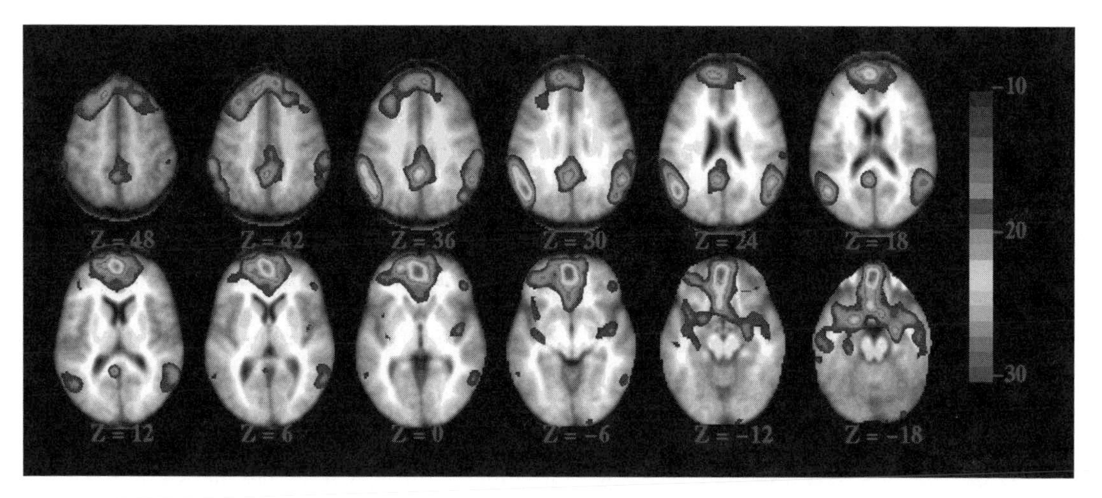

Figure 3.1
Transverse images of positron emission tomographic data averaged across nine cognitive tasks (n = 134) show the pattern of the common (impersonal task-induced) decreases. See plate 2 for color version.

the processing of information expected to carry behavioral significance by filtering out unattended sensory input.

In addition to such signal decreases within sensory cortices, however, a special set of areas in association cortices have been observed to exhibit decreases in activity in averaged data obtained relative to a passive resting state condition across a wide variety of goal-directed tasks (figure 3.1, plate 2). This set of areas is of particular note in the context of this discussion, and its significance is detailed below. What are important to note here, however, are physiological characteristics of imaging signal decreases.

In particular, such signal decreases are not at all likely to be directly attributable to local inhibitory processes. This is because the latter are known to require energy and appear as likely to be associated with signal increases as excitatory processes are (Ackerman et al., 1984; Lauritzen, 2001; Lauritzen & Gold, 2003), although details of the neurovascular coupling involved in inhibitory processes are less well understood. Also, and most important, recent neurophysiological observations (Gold & Lauritzen, 2002; Shmuel et al., 2003a, b) indicate that these signal decreases are the result of actual decreases in neuronal activity, particularly in local field potentials, which reflect input to and local processing within neuronal ensembles.

Thus, imaging signal changes (and underlying neural activity) may consist of both increases and decreases, as the information-processing landscape is organized and reorganized to address particular task demands.

The Physiological Baseline

In the average adult human, the brain represents about 2 percent of total body weight, but accounts for about 20 percent of oxygen consumed and, thus, calories (energy) consumed by the body (Clark & Sokoloff, 1999). In relation to this very high rate of baseline metabolism, regional imaging signals are remarkably small, in metabolic terms often less than 5 percent of the ongoing metabolic activity of the brain in that particular area. These are modest modulations in ongoing or baseline activity and do not appreciably affect the overall metabolic rate of the brain (Sokoloff et al., 1955; Fox, Burton, & Raichle, 1987; Fox et al., 1985, 1987).[1]

Of importance, not only does this physiological baseline account for the majority of the brain's metabolic requirements, but evidence suggests that it may inhere functionally significant signaling processes. Several lines of evidence strongly suggest this. These include measurements of brain energy metabolism using magnetic resonance spectroscopy (Sibson et al., 1998; Shulman, Hyder, & Rothman, 2001; Hyder, Rothman, & Shulman, 2002) that indicate that up to 75 percent to 80 percent of the entire energy consumption of the brain is devoted to glutamate cycling and hence, signaling processes. There are also neurophysiological studies that have noted spontaneous electrical activity that does not bear an obvious relationship to specific sensory or motor tasks (Arieli et al., 1996; McCormick, 1999; Tsodyks et al., 1999; Sanchez-Vives & McCormick, 2000; Shu, Hasenstaub, & McCormick, 2003), and is thought not simply to represent noise (see Ferster, 1996; McCormick, 1999). This spontaneous spiking activity is relatively low in frequency compared with vigorous bursts of activity seen during task performance, which may be related to offering different forms of representation or neural coding of information and differential economies of energy use as well (Attwell & Laughlin, 2001). Such considerations raise the possibility that in the physiologic reservoir of baseline activity, where oxidative metabolism is the primary source of energy, a very efficient strategy has developed to manage large amounts of information on a sustained or long-term basis. This would complement the situation for activations, where glycolysis assumes an important role as a source of energy (Fox et al., 1988), which may have the potential for supporting more specific, rapidly varying, and time-limited information processing required for immediate purposes.

How might one think about this ongoing and so-called baseline neuronal activity? One view suggests that it may serve a preparatory or facilitatory processing role (Salinas & Sejnowski, 2001). From this perspective it might be argued that such a process is a necessary and expensive component of brain function but it is not associated with information processing with any inherent functionality. However, others have proposed that information processing may, in fact, be a property of such a state (Tononi & Edelman, 1998; Shu et al., 2003), and here functional brain imaging with both PET and fMRI provide a potentially unique perspective. This emanates directly from the observation of the previously mentioned imaging signal decreases and, in particular, those that occur in a specific set of association cortices during the performance of a wide variety of goal-directed tasks.

The set of brain areas exhibiting these decreases consists, in large part, of the medial prefrontal and parietal cortices as well as lateral parietal cortices bilaterally (Shulman et al., 1997; Mazoyer et al., 2001; figure 3.1). Current evidence indicates that these specific imaging decreases do not correspond to activations in the resting state as has been suggested (Mazoyer et al., 2001), but do, in fact, arise from the physiologic baseline (for a more detailed defense of this assertion, see Raichle et al., 2001). This suggests that they might more appropriately be referred to as areas that, in the resting state, are active rather than activated. This would be consistent with functions that are spontaneous and virtually continuous, being attenuated only when subjects engage in certain goal-directed behaviors.

Two observations using novel approaches to the study of the resting state (awake, relaxed with eyes closed, unconstrained cognition) provide information on how activity in these regions may relate to resting state cognition.

Using an imaging strategy that has been employed in several laboratories (Biswal et al., 1995; Coren, 1969; Lowe, Mock, & Sorenson, 1998; Xiong et al., 1999), investigators (Greicius et al., 2003) explored the interregional temporal correlations of spontaneous BOLD signal fluctuations in the resting state in fourteen subjects using regions-of-interest in medial parietal cortex as well as ventral anterior cingulate cortex, regions that are among those regularly exhibiting the commonly observed decreases in association cortices. What emerged was evidence for significant correlations in the spontaneous activity among a group of areas virtually identical with those that have been identified with these common decreases (figure 3.2, plate 2). While up to this point the evidence for a network of interrelated areas exhibiting coordinated activity in the resting state had been indirect, being based on repeated demonstrations of the common decreases (areas attenuated in their activity relative to the resting state condition), this study was the first to provide direct evidence of coordinated activity in the same network of areas *in* the resting state.

The second study (Laufs et al., 2003) used an approach to resting state fMRI data employing an analysis driven by variations in simultaneously acquired electroencephalographic (EEG) frequency bands. Changes in power within these bands were correlated with magnitude variations in regional BOLD within the brain in normal

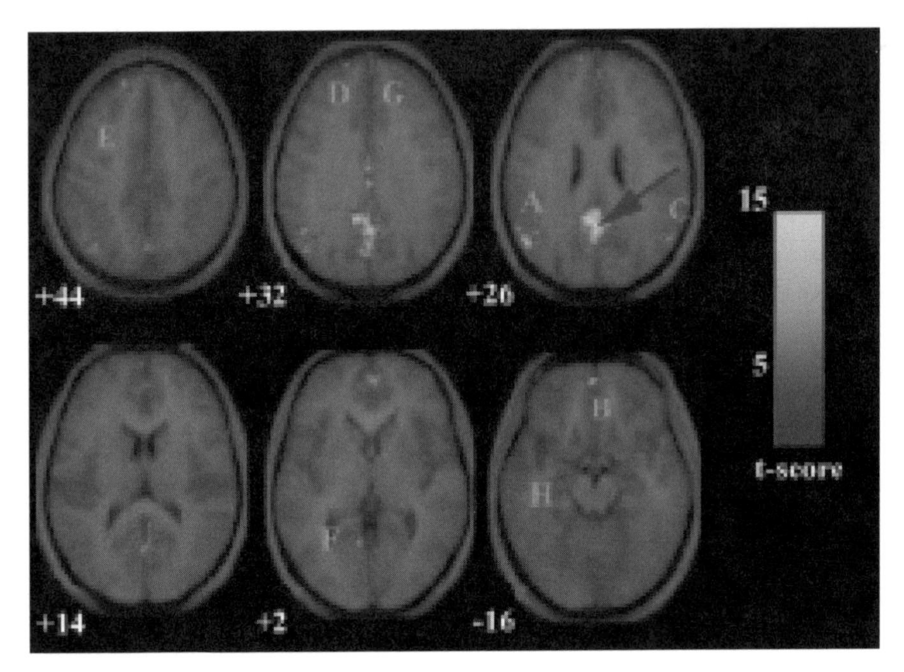

Figure 3.2
Map of resting-state neural connectivity for the posterior cingulate cortex (PCC) (blue arrow indicates the PCC peak used for the analysis). Labels A → H designate the significant clusters in order of descending *t* score (Greicius et al., 2003). Note correspondence of this map with the pattern of the common decreases (figure 3.1). See plate 3 for color version.

subjects engaged for twenty minutes in the eyes-closed resting state. Areas in dorsolateral parietal and prefrontal cortices (areas thought to be important in the support of directed visual attention) varied inversely with alpha power (8–12 Hz), whereas a significant number of areas associated with the common decreases varied directly with power in a beta band (17–23 Hz), an EEG band thought to be associated with cognitive processing. The authors thus concluded that the resting state appears to be characterized by temporal fluctuations in functional activity ranging from inattention to spontaneous cognition, and that these functional states are associated with changes in activity in the noted brain systems.

Self-Awareness

Data that support the view that circumstances exist in which this set of brain regions may be regarded as constituting a coordinated network suggest the need for a functional description of this network that subsumes those that have been offered for individual regions when they are considered separately (which can occur when such individual regions are dissociated in their activation in the context of specific imaging paradigms). Accumulating evidence suggests, for example, that processes in these various areas may be regarded as participating significantly in enabling various aspects of self-awareness. Such data come from both positron emission tomography (PET) and block-design fMRI studies.

Several PET studies, for example, have contrasted particular altered states of awareness with the normal awake state. Levels of awareness or consciousness have been investigated by varying levels of anesthesia and, in one study, by means of the general anesthetic propofol consciousness in five volunteers was varied from the completely awake alert state to unconsciousness (Fiset et al., 1999). In addition to global depression of cerebral blood flow that is well known to occur with anesthesia, regional decreases in blood flow were observed that, in the cortex, were most prominent in medial parietal and ventral prefrontal areas, areas among the set of common decreases.

Another group used PET to assess brain metabolism in patients during and after recovery from a vegetative state, which is defined as impaired awareness in the setting of normal arousal (Laureys et al., 1999). Areas that resumed normal metabolism after patients' recovery of consciousness largely consisted of medial parietal and lateral parietal areas, which again significantly overlap some of those engaged by the common decreases. This group also performed functional connectivity analyses on imaging data from such patients and reported an altered correlation in activity between these medial parietal cortices and dorsomedial and lateral prefrontal cortices compared with normal controls, which returned to normal after patients' recovery (Laureys et al., 2000). In PET studies of normal adult subjects during various stages of sleep, medial parietal and medial prefrontal cortices together with portions of dorsolateral prefrontal cortex exhibited attenuated activity relative to the awake state (Maquet, 2000). In addition, the areas involved in the set of common decreases predominate in the reduced cortical activity exhibited during spike and wave discharges in patients with idiopathic generalized epilepsy (Aghakhani et al., 2004; personal communication, J. Gotman, Montreal Neurological Institute).

In addition to such studies, which suggest an important relationship between activity in these brain regions and alterations in consciousness or levels of awareness, significant amounts of PET and fMRI data have also emerged from studies addressing what might be termed various self-processes, which also implicate these same brain regions. Such studies usually involve the presentation of visual or auditory stimuli (typically words or pictures) and require subjects to perform some task. These tasks range from asking subjects to make explicit evaluative judgments about themselves, for example, to having them explicitly take an interpersonal or spatial perspective, or endorse the experience of themselves versus another as being the cause of an action. In all of the cited studies, subjects made explicit reference (typically by a graded or yes or no categorical judgment) to some aspect of their own mental or somatic states, to some aspect of themselves (or their construal of some aspect of another agent) in a relational context, or to some conceptual knowledge of themselves (or another agent). Neural correlates of processes so targeted may be regarded in that sense as subserving some process of explicit self-focus or self-awareness (and, in the case of "other" person assessments, some computational transform fundamentally based on some aspect of self-awareness).

Mental state considerations, for example, whether explicitly reflecting on aspects of one's own mental state, with or without an affective component, or attributing mental states to others, have generally engaged the dorsal anterior paracingulate and dorsomedial prefrontal regions (Frith & Frith, 2003; Gusnard et al., 2001; Lane et al., 1997; Shallice, 2001). Other imaging studies targeting retrieval of personal or episodic memories involving verbal and nonverbal material (Cabeza & Nyberg, 2000; Cabeza et al., 2004) have often tended to include this same prefrontal region.

Such episodic memory processes as well as explicit spatial perspective taking have also been shown to engage portions of the medial parietal region comprising posterior cingulate, precuneus, and retrosplenial cortex (Maguire, 2001; Vogeley & Fink, 2003). This region has also been implicated in numerous tasks involving orienting oneself in large-scale space (Maguire, 2001). In a summary of work considering a variety of empirical data, including functional imaging data (Vogeley & Fink, 2003), it was posited that tasks involving taking a first-person perspective in space, in action as well as social interaction, tend to implicate these medial prefrontal and parietal regions, along with a lateral parietal region, particularly on the right side.

A long history of neuropsychological lesion data indicates that this right lateral parietal region plays an important role in attending to aspects of one's personal space (Behrmann, 1999; Farrell & Robertson, 2000) as well as one's extensions (e.g., by means of tool use) into that space (Ackroyd et al., 2002; Maravita, Spence, & Driver, 2003). In the functional imaging literature, this region has also been implicated in attention to the experience of movement and in effecting attributions of agency (Farrer et al., 2003; Farrer & Frith, 2002; Jackson & Decety, 2004).

Accessing semantic information or making semantic judgments about persons, including oneself, have also been shown to engage portions of the medial prefrontal region (Craik et al., 1999; Kelley et al., 2002; Kjaer, Nowak, & Lou, 2002; Mitchell,

Heatherton, & Macrae, 2002) as well as the medial parietal (Kelley et al., 2002; Kjaer et al., 2002) and lateral parietal regions in both hemispheres (Lou et al., 2004; Mitchell et al., 2002). Of interest, a study involving PET and transcranial magnetic stimulation (TMS) reported selective disruption of accessing semantic knowledge about oneself when TMS was applied to the medial parietal area (Lou et al., 2004).

Those portions of the medial prefrontal cortex implicated in the cited studies (which involve explicit reference to some personal standard or reference frame) largely consist of areas that are located either relatively centrally at the level of the genu of the corpus callosum or more dorsally. By contrast, to date, little evidence suggests that portions of the medial prefrontal and adjacent cingulate cortex located more ventrally are activated by such explicit assessments (at least when such task conditions are compared with a low-level control condition such as visual fixation). Rather, these ventral areas are generally modulated by tasks that elicit a measurable emotional or arousal response on the part of the subject (Bush, Luu, & Posner, 2000; Nagai et al., 2004; Simpson et al., 2001a, b), so that the degree to which a behavior or task may be perceived as (potentially) challenging, rewarding, or threatening by that individual, for example, appears to have an influence on the degree of activity (typically ranging from near baseline to some degree of deactivation) in these areas. Thus, while this ventral medial prefrontal region is among those involved in the set of common decreases, its functionality appears to have less to do with explicit aspects of self-awareness and more with implicit modulations of the workings of the first-person perspective or core self (Damasio, 1999).

One area that has sometimes been reported in this context is not among the common set of decreases. This region, the insula, has been observed to be engaged in studies targeting interoceptive awareness (Critchley et al., 2004) as well as attributions of agency (Farrer et al., 2003; Farrer & Frith, 2002). This apparent discrepancy may be explained by some of our own data (unpublished), based on having subjects go from an eyes-closed resting condition to simply opening their eyes and fixating on a crosshair. With this maneuver, typical activations were seen in the visual cortex as well as in parietal and frontal areas involved in visual attention. At the same time, however, there was evidence of several of the common decreases, in this case involving medial prefrontal and parietal regions. Of interest, in this setting, decreases were also observed in the insula. It is thus possible that such decreases in the insula were observed in this study and are not observed more routinely in other studies showing common decreases because they are already subtracted away in cases where a visual fixation condition is used as the lowest level control condition; and a resting state condition is not included as it was in our study.

One could argue that there are functional imaging studies that have yielded results engaging some of these same areas and have attributed other information-processing functions to them, using terminology such as emotion-processing, which makes no mention of the self or self-reference. Some have even suggested that current evidence does not point to a unitary common system supporting the subjective experience of a unified self (Gillihan & Farah, 2005). The position taken here, however, is that the presence of such a system is suggested if one's conceptual view takes into account imaging data that spans consideration of altered states of awareness as well as dynamics (deactivation as well as activation) of information-processing functionality of personal salience to the individual subject.

The argument, in sum, is that such data suggest that there are a set of brain regions involved in supporting an internal self-model whose activity is significant in influencing whether awareness of aspects of the self (the subject/agent) are more or less within the purview of attention. When such information is salient, some of these areas exhibit greater activation. But when such information is behaviorally less salient and could interfere with the performance of goal-directed tasks, their activity is attenuated. Notably, the magnitudes of some of these regions' decreases in activity have been observed to vary with such things as task difficulty (McKiernan et al., 2003) and the subject's emotional state (Simpson et al., 2001b).

Such a theoretical model would be commensurate with the experimental social psychological view that there are manipulations by which we may be made to become more or less aware of aspects of our selves, whether these be our own mental states, emotions, percepts, or other visceral experiences, in a particular behavior context. The diversity of the brain regions that appear to be involved, as well as the moderating effects of such factors as perceived task difficulty and engendered emotional response, also suggest possible sources of differences in process and reactivity that individuals display in the context of experimental manipulations of self-awareness.

We are, in fact, beginning to employ such a model in investigating effects of manipulations of objective self-awareness, as suggested by Duval and Wicklund (1972) and other experimental social psychologists, in a functional imaging setting.

In a pilot study, normal undergraduates underwent fMRI scanning while performing a novel and moderately difficult lexical decision task (alternating with visual fixation) in a mixed-blocked and event-related design. The self-awareness manipulation consisted of informing the subjects that they were being observed and exposing them to a camera during task performance at some times and not others, the timing of which varied between conditions and across runs. Effect of the camera manipulation was assessed by subjects' self-report (verbal thought listing) (Cacioppo, Von Hippel, &

Ernst, 1997) at the end of each run, as well as by reaction times during performance of the task.

The camera manipulation was successful in enhancing subjects' self-consciousness and promoting a more discriminating and evaluative stance with regard to their own appearance and/or behavior. Subjects' self-reports included content such as "wondering how I look," "feel like people are watching me," and "feel like I'm being evaluated" when referring to their experiences while performing the task with the camera on, which was not the case with the camera off. On average, reaction time while performing the lexical task was somewhat longer when the camera was on compared with when it was not. Subjects differed in the extent to which they expressed self-consciousness and were affected in their performance, however. Also, of particular interest were the preliminary imaging findings in those expressing the most self-consciousness (figure 3.3, plate 4), where significant activations included portions of medial prefrontal and medial parietal regions. This is consistent with the hypothesis

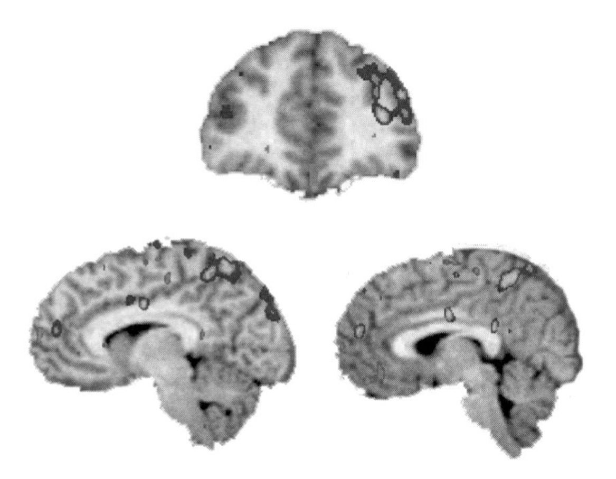

Figure 3.3
In the single subject whose image data is displayed here (contrasting the condition of task with camera on to that with camera off), one observes increased activity in portions of medial prefrontal and medial parietal (precuneus) cortices, consistent with enhanced attention to self-evaluative processes in this behavioral context. Activations are also seen in the dorsal anterior cingulate, which has been implicated in the affective processing of physical pain (Rainville et al., 1997) and social pain (Eisenberger, Lieberman, & Williams, 2003), as well as in the left dorsolateral prefrontal cortex (typically associated with aspects of working memory) and left ventral prefrontal cortex, both of which are implicated in control strategies, including those involved in emotional self-regulation (Ochsner et al., 2002), which are presumably engaged in this context of becoming more self-conscious, as the subject tries to persist in successful task performance. See plate 4 for color version.

that these areas may be enhanced in their activity as individuals are motivated to recruit information-processing resources in the service of attention to more proximal aspects (personal details and concerns) of their performance.

Conclusion

I believe that we now have data that are beginning to provide clues to an understanding of brain regions and neural dynamics significantly involved in enabling self-awareness. The implicated brain regions not only exhibit a coordinated functionality and apparently spontaneous, although fluctuating, activity in the passive resting state that appears likely to be related to ongoing spontaneous cognition or stream of consciousness, but they also show attenuated activity when subjects perform impersonal tasks (having either little inherent self-relevance or requiring little attention to details of one's performance). When personally relevant phenomena (interoceptive sense-data, one's orientation in space, one's mental states, episodic memories, personal semantic knowledge) are explicitly targeted in a functional imaging experiment, however, such targeting typically results in a dissociation of the activity in these brain regions from each other, and either little or no deactivation or actual activation of one or more of the regions in this set. Activity in some of these brain regions has also been correlated with phenomenology of conscious states of awareness, although the specific means by which they might be involved in supporting such states remains unclear. I also contend that many of these data, particularly those regarding the common decreases and their regions' coordinated fluctuations in activity in the resting state, underscore the dynamic character of self-awareness—awareness of ourselves (or, rather, a representation of some aspect of our so-called selves)—often only fleetingly in the cognitive foreground, depending on the attentional resources that we are motivated to devote to aspects of our selves in any behavioral circumstance.

Social psychologists have noted that, under certain circumstances, we may be encouraged or cued to devote more attentional resources to aspects of our selves and that this may be associated with destabilization of our performance as we become preoccupied with excessive execution focus or if we discern discrepancies between our current behavior and desired goal states. Clearly, however, competition between motivations to attend to more self-related versus more goal-related features of task performance may, at other times, be adjudicated smoothly and yield the ready achievement of intended results.

Many details regarding the means by which relevant brain regions are so engaged and temporally coordinated, as well as factors accounting for individual differences in sensitivity to cues or dispositions likely to be associated with greater or lesser degrees of self-awareness in any such circumstances, remain to be elucidated. Such details and the accompanying increase in our self-knowledge may permit a more mature understanding of the fundamentally distributed and dynamic nature of our selves, as well as offer clues to means by which we may more flexibly and positively confront the adaptive challenges of many demanding personal and social circumstances.

Note

1. For purposes of clarity, it is useful to define more precisely what we mean by a physiological baseline and how it is to be distinguished from activations and deactivations. It should be noted that this definition has arisen in the context of a functional imaging framework, but that converging evidence using other methodologies suggests that its conceptual foundations rest on aspects of brain function of broader relevance.

Fundamental to this notion of a physiological baseline is the observation that, in normal adult humans lying quietly in a PET scanner awake and alert with eyes closed (Raichle et al., 2001), there is a close match between local blood flow and oxygen utilization averaged over time, usually over about 30 minutes (Mintun et al., 1984), a relationship that is often expressed in terms of the ratio of oxygen consumption to oxygen delivery. This ratio (~0.40 in the adult human) has come to be known as the oxygen extraction fraction (OEF) and is characterized by its spatial uniformity across the brain despite marked regional differences in both blood flow and oxygen consumption between gray matter and white matter, as well as among gray matter areas.

In contrast, activations are characterized by a local fall in the average resting OEF, which results from a local increase in blood flow that is not accompanied by a commensurate increase in oxygen consumption (Fox & Raichle, 1986; Fox et al., 1988), leading to increased local blood oxygen content and the fMRI BOLD signal (Ogawa et al., 1990).

It follows that a physiological baseline can be defined as the average OEF obtained across individuals or within an individual across time while such individuals rest quietly awake and alert with eyes closed. It is thus constituted by an absence of activation in terms of an average OEF. (For more details regarding derivation of the concept of the physiologic baseline, see Raichle et al., 2001).

The uniformity of the OEF in averaged data obtained at rest in the normal brain suggests that equilibrium exists between local metabolic requirements necessary to sustain a continuing level of neural activity and blood flow in that region. We have suggested that this equilibrium state identifies a baseline level of neuronal activity (Gusnard & Raichle, 2001; Raichle et al., 2001). Consequently, areas with a reduction in this equilibrium OEF are regarded as activated (neural activity is increased above the baseline level), while those areas not different from the brain mean OEF are considered to be at baseline. In this scheme, increases in the OEF (decreases in the BOLD signal) define areas of deactivation (where neural activity is decreased below the baseline level). Of importance, obvious decreases in OEF from the brain mean, which would indicate areas of activation, are not present in averaged data from subjects resting quietly awake and alert with eyes closed (Raichle, 1998).

References

Ackerman, R. F., Finch, D. M., Babb, T. L., & Engel, J. J. (1984). Increased glucose utilization during long-duration recurrent inhibition of hippocampal pyramidal cells. *Journal of Neuroscience, 4,* 251–264.

Ackroyd, K., Riddoch, M. J., Humphreys, G. W., Nightingale, S., & Townsend, S. (2002). Widening the sphere of influence: Using a tool to extend extrapersonal visual space. *Neurocase, 8,* 1–12.

Aghakhani, Y., Bagshaw, A. P., Benar, C. G., Hawco, C., Andermann, F., Dubeau, F., & Gotman, J. (2004). fMRI activation during spike and wave discharges in idiopathic generalized epilepsy. *Brain, 127,* 1127–1144.

Arieli, A., Sterkin, A., Grinvald, A., & Aertsent, A. (1996). Dynamics of ongoing activity: Explanation of the large variability in evoked cortical responses. *Science, 273,* 1868–1871.

Attwell, D. & Laughlin, S. B. (2001). An energy budget for signaling in the grey matter of the brain. *Journal of Cerebral Blood Flow and Metabolism, 21,* 1133–1145.

Behrmann, M. (1999). Spatial reference frames and hemispatial neglect. In M. Gazzaniga (Ed.), *The New Cognitive Neurosciences*. Cambridge: MIT Press.

Biswal, B., Yetkin, F., Haughton, V., & Hyde, J. (1995). Functional connectivity in the motor cortex of resting human brain using echo-planar MRI. *Magnetic Resonance in Medicine, 34,* 537–541.

Boynton, G. M., Engel, S. A., Glover, G. H., & Heeger, D. J. (1996). Linear systems analysis of functional magnetic resonance imaging in human V1. *Journal of Neuroscience, 16,* 4207–4221.

Bush, G., Luu, P., & Posner, M. I. (2000). Cognitive and emotional influences in anterior cingulate cortex. *Trends in Cognitive Sciences, 4,* 215–222.

Cabeza, R. & Nyberg, L. (2000). Imaging cognition. II. An empirical review of 275 PET and fMRI studies. *Journal of Cognitive Neuroscience, 12,* 1–47.

Cabeza, R., Prince, S. E., Daselaar, S. M., Greenberg, D. L., Budde, M., Dolcos, F., et al. (2004). Brain activity during episodic retrieval of autobiographical and laboratory events: An fMRI study using a novel photo paradigm. *Journal of Cognitive Neuroscience, 16,* 1583–1594.

Cacioppo, J. T., Von Hippel, W., & Ernst, J. M. (1997). Mapping cognitive structures and processes through verbal content: The thought-listing technique. *Journal of Consulting and Clinical Psychology, 65,* 928–940.

Carver, C. S. (2003). Self-awareness. In J. P. Tangney (Ed.), *Handbook of Self and Identity* (pp. 176–196). New York: Guilford Press.

Clark, D. D. & Sokoloff, L. (1999). Circulation and energy metabolism of the brain. In G. J. Siegel, B. W. Agranoff, R. W. Albers, S. K. Fisher, & M. D. Uhler (Eds.), *Basic Neurochemistry. Molecular, Cellular and Medical Aspects,* 6th ed. (pp. 637–670). Philadelphia: Lippincott–Raven.

Coren, S. (1969). Brightness contrast as a function of figure-ground relations. *Journal of Experimental Psychology, 80,* 517–524.

Craik, F. I. M., Moroz, T. M., Moscovitch, M., Stuss, D. T., Winocur, G., Tulving, E., et al. (1999). In search of the self: A positron emission tomography study. *Psychological Science, 10,* 26–34.

Critchley, H. D., Wiens, S., Rotshtein, P., Ohman, A., & Dolan, R. J. (2004). Neural systems supporting interoceptive awareness. *Nature Neuroscience, 7,* 189–195.

Damasio, A. R. (1999). *The Feeling of what Happens: Body and Emotion in the Making of Consciousness.* New York: Harcourt Brace.

Dehaene, S. & Naccache, L. (2001). Towards a cognitive neuroscience of consciousness: Basic evidence and a workspace framework. *Cognition, 79,* 1–37.

Drevets, W. C., Burton, H., Videen, T. O., Snyder, A. Z., Simpson, J. R., Jr., & Raichle, M. E. (1995). Blood flow changes in human somatosensory cortex during anticipated stimulation. *Nature, 373*(6511), 249–252.

Duval, S. & Wicklund, R. A. (1972). *A Theory of Objective Self-Awareness.* New York: Academic Press.

Edelman, G. M. (1989). *The Remembered Present: A Biological Theory of Consciousness.* New York: Basic Books.

Edelman, G. M. & Tononi, G. (2000). *A Universe of Consciousness: How Matter Becomes Imagination.* New York: Basic Books.

Eisenberger, N. I., Lieberman, M. D., & Williams, K. D. (2003). Does rejection hurt? An fMRI study of social exclusion. *Science, 302,* 290–292.

Farrell, M. J. & Robertson, I. H. (2000). The automatic updating of egocentric spatial relationships and its impairment due to right posterior cortical lesions. *Neuropsychologia, 38,* 585–595.

Farrer, C., Franck, N., Georgieff, N., Frith, C. D., Decety, J., & Jeannerod, M. (2003). Modulating the experience of agency: A positron emission tomography study. *Neuroimage, 18,* 324–333.

Farrer, C. & Frith, C. D. (2002). Experiencing oneself vs another person as being the cause of an action: The neural correlates of the experience of agency. *Neuroimage, 15,* 596–603.

Ferster, D. (1996). Is neural noise just a nuisance? *Science, 273,* 1812.

Fiset, P., Paus, T., Daloze, T., Plourde, G., Meuret, P., Bonhomme, V., et al. (1999). Brain mechanisms of propofol-induced loss of consciousness in humans: A positron emission tomographic study. *Journal of Neuroscience, 19,* 5506–5513.

Fox, P. T., Burton, H., & Raichle, M. E. (1987). Mapping human somatosensory cortex with positron emission tomography. *Journal of Neurosurgery, 67,* 34–43.

Fox, P. T., Fox, J. M., Raichle, M. E., & Burde, R. M. (1985). The role of cerebral cortex in the generation of voluntary saccades: A positron emission tomographic study. *Journal of Neurophysiology, 54,* 348–369.

Fox, P. T., Miezin, F. M., Allman, J. M., Van Essen, D. C., & Raichle, M. E. (1987). Retinotopic organization of human visual cortex mapped with positron emission tomography. *Journal of Neuroscience, 7*, 913–922.

Fox, P. T. & Raichle, M. E. (1986). Focal physiological uncoupling of cerebral blood flow and oxidative metabolism during somatosensory stimulation in human subjects. *Proceedings of the National Academy of Sciences of the United States of America, 83*, 1140–1144.

Fox, P. T., Raichle, M. E., Mintun, M. A., & Dence, C. (1988). Nonoxidative glucose consumption during focal physiologic neural activity. *Science, 241*, 462–464.

Frith, U. & Frith, C. (2003). Development and neurophysiology of mentalizing. *Philosophical Transactions of the Royal Society of London Series B: Biological Sciences, 258*, 459–473.

Ghatan, P. H., Hsieh, J.-C., Petersson, K. M., Stone-Elander, S., & Ingvar, M. (1998). Coexistence of attention-based facilitation and inhitibion in human cortex. *Neuroimage, 7*, 23–29.

Gillihan, S. J. & Farah, M. J. (2005). Is self special? A critical review of evidence from experimental psychology and cognitive neuroscience. *Psychological Bulletin, 131*, 76–97.

Gold, L. & Lauritzen, M. (2002). Neuronal deactivation explains decreased cerebellar blood flow in response to focal cerebral ischemia or suppressed neocortical function. *Proceedings of the National Academy of Sciences of the United States of America, 99*, 7699–7704.

Greicius, M. D., Krasnow, B., Reiss, A. L., & Menon, V. (2003). Functional connectivity in the resting brain: A network analysis of the default mode hypothesis. *Proceedings of the National Academy of Sciences of the United States of America, 100*, 253–258.

Gusnard, D. A., Akbudak, E., Shulman, G. L., & Raichle, M. E. (2001). Medial prefrontal cortex and self-referential mental activity: Relation to a default mode of brain function. *Proceedings of the National Academy of Sciences of the United States of America, 98*, 4259–4264.

Gusnard, D. A. & Raichle, M. E. (2001). Searching for a baseline: Functional imaging and the resting human brain. *Nature Reviews Neuroscience, 2*, 685–694.

Harris, J. (1998). *Developmental Neuropsychiatry: Fundamentals* (vol. I). New York: Oxford University Press.

Haxby, J. V., Horwitz, B., Ungerleider, L. G., Maisog, J. M., Pietrini, P., & Grady, C. L. (1994). The functional organization of human extrastriate cortex: A PET-rCBF study of selective attention to faces and locations. *Journal of Neuroscience, 14*, 6336–6353.

Hyder, F., Rothman, D. L., & Shulman, R. G. (2002). Total neuroenergetics support localized brain activity: Implications for the interpretation of fMRI. *Proceedings of the National Academy of Sciences of the United States of America, 99*, 10771–10776.

Jackson, P. L. & Decety, J. (2004). Motor cognition: A new paradigm to study self-other interactions. *Current Opinion in Neurobiology, 14*, 259–263.

James, W. (1890). *Principles of Psychology* (pp. 97–99). New York: Henry Holt.

James, W. (1977). Does consciousness exist? In J. McDermott (Ed.), *Writings of Williams James* (pp. 169–183). Chicago: University of Chicago Press.

Kawashima, R., O'Sullivan, B. T., & Roland, P. E. (1995). Positron-emission tomography studies of cross-modality inhibition in selective attentional tasks: Closing the "mind's eye." *Proceedings of the National Academy of Sciences of the United States of America, 92*, 5969–5972.

Kelley, W. M., Macrae, C. N., Wyland, C. L., Caglar, S., Inati, S., & Heatherton, T. F. (2002). Finding the self? An event-related fMRI study. *Journal of Cognitive Neuroscience, 14*(5), 785–794.

Kjaer, T. W., Nowak, M., & Lou, H. C. (2002). Reflective self-awareness and conscious states: PET evidence for a common midline parietofrontal core. *Neuroimage, 17*, 1080–1086.

Kwong, K. K., Belliveau, J. W., Chesler, D. A., Goldberg, I. E., Weiskoff, R. M., Poncelet, B. P., et al. (1992). Dynamic magnetic resonance imaging of human brain activity during primary sensory stimulation. *Proceedings of the National Academy of Sciences of the United States of America, 89*, 5675–5679.

Lane, R. D., Fink, G. R., Chau, P. M.-L., & Dolan, R. J. (1997). Neural activation during selective attention to subjective emotional responses. *Neuroreport, 8*, 3969–3972.

Laufs, H., Krakow, K., Sterzer, P., Egger, E., Beyerle, A., Salek-Haddadi, A., et al. (2003). Electroencephalographic signatures of attentional and cogntive default modes in spontaneous brain activity fluctuations at rest. *Proceedings of the National Academy of Sciences of the United States of America, 100*, 11053–11058.

Laureys, S., Faymonville, M. E., Luxen, A., Lamy, M., Franck, G., & Maquet, P. (2000). Restoration of thalamocotical connectivity after recovery from persistent vegetative state. *Lancet, 355*, 1790–1791.

Laureys, S., Lemaire, C., Maquet, P., Phillips, C., & Franck, G. (1999). Cerebral metabolism during vegetative state and after recovery to consciousness. *Journal of Neurology Neurosurgery and Psychiatry, 67*, 121–122.

Lauritzen, M. (2001). Relationship of spikes, synaptic activity and local changes of cerebral blood flow. *Journal of Cerebral Blood Flow and Metabolism, 21*, 1367–1383.

Lauritzen, M. & Gold, L. (2003). Brain function and neurophysiological correlates of signals used in functional neuroimaging. *Journal of Neuroscience, 23*, 3972–3980.

Logothetis, N. K., Pauls, J., Augath, M., Trinath, T., & Oeltermann, A. (2001). Neurophysiological investigation of the basis of the fMRI signal. *Nature, 411*.

Lou, H. C., Luber, B., Crupain, M., Keenan, J. P., Nowak, M., Kjaer, T. W., et al. (2004). Parietal cortex and representation of the mental self. *Proceedings of the National Academy of Sciences of the United States of America, 101*, 6827–6832.

Lowe, M. J., Mock, B. J., & Sorenson, J. A. (1998). Functional connectivity in single and multislice echoplanar imaging using resting-state fluctuations. *Neuroimage, 7*, 119–132.

Maguire, E. A. (2001). The retrosplenial contribution to human navigation: A review of lesion and functional imaging findings. *Scandinavian Journal of Psychology, 42,* 225–238.

Maquet, P. (2000). Functional neuroimaging of normal human sleep by positron emission tomography. *Journal of Sleep Research, 9,* 207–231.

Maravita, A., Spence, C., & Driver, J. (2003). Multisensory integration and the body schema: Close to hand and within reach. *Current Biology, 13,* R531–R539.

Mazoyer, B., Zago, L., Mellet, E., Bricogne, S., Etard, O., Houde, O., et al. (2001). Cortical networks for working memory and executive functions sustain the conscious resting state in man. *Brain Research Bulletin, 54,* 287–298.

McCormick, D. A. (1999). Spontaneous activity: Signal or noise? *Science, 285,* 541–543.

McKiernan, K. A., Kaufman, J. N., Kucera-Thompson, J., & Binder, J. R. (2003). A parametric manipulation of factors affecting task-induced deactivation in functional neuroimaging. *Journal of Cognitive Neuroscience, 15,* 394–408.

Mintun, M. A., Raichle, M. E., Martin, W. R., & Herscovitch, P. (1984). Brain oxygen utilization measured with O-15 radiotracers and positron emission tomography. *Journal of Nuclear Medicine, 25,* 177–187.

Mitchell, J. P., Heatherton, T. F., & Macrae, C. N. (2002). Distinct neural systems subserve person and object knowledge. *Proceedings of the National Academy of Sciences of the United States of America, 99,* 15238–15243.

Nagai, Y., Critchley, H. D., Featherstone, E., Trimble, M. R., & Dolan, R. J. (2004). Activity in ventromedial prefrontal cortex covaries with sympathetic skin conductance level: A physiological account of a "default mode" of brain function. *Neuroimage, 22,* 243–251.

Ochsner, K. N., Bunge, S. A., Gross, J. J., & Gabrieli, J. D. E. (2002). Rethinking feelings: An fMRI study of the cognitive regulation of emotion. *Journal of Cognitive Neuroscience, 14,* 1215–1229.

Ogawa, S., Lee, T. M., Kay, A. R., & Tank, D. W. (1990). Brain magnetic resonance imaging with contrast depedent on blood oxygenation. *Proceedings of the National Academy of Sciences of the United States of America, 87,* 9868–9872.

Povinelli, D. J. & Cant, J. G. H. (1995). Arboreal clambering and the evolution of self-conception. *Quarterly Review of Biology, 70,* 393–421.

Premack, D. & Premack, A. J. (1983). *The Mind of an Ape.* New York: Norton.

Raichle, M. E. (1998). Behind the scenes of functional brain imaging: A historical and physiological perspective. *Proceedings of the National Academy of Sciences of the United States of America, 95,* 765–772.

Raichle, M. E., MacLeod, A. M., Snyder, A. Z., Powers, W. J., Gusnard, D. A., & Shulman, G. L. (2001). A default mode of brain function. *Proceedings of the National Academy of Sciences of the United States of America, 98,* 676–682.

Rainville, P., Duncan, G. H., Price, D. D., Carrier, B., & Bushnell, M. C. (1997). Pain affect encoded in human anterior cingulate but not somatosensory cortex. *Science, 277*, 968–971.

Roessler, J. & Eilan, N. (Eds.). (2003). *Agency and Self-Awareness: Issues in Philosophy and Psychology*. New York: Oxford University Press.

Salinas, E. & Sejnowski, T. J. (2001). Correlated neuronal activity and the flow of neural information. *Nature Reviews Neuroscience, 2*, 539–550.

Sanchez-Vives, M. V. & McCormick, D. A. (2000). Cellular and network mechanisms of rhythmic recurrent activity in neocortex. *Nature Neuroscience, 3*, 1027–1034.

Shallice, T. (2001). Theory of mind and the prefrontal cortex. *Brain, 124*, 279–286.

Shmuel, A., Augath, M., Oeltermann, A., Pauls, J., & Logothetis, N. K. (2003a). The negative BOLD response in monkey V1 is associated with decreases in neuronal activity. Presented at the Thirty-Third Annual Meeting of the Society for Neuroscience, New Orleans, LA, November 7–12, 2003.

Shmuel, A., Augath, M., Rounis, E., Logothetis, N. K., & Smirnakis, S. (2003b). Negative BOLD response ipsi-lateral to the visual stimulus: Origin is not blood stealing. Presented at the Ninth Annual Meeting of the Organization for Human Brain Mapping, New York, NY, June 18–22, 2003.

Shmuel, A., Yacoub, E., Pfeuffer, J., Van De Moortele, P.-F., Adriany, G., Hu, X., et al. (2002). Sustained negative BOLD, blood flow and oxygen consumption response and its coupling to the positive response in the human brain. *Neuron, 36*, 1195–1210.

Shmuel, A., Yacoub, E., Pfeuffer, J., Van De Moortele, P.-F., Adriany, G., Ugurbil, K., et al. (2001). Negative BOLD response and its coupling to the positive response in the human brain. *Neuroimage, 13*, S1005.

Shu, Y., Hasenstaub, A., & McCormick, D. A. (2003). Turning on and off recurrent balanced cortical activity. *Nature, 423*, 288–293.

Shulman, G. L., Fiez, J. A., Corbetta, M., Buckner, R. L., Miezin, F. M., Raichle, M. E., et al. (1997). Common blood flow changes across visual tasks. II. Decreases in cerebral cortex. *Journal of Cognitive Neuroscience, 9*, 648–663.

Shulman, R. G., Hyder, F., & Rothman, D. L. (2001). Cerebral energetics and the glycogen shunt: Neurochemical basis of functional imaging. *Proceedings of the National Academy of Sciences of the United States of America, 98*, 6417–6422.

Sibson, N. R., Dhankhar, A., Mason, G. F., Rothman, D. L., Behar, K. L., & Shulman, R. G. (1998). Stoichiometric coupling of brain glucose metabolism and glutamatergic neuronal activity. *Proceedings of the National Academy of Sciences of the United States of America, 95*, 316–321.

Simpson, J. R. Jr., Drevets, W. C., Snyder, A. Z., Gusnard, D. A., & Raichle, M. E. (2001a). Emotion-induced changes in human medial prefrontal cortex. II. During anticipatory anxiety. *Proceedings of the National Academy of Sciences of the United States of America, 98*, 688–691.

Simpson, J. R. Jr., Snyder, A. Z., Gusnard, D. A., & Raichle, M. E. (2001b). Emotion-induced changes in human medial prefrontal cortex. I. During cognitive task performance. *Proceedings of the National Academy of Sciences of the United States of America, 98*, 683–687.

Smith, A. T., Singh, K. D., & Greenlee, M. W. (2000). Attentional suppression of activity in the human visual cortex. *NeuroReport, 11*, 271–277.

Sokoloff, L., Mangold, R., Wechsler, R., Kennedy, C., & Kety, S. S. (1955). The effect of mental arithmetic on cerebral circulation and metabolism. *Journal of Clinical Investigation, 34*, 1101–1108.

Tononi, G. & Edelman, G. M. (1998). Consciousness and the integration of information in the brain. In H. H. Jasper, J. L. Descarries, V. F. Castellucci, & S. Rossignol (Eds.), *Consciousness: At the Frontiers of Neuroscience* (vol. 77). Philadelphia: Lippincott–Raven.

Tootell, R. B. H., Hadjikhani, N., Hall, E. K., Marrett, S., Vanduffel, W., Vaughn, J. T., et al. (1998). The retinotopy of visual spatial attention. *Neuron, 21*, 1409–1422.

Tootell, R. B. H., Tsao, D., & Vanduffel, W. (2003). Neuroimaging weighs in: Humans meet macaques in primate visual cortex. *Journal of Neuroscience, 23*, 3981–3989.

Tsodyks, M., Kenet, T., Grinvald, A., & Arieli, A. (1999). Linking spontaneous activity of single cortical neurons and the underlying functional architecture. *Science, 286*, 1943–1946.

Ueki, M., Linn, F., & Hossmann, K.-A. (1988). Functional activation of cerebral blood flow and metabolism before and after global ischemia of rat brain. *Journal of Cerebral Blood Flow and Metabolism, 8*, 486–494.

Vogeley, K. & Fink, G. R. (2003). Neural correlates of the first-person perspective. *Trends in Cognitive Sciences, 7*, 38–42.

Woolsey, T. A., Rovainen, C. M., Cox, S. B., Henegar, M. H., Liang, G. E., Liu, D., et al. (1996). Neuronal units linked to microvascular modules in cerebral cortex: Response elements for imaging the brain. *Cerebral Cortex, 6*, 647–660.

Xiong, J., Parsons, L. M., Gao, J. H., & Fox, P. T. (1999). Interregional connectivity to primary motor cortex revealed using MRI resting state images. *Human Brain Mapping, 8*, 151–156.

4 Thinking about Others: The Neural Substrates of Social Cognition

Jason P. Mitchell, Malia F. Mason, C. Neil Macrae, and Mahzarin R. Banaji

Humans are set apart from other animals by a number of behaviors and the cognitive faculties that give rise to them. For example, *Homo sapiens* is the only species that has harnassed arbitrary symbols into languages capable of expressing an infinite number of ideas; that routinely develops and improves upon tools for augmenting our natural abilities (including mathematics); and that can suppress immediate desires or prepotent tendencies indefinitely in pursuit of abstract goals that may be realized only in a distant future (a fact to which any academic can attest). These faculties have no doubt contributed critically to the vast global changes wrought by humans.

This list can safely be expanded to include another special feature of human behavior, namely, social behavior. Although obvious examples of sociability are to be found in other animals throughout the phylogenetic tree, from other primates to social insects such as bees and ants, the scale and complexity of human social abilities far outstrip those of even our closest primate relatives (Byrne & Whiten, 1988; Parker & Gibson, 1990; Ristau, 1991). Moreover, many of these abilities, such as recognizing oneself as a mental agent and inferring the psychological states of other such agents (even when their beliefs conflict with one's own), do not appear to have ready homologues among other animals (Gallup, 1985; Povinelli, Parks, & Novak, 1991; Tomasello & Call, 1997), suggesting that humans may have their own adaptation for particular aspects of social cognition. Indeed, this apparent lack of comparable social-cognitive skills in other animals, together with the centrality of social interaction to human life, has prompted some observers to suggest that social cognition represents the primary focus of evolutionary change in humans (Kamil, 2004; Tomasello, 1999).

Recently, researchers in the neurosciences have turned their attention to understanding the ways in which the brain gives rise to human social abilities. Of particular interest to both traditional social-cognitive psychologists and cognitive neuroscientists is whether the processes that give rise to social cognition are a subset of more general cognitive processes, or whether specific social-cognitive processes exist (Adolphs, 1999,

2001; Blakemore, Winston, & Frith, 2004; Ostrom, 1984). As Blakemore et al. (2004) asked, are the "general cognitive processes involved in perception, language, memory and attention . . . sufficient to explain social competence or, over and above these general processes, are there specific processes that are special to social interaction?" (p. 216).

The notion that social cognition may rely on a set of specific mental processes is supported by observations that the cognitive challenges posed by social interaction appear to be distinct from those presented by physical (nonhuman) objects. That is, a successful encounter with another person demands certain kinds of cognitive skills that are not generally required in the rest of everyday life. For example, appropriate interaction with another person requires that we first recognize that the other entity is indeed another mental agent, possessing internal psychological states not unlike our own. Having done so, we must accurately and rapidly intuit the motivations, feelings, and beliefs underlying that individual's behavior (Baron-Cohen, 1995), while keeping in mind that in addition to moment-to-moment mental states, people possess stable dispositional characteristics (personalities) that influence their actions. Finally, we must compute how our own behavior will influence the other person, both in order to act in a socially appropriate manner as well as to manipulate the other's mental states and concomitant behavior (such as when attempting to convey a complicated idea or to persuade someone to act in a certain way).

At a minimum, these informal observations of the special problems posed by social behavior provide probable cause for hypothesizing that social cognition may rely on a distinct set of mental processes. How, then, does one proceed to gather evidence in support of this possibility? Within several other fields of psychology, neuroimaging and neuropsychological research have proved critical to the resolution of similar theoretical debates. In particular, these techniques have contributed significantly to several entrenched controversies when the central theoretical question could be specified in the following way: do two psychological phenomena result from a single set of mental processes or from multiple processes? For example, ambiguities regarding the relation between implicit and explicit memory (whether implicit memory should be considered a degraded form of explicit memory or the product of an entirely different memory system) persisted even after years of accumulated behavioral data because they were explainable within a number of different theoretical frameworks. Resolution of this controversy in favor of the multiple-systems view came about after observations of patients with neuropsychological disorders (Gabrieli et al., 1995), and neuroimaging (see Schacter & Buckner, 1998) showed that the basis of implicit memory was both neuroanatomically and functionally distinct from explicit memory.

In a different controversy, neuroimaging data were crucial in showing that visual imagery relies on the very same neural mechanisms as actual visual perception (Kosslyn, 1994), thus providing compelling evidence against the opposing theoretical account that imagery relies on nonsensory, propositional knowledge (Pylyshyn, 1981, 2003).[1]

Based on such results, we propose that data from neuroimaging and patient studies will also provide an efficient means for addressing the question of whether social cognition relies on its own set of mental processes or instead piggybacks on other, more general processes of memory, inference, executive planning, and so on. By demonstrating that social cognition relies on a discrete set of brain regions, extant neuroscience research has generally supported a view of social cognition as distinct from other types of mental processes. The brain region most frequently implicated in social cognition is the medial prefrontal cortex (PFC), although research also suggests that a number of other regions contribute critically to social-cognitive processing, including the temporoparietal junction, orbitofrontal cortex, amygdala, superior temporal sulcus, and temporal poles (for reviews, see Adolphs, 1999, 2001; Blakemore et al., 2004; Frith & Frith, 1999; Gallagher & Frith, 2003).

Knowledge about the Characteristics of Other People

In being governed by complex mental states, other people are a unique kind of stimulus. Whereas people's behaviors are understood through consideration of their underlying motivations and feelings (intellectual curiosity, teenage angst), objects such as jackhammers, avocados, and Jeeps—indeed everything other than people—are governed by external forces. Because appropriate social behavior is predicated on the recognition that other people can be described in such mental terms, one fundamental social-cognitive challenge is to distinguish person-relevant from person-irrelevant semantic knowledge.

Until recently, studies of how the brain represents semantic knowledge focused on dissociations among different classes of inanimate objects. An interesting, and somewhat unexpected, finding from this research is that the brain appears to organize some types of semantic knowledge in a category-specific manner (Warrington, 1975). That is, knowledge about various kinds of objects (animals, tools) appears to be subserved by different brain regions. Although some controversy exists regarding the precise organizing principles underlying such category specificity, several influential theories suggested that the brain's semantic representations of a class of object center around the features that are specific to that class of object (Caramazza & Shelton, 1998). For instance, because most tools are defined by their function and not some arbitrary physical property, such as color, motor regions—such as left premotor cortex—are involved in the representation of knowledge about tools (Martin, 2001). In contrast, because animals are differentiated from one another primarily on the basis of their visual features rather than on the basis of function, semantic knowledge of animals appears to be represented by brain regions involved in the visual perception of animate objects and biological motion—lateral fusiform gyrus and superior temporal sulcus (Chao, Haxby, & Martin, 1999). Moreover, recent work suggests that regions of the motor cortex that support movement of various body parts (foot, arm, mouth) are also recruited when people read action words that are associated with these body parts, such as *kick*, *throw*, or *chew* (Hauk, Johnsrude, & Pulvermuller, 2004).

Is semantic knowledge about other people also represented in such a category-specific manner? That is, given that people differ from other types of entities by virtue of having mental states, do discrete brain regions subserve knowledge about people as mental agents? Alternately, could our understanding about the characteristics of other people simply rely on the same brain regions known to subserve semantic processing more generally (perceptual and functional representations)?

Using functional magnetic resonance imaging (fMRI), a series of studies suggested that the brain may indeed represent semantic knowledge about the mental states of other people in a discrete manner. In an initial study, Mitchell, Heatherton, and Macrae (2002) presented participants with items from three different categories: people (denoted by common American forenames, e.g., John, Mary), fruits (banana, grape), and articles of clothing (mitten, socks). Each item was presented beside an adjective (curious, pitted, woolen), and participants were asked to indicate whether the adjective could ever be used to describe the target item (which it could on half the trials). Replicating earlier work on the neural basis of semantic processing, event-related fMRI analyses indicated that semantic judgments about inanimate objects engaged regions previously implicated in object-knowledge tasks, specifically, left-lateralized inferotemporal cortex and ventrolateral PFC. In stark contrast, and despite the similarity between object and person trials, judgments about other people were associated with modulations in a qualitatively different set of brain regions. Dovetailing with earlier work on the potential neural basis of social cognition, these regions consisted of medial PFC, right temporoparietal junction, superior temporal sulcus, and fusiform gyrus (figure 4.1). In other words, making semantic decisions about characteristics of other people appeared to engage a qualitatively different set of brain regions, and, presumably, a concomitantly different set of cognitive processes, than did similar decisions about inanimate objects.

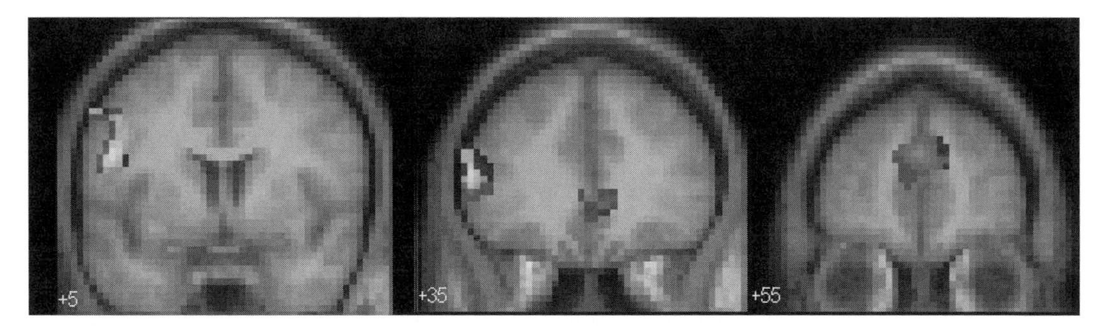

Figure 4.1
Regions of the prefrontal cortex that were differentially engaged by semantic judgments of objects and of people (Mitchell, Heatherton, & Macrae, 2002). Replicating earlier studies of neural systems that subserve semantic knowledge, object judgments engaged an extensive region of left ventrolateral PFC (red-orange scale). In contrast, person judgments were associated with activity in qualitatively distinct regions of the prefrontal cortex, specifically, medial PFC (blue-green scale). T-maps from comparisons of object and person judgments are overlaid on coronal slices (y values = 5, 35, and 55, respectively) of participants' mean normalized brain.

While suggesting that distinct neural representations may subserve knowledge about other people, this study raises several important questions about the precise nature of these representations. First, do observed functional differences between person and object knowledge extend to all aspects of other people, or just to information about mental states? Because people differ from other stimuli by virtue of having such mental states, one might expect special mechanisms for representing these person characteristics, but not aspects of people shared with other stimuli, such as physical descriptors. Second, do these observed differences extend to knowledge about the psychological states of any stimulus or are they specific to understanding those of other people? Many nonhuman animals can be anthropomorphized as having mental states, such as curious or afraid; perhaps social-cognitive representations extend to the internal "mental" states of other animals.

To examine directly the extent to which brain regions observed in our earlier work (Mitchell et al., 2002) are selective for understanding the mental states of other people, we scanned participants while they made semantic judgments about two different kinds of targets, people and dogs (Mitchell, Banaji, & Macrae, in press-a). As in our earlier work, participants judged whether a word could ever be used to describe the presented target. In this experiment, words could refer to one of two aspects of the targets—their potential psychological states (e.g., curious, frightened, angry) or their unobservable physical parts (e.g., lung, heart, liver). An equal number of words could not serve as potential descriptors (e.g., onic, metallic) and parts (nozzle, rudder). Critically, the words that could be used to describe people were pretested to be equally applicable to dogs, and the same response was made to each word regardless of whether the target was a person or a dog (i.e., curious would require a "yes" response regardless of whether the target was a person or a dog). In examining the brain response to each of these four types of trials, we observed that activity in the medial PFC region previously implicated in person knowledge was selective for judging words that referred to mental states, regardless of whether the target was another person or a dog. That is, the medial PFC was not simply engaged when participants made any judgment about a person (there was relatively low activity in this region when responding to "parts" trials). However, the medial PFC did appear to generalize to making mental state judgments regardless of whether the target of those judgments was another person or another mental agent (a dog).

Another fMRI experiment addressed a similar question by examining the neural basis of action knowledge (Mason, Banfield, & Macrae, 2004). Participants judged whether a series of actions (denoted by verbs such as run, sit, and bite) could be performed by a target, which again could be a person or a dog. As before, action words were pretested to ensure that they could apply equally to dogs and people. Whereas action-related judgments about dogs were associated with activity in regions involved with mental imagery (occipital and parahippocampal gyri), identical judgments about people yielded activity in the medial PFC. Together, these data suggest that activity in some brain regions, such as the medial PFC, specifically subserves social-cognitive representations about specific aspects of other people, including their mental and behavioral characteristics.

Inferring the Current Mental States of Another Person

Arguably the most important social-cognitive challenge is understanding the forces that govern other people's behavior. Unlike inanimate objects, the behavior of people can often be attributed to unobservable mental states. According to Dennett (1987), perceivers understand other people and predict how they will act by adopting the "intentional stance"—assuming that people are motivated by their current beliefs, desires, feelings, and goals. As such, a fundamental challenge to understanding other people is the ability to infer what these underlying mental states might be.

A fair proportion of neuroimaging and neuropsychological research on social cognition has focused on understanding brain mechanisms that subserve the capability to take others' perspectives or infer their mental states. This enterprise has generated an extraordinarily rich diversity of paradigms designed to manipulate the extent to which perceivers must infer the mental states of others. For example, in some of the first work on this topic, Goel and colleagues (1995) asked participants to indicate whether a historical figure (Christopher Columbus) would know the function of various artifacts (such as a compact disk), and compared neural activity during this task with that in one in which participants considered semantic or visual knowledge about those objects. Around the same time, Fletcher et al. (1995) presented participants with stories that were understandable only if one considered the mental states of characters, as well as stories that instead required understanding physical causality. Similarly, participants in a later study by Gallagher et al. (2002) were presented with the same mental-state stories as well as cartoons that also required understanding the minds of the characters in them. Finally, more recent work has had participants playing interactive games that require second-guessing one's opponent, such as the children's game "rock, paper, scissors" (Gallagher et al., 2002), and compared activations when subjects thought they were playing against a human opponent versus a computer.

Despite the wide diversity of tasks used to prompt mental state attribution (verbal stories, cartoons, competetive games, etc.), a remarkable empirical consensus has emerged regarding the underlying brain regions important for understanding the mind of another person. In each of these studies, greater activity was observed in medial PFC during tasks that required participants to infer the mental state of another person. Of particular interest is the gaming study of Gallagher et al. (2002), in which the only manipulation was whether participants believed they were playing against a human or a computer; despite identical visual input and task requirements, activity in medial PFC differed as a function of whether or not mental state attribution was required.

Extracting and Encoding Personality Information

Although rapid understanding of others' transient mental states must surely be a central capacity of any successful social agent, human interaction can also be marked by repeated encounters with the same person, and, as such, the opportunity to extract information about a person's stable, idiosyncratic characteristics. For better or worse, we share an apartment with the same roommate, ride the elevator with the same colleagues at work, see the same family members on holidays. Whether or not our social encounters with these individuals are successful hinges in part upon our ability to form theories of their stable qualities. Decades of research in social psychology suggest that dispositional aspects of another person can help provide an efficient famework for explaining the behavior of others. Indeed, humans appear to have a bias toward attributing the behavior of others to their stable dispositions, while often ignoring important additional influences on behavior, such as situational constraints (Gilbert & Malone, 1995).

Given the utility of extracting and remembering regularities that may guide another person's behavior, as well as our tendency to explain others' behaviors in terms of their dispositional traits (rather than situational constraints), it seems important to understand the underlying cognitive processes that give rise to such abilities. In the late 1970s, researchers began examining mechanisms underlying perceivers' abilities to infer the dispositional regularities that define other people (Hamilton, Driscoll, & Worth, 1989; Hamilton, Katz, & Leirer, 1980; Hastie & Kumar, 1979; Srull & Wyer, 1989; Wyer, Bodenhausen, & Srull, 1984). Participants were typically given information about a series of novel target individuals. Some participants would be asked to use the information to form an impression of the target (to attend to the stable, dispositional aspects of the person being described), whereas others would simply be asked to memorize the information. Subsequently, participants were asked to retrieve all the information that had been presented. Researchers observed a number of intriguing dissociations in memory performance between the social-cognitive (impression formation) task and the nonsocial (memorize) task. First, and somewhat a surprise, participants' memory was typically better after impression formation than after intentionally memorizing. Second, *patterns* of memory performance differed across the two conditions. For example, participants directed to form an impression were likely to recall items in clusters, suggesting that they had spontaneously organized information around implied traits (consecutively recalling many of the items that implied someone was honest, then recalling many of the items that implied intelligence, etc.). Moreover, impression formation often led to increased memory for information that was inconsistent with participants' expectations of the target's personality (Hastie & Kumar, 1979; Srull, 1981). For example, if a target was first described as honest, information that implied dishonesty would be particularly well remembered, but only if the participant was trying to form an impression of the person (Hartwick, 1979; Hastie & Kumar, 1979).

In making sense of such differences, researchers generally suggested that social-cognitive processing prompts deeper, more elaborative encoding of the sort that generally supports episodic memory (Craik & Lockhart, 1972), such as generation of schemas (Cohen & Ebbesen, 1979) or formation of a particularly rich network of inter-item associations (Hastie & Kumar, 1979; Srull, 1981). However, a second possibility suggests that rather than simply engaging deeper processing, attempts to form an impression actually engage different processing operations. In other words, introducing social-cognitive goals such as impression formation may prompt deployment of qualitatively distinct cognitive mechanisms. Functional brain imaging may shed light on these issues by distinguishing between the circumstances under which (1) processing information about social objects, such as people, overlaps with processing information about nonsocial objects, as well as (2) conditions under which these processes diverge and distinct neural networks are recruited for social and nonsocial cognition.

Neuroimaging research suggest that impression formation does indeed prompt distinct kinds of cognitive processing (Mitchell, Macrae, & Banaji, 2004). In one study, participants were scanned while incidentally encoding information that described a series of unfamiliar people. Each person (denoted by photographs of his face) was paired with ten statements that described various activities ostensibly performed by the person (e.g., "studied for his calculus final on the flight home for the holidays"). Each pair was accompanied by one of two cues (form impression, remember order) that indicated which of two orienting tasks was to be performed. For impression formation trials, participants were instructed to use the statement to infer the personal characteristics and traits of the target person, integrating across the entire set of statements for each person. For sequencing trials, participants were instructed to encode the order in which statements were paired with each face. (Although earlier cognitive work generally compared impression formation with explicit attempts to memorize the information, the cognitive processes engaged during memorization may vary from participant to participant, or even from trial to trial, making intentional encoding too underconstrained for event-related neuroimaging.) Subsequently, participants performed an associative memory task during which they were asked to match statements to the face with which it was originally presented.

Initial fMRI analyses revealed expected neural differences between impression formation and sequencing; specifically, greater activation for impression formation along a wide extent of the medial PFC. Of course, given that participants' task during impression formation was very different from that during sequencing, brain differences between the tasks are somewhat unsurprising. However, the design of this experiment allowed an additional analysis based on conditionalizing encoding data as a function of subsequent memory. Specifically, encoding trials were retroactively conditionalized ("binned") on the basis of both what orienting task had been performed (impression formation, sequencing) as well as subsequent memory success; that is, whether an item went on to be correctly remembered (hits) or to be forgotten (misses), resulting in four types of trials: impression hits, impression misses, sequencing hits, sequencing misses. Results indicated that, for trials encoded as part of the impression formation task, only a single region—dorsomedial PFC—had higher activity for subsequent hits than for subsequent misses. Of importance, no significant difference was observed between hits and misses in this region for items that were initially encoded as part of the sequencing task. That is, whereas encoding activity in dorsomedial PFC was greater for impression hits than impression misses, activity in this region did not differentiate between sequencing hits and sequencing misses. In contrast, for trials encoded as part of the sequencing task, subsequent memory success was correlated only with activity in the right hippocampus. It was again important that no significant difference was observed between impression hits and impression misses in this region (figure 4.2). That is, encoding activity in right hippocampus was selectively correlated with subsequent memory for sequencing, but not for impression formation. By showing the distinct neural basis of impression formation, these results suggest that not only are specific cognitive processes engaged by impression formation tasks, but

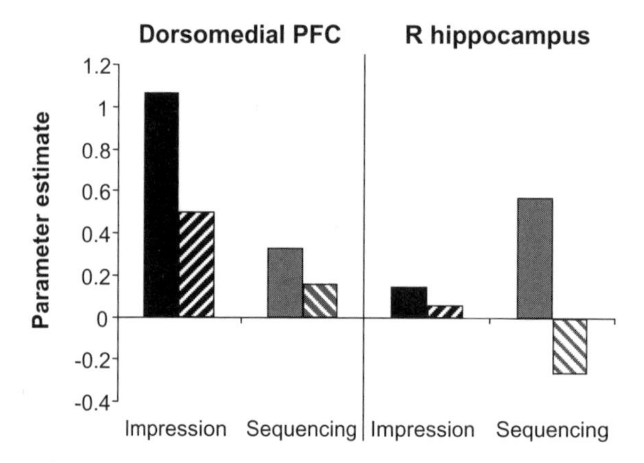

Figure 4.2

Participants encoded face-statement pairs in either a socially relevant (impression formation) or irrelevant (sequencing) manner (Mitchell, Macrae, & Banaji, 2004). Distinct neural correlates of subsequent memory success were observed as a function of the orienting task performed during encoding. For impression formation trials, activity in dorsomedial PFC (left panel) was higher for subsequent hits (solid black bars) than misses (striped black bars); no significant difference was observed between hits and misses for sequencing trials (gray solid and striped bars) in this area. In contrast, for sequencing trials, activity in right hippocampus (right panel) was higher for subsequent hits than misses, and no difference was observed in this region for impression formation trials.

memory differences after social and nonsocial tasks result from the operation of these separate cognitive processes.

However, just as for research on the neural systems that subserve semantic knowledge about other mental agents, the demonstration that distinct processes are engaged for social cognition requires that tasks be held constant across targets that are both social and nonsocial. Perhaps these earlier results reflect the particular demands of forming an impression about any stimulus, whether another person or an inanimate object. To address this possibility, a follow-up study examined the functional neuroanatomy associated with impression formation for both other people as well as inanimate objects (Mitchell, Macrae, & Banaji, in press). As before, participants saw a series of statements that ostensibly described an associated target. However, only half the targets were other people; the remaining half were inanimate objects (cars, computers). For half of the people and half of the inanimate objects, participants were once again asked to use the statement to form an impression of the target, and for the other half they were asked to remember the order in which statements were paired with the target. When brain regions were defined in the same way as in the initial study (impression formation > sequencing for person trials only), a very similar region of dorsomedial PFC was observed, providing a direct replication of earlier findings. However, of critical interest in this experiment was whether or not the medial PFC was generally engaged by attempts to form an impression, regardless of social aspects of the targets. To address this question, the neural response to forming an impression of inanimate objects was examined. Consistent with the notion that medial PFC specifically subserves social-cognitive processing, results revealed a weak response in this region when participants formed an impression of inanimate objects, similar to the response observed for the sequencing task. In other words, although the dorsomedial PFC was significantly activated during attempts to form an impression of other people, its activity was not increased by putatively similar attempts to form an impression of inanimate objects. Activity in this region appeared to track specifically with the social-cognitive demands of the orienting task.

Guiding Social Behavior

Although understanding certain aspects of the minds of other people is generally necessary for appropriate social behavior, the ultimate output of a system for social cognition must be the direct guidance of such behavior. As Fiske (1992) pointed out, "thinking is for doing"; a system that apprehends something (such as the state of another's mind), but is unable to act appropriately based on that information, will have little adaptive value. As such, social cognition must be able to direct social behavior, including the selection and initiation of appropriate actions (is it permissible to start the wave during a particularly inspiring lecture?), speech (how do I effectively communicate a complex scientific concept to a novice?), and reciprocal responses (is it a good idea to snort derisively at the search committee member's uninformed question?).

However, relative to the amount of extant work on the intrapersonal aspects of social cognition, little neuroscience research has examined the demands of *inter*personal social behavior. This state of affairs is due at least in part to the strict contraints of neuroimaging techniques, including restriction of movement, limit to one participant at a time (although recent attempts have been made to use coordinated scanners to image two or more people as they interact in real time (Montague et al., 2002), and difficulty recording verbal output (at least for fMRI, where scanner noise is often prohibitively loud). As such, what little is known about the systems that guide social behavior has tended to come from research in patients with neuropsychological disorders. Damage to the orbitofrontal cortex and medial aspects of the PFC, especially to its more ventral aspects, is associated with a wide range of social deficits, including selective failure on theory of mind tasks (including second-order false belief and faux pas tasks). Although individuals with orbitofrontal-medial PFC damage typically come to the attention of neurologists because of marked changes in social behavior, including personality changes, lack of empathy, and inappropriate social interactions, no detailed account of their behavioral deficits has yet emerged. Perhaps the fullest sketch of social changes after damage to these regions can be found in the description of a patient known as Elliot (Damasio, 1994). After resection of an area of his frontal lobes to remove a tumor (mainly confined to orbitofrontal cortex and ventromedial aspects of the medial PFC), Elliot suffered severe changes to his ability to negotiate the social world. Formerly a reliable professional with a rewarding family life, he became unable to hold a job, lost his life savings in a series of bad business ventures, and was divorced by his wife; he married another woman of whom his friends and family disapproved, and divorced a second time.

Of interest, as is typical in cases of individuals with damage to orbitofrontal cortex and medial PFC, Elliot had unimpaired intelligence, language, and working memory. Such cases provide an important clue that social behavior results from the operation of cognitive processes that are distinct from those guiding behavior in other domains. Despite severe social deficits, patients with orbitofrontal and medial PFC damage are often unimpaired on tasks outside of the social domain, even when those tasks are quite challenging.

The distinctiveness of social behavior is also underscored by research on the autistic syndrome (Baron-Cohen, 1995). Although little is understood about the brain basis of the disorder, one characteristic of autism is a profound disengagement from the social world. Autistic individuals may shun social interaction and appear unable to learn its basic rules, often despite relative sparing of abilities in other domains. Of interest, a disorder known as Williams syndrome appears in many ways to present the converse pattern of deficits: these individuals often appear hypersocial and verbal, but are profoundly impaired in nonsocial domains (Tager-Flusberg, Boshart, & Baron-Cohen, 1998).

In another interesting line of research, patients with frontal variant frontotemporal dementia (FTD) were examined on a series of mental tasks of varying degrees of difficulty (Gregory et al., 2002). FTD is a progressive disorder that results in degeneration to areas of frontal and temporal lobe. Individuals with the frontal variant of the disorder typically experience various changes in social behavior similar to those in patients with lesions to medial PFC, including personality changes, lack of empathy, and socially inappropriate behavior. Meta-analytic procedures over large numbers of individuals diagnosed with frontal variant FTD (Salmon et al., 2003) suggested that neural degeneration is most severe in a circumscribed region of PFC highly similar to medial regions observed in neuroimaging studies of social cognition. Gregory et al. compared performance of such patients with that of individuals with Alzheimer's disease (AD) on a series of mental tasks: first-order false belief, second-order false belief, reading the minds in the eyes (Baron-Cohen et al., 2001), and faux pas detection. Compared with AD controls, patients with frontal variant FTD were impaired on all the tasks, including the ability to detect socially inappropriate behavior on the faux pas task.

One question regarding such results is whether the deficits of social behavior of patients with neuropsychological disorders are a direct result of impairments in the ability to mentalize about the minds of other people. Future research is necessary to examine whether separate systems exist for such guidance of social behavior, or whether social interaction is predicated exclusively on the intact ability to understand the minds of those with whom one is engaging.

Thinking about Oneself and Others

Although the data reviewed above suggest that social cognition may indeed rely on distinct mental processes, a central question remains regarding precisely of what those processes consist. Given the fact that understanding people is not the same as understanding inanimate objects, how is it that one goes about making sense of the mental states and behavior of other people? One influential theory suggests that knowledge of one's own mind can be used successfully to help infer the mind of another person (Davis & Stone, 1995a, b). This account, broadly known as simulation theory, proposes that one valuable source of information about the thoughts, feelings, or potential behavior of another person is a first-person prediction about what I myself might think, feel, or do in a similar situation. Although clues about what is going on in another's mind can certainly be gleaned from a variety of sources, (emotional expression, direction of eye gaze, folk theories about how other people work), simulation could provide particularly useful information in this regard, especially in complex or novel social situations.

A good deal of overlap appears to exist between brain regions that subserve thinking about other people and those that subserve thinking about oneself. Specifically, in addition to its general role in social cognition, the medial PFC appears to be an integral component of tasks that require participants to assess one's own qualitities or current feelings (Johnson et al., 2002; Kelley et al., 2002; Macrae et al., 2004; Schmitz, Kawahara-Baccus, & Johnson, 2004; Zysset et al., 2002). For example, a recent study examined the neural basis of the self-relevance effect in memory, whereby participants typically demonstrate enhanced episodic memory for information that has been related to oneself (Rogers, Kuiper, & Kirker, 1977). Participants incidentally encoded a series of adjectives by making one of three judgments about each: whether the word described their own personality, described the personality of current president George W. Bush, or appeared in uppercase or lowercase letters. Results indicated that self-judgments were associated with additional activity in medial PFC compared with other or case judgments, suggesting a role for the structure in self-relevant processing (Kelley et al., 2002).[2] Later research revealed that activity in this region of medial PFC also correlates with subsequent memory for items that were encoded in a self-referential manner (Macrae et al., 2004). These results dovetail with those of Gusnard and colleagues (2001), that a region of dorsomedial PFC was activated during judgments of photographs in a way that required participants to refer to their own affective experience.

Together, these observations that medial PFC appears to subserve thinking about self as well as thinking about others provide initial support for simulation accounts of social cognition. By suggesting that social cognition and self-referential thought may rely on a common set of cognitive processes, these disparate lines of research converge on the notion that understanding oneself is an integral component in understanding other people. Mitchell et al. (in press-b) performed an fMRI study to test two predictions that derive from such simulation accounts of social cognition. First, one should engage in simulation only when one's task is to infer the current mental states of another person. When interacting with a person in a way that does not require mental state attribution (e.g., when looking for a friend's familiar face in a crowded bar), one need not simulate the minds of others based on one's own. Second, simulation should be useful only when one has reason to believe that it will be applicable to the person in question. If a person thinks very differently (perhaps because of cultural or interpersonal differences), it is unclear that understanding his mental states or predicting his behavior can be achieved through consideration of one's own.

During event-related fMRI scanning, participants saw a series of faces for which they were asked to perform either a mentalizing or nonmentalizing task. For half the faces, participants were asked to mentalize about the target's internal states by judging how pleased the person seemed to be to have his or her photograph taken. For the other half, participants were asked to make judgments that did not include a mentalizing component, namely, indicating how symmetrical (left to right) each face appeared. After scanning, participants saw each of the faces a second time and were asked to judge each for how similar they felt the person was to themselves. The event-related nature of the design allowed us to conditionalize items retroactively as a function of the task performed during the initial presentation (mentalize, nonmentalize) as well as how similar each participant felt he or she was to each target (similar, dissimilar). As predicted on the basis of extant research on social cognition, the contrast of mentalizing > nonmentalizing yielded a region of medial PFC similar to that in our earlier work on impression formation. More important for simulation accounts of social cognition, activity in medial PFC differentiated between similar and dissimilar targets, but only for those for whom the mentalizing task was performed. Specifically, for faces in the mentalizing task, activity in the medial PFC was higher for faces that participants judged to be similar than for faces judged to be dissimilar; no such dissociation was observed for faces in the nonmentalizing task. These results are consistent with the prediction that simulation should occur only when a target is similar enough to the perceiver to make simulation an appropiate basis for understanding that person's mind.

Conclusion

Social cognition—thinking about the minds of other people—poses a set of distinct challenges that may not have ready parallels in the physical world. Other individuals are complex, dynamic entities who have properties (mental states) that are not possessed by any other class of stimuli. To make sense of another's behavior we must accurately identify these mental states and consider how, in combination with stable dispositional traits, they may influence how people interact with the world around them.

The centrality of these abilities to human life is demonstrated most clearly by individuals who have selective social deficits, such as autism or particular kinds of brain damage. Indeed, the fundamental importance of human sociability both to everyday life as well as to cultural achievements of the species prompted some observers to suggest that social cognition may have been one of the primary engines of human evolution. For example, Tomasello (1999) argued convincingly that what sets *Homo sapiens* apart from other primates is the ability to represent the mind of conspecifics.

Given both the uniqueness of challenges posed by the social world and the importance of sociability to human life, one might expect distinct cognitive processes to be dedicated to thinking about and interacting with other individuals. Using neuroimaging and neuropsychological methods, researchers have begun to provide triangulating support for the notion that separate systems do indeed subserve social and nonsocial cognition. The promise of future research in this domain is to begin to specify the precise computations that allow us to perform the remarkable feats of social gymnastics of which humans are routinely capable.

Notes

1. Of course, the ability to use neuroimaging and patients with neuropsychological disorders to address issues of theoretical relevance in this way relies on the following two assumptions: first, that two different brain regions cannot give rise to precisely the same cognitive process, and second, that no single brain region instantiates multiple such processes. Although neither of these assumptions is compelled philosophically or empirically, nor has serious challenge been posed to these two foundations of the cognitive neuroscience enterprise.

2. A somewhat counterintuitive finding from this study was that the medial PFC was not engaged by judging the personality characteristics another person (George W. Bush), despite consistent observations that this region accompanies similar social-cognitive tasks. One possibility is that the personality of famous figures can be judged in an abstract, semantic manner that does not involve self-referencing. For instance, one may know that a well-known politician is dishonest or unintelligent in much the same way that one has semantic knowledge that grapes grow on vines and can be pressed into wine. Recent research suggested that the region of medial PFC implicated in self-referencing is also engaged when making comparable judgments about the personality of a significant other (Schmitz et al., 2004).

References

Adolphs, R. (1999). Social cognition and the human brain. *Trends in Cognitive Science, 3*(12), 469–479.

Adolphs, R. (2001). The neurobiology of social cognition. *Current Opinion in Neurobiology, 11*, 231–239.

Baron-Cohen, S. (1995). *Mindblindness: An Essay on Autism and Theory of Mond.* Cambridge: MIT Press.

Baron-Cohen, S., Wheelwright, S., Hill, J., Raste, Y., & Plumb, I. (2001). The "reading the mind in the eyes" test revised version: A study with normal adults, and adults with Asperger syndrome or high-functioning autism. *Journal of Child Psychology and Psychiatry, 42*(2), 241–251.

Blakemore, S. J., Winston, J., & Frith, U. (2004). Social cognitive neuroscience: Where are we heading? *Trends in Cognitive Science, 8*(5), 216–222.

Byrne, R. E. & Whiten, A. (1988). *Machiavellian Intelligence: Social Expertise and the Evolution of Intellect in Monkeys, Apes, and Humans.* Oxford: Oxford University Press.

Caramazza, A. & Shelton, J. R. (1998). Domain-specific knowledge systems in the brain: The animate-inanimate distinction. *Journal of Cognitive Neuroscience, 10*, 1–34.

Chao, L. L., Haxby, J. V., & Martin, A. (1999). Attribute-based neural substrates in temporal cortex for perceiving and knowing about objects. *Nature Neuroscience, 2*, 913–919.

Cohen, C. E. & Ebbesen, E. B. (1979). Observational goals and schema activation: A theoretical framework for behavior perception. *Journal of Experimental Social Psychology, 15*(4), 305–329.

Craik, F. I. M. & Lockhart, R. S. (1972). Levels of processing: A framework for memory research. *Journal of Verbal Learning and Verbal Behavior, 11*, 617–684.

Damasio, A. R. (1994). *Descartes' Error.* New York: Grosset/Putnam.

Davis, M. & Stone, T. (1995a). *Folk Psychology: Readings in Mind and Language.* Oxford: Blackwell.

Davis, M. & Stone, T. (1995b). *Mental Simulation: Readings in Mind and Language.* Oxford: Blackwell.

Dennett, D. C. (1987). *The Intentional Stance.* Cambridge: MIT Press.

Fiske, S. T. (1992). Thinking is for doing: Portraits of social cognition from daguerreotype to laser-photo. *Journal of Personality and Social Psychology, 63*(6), 877–889.

Fletcher, P. C., Happe, F., Frith, U., Baker, S. C., Dolan, R. J., Frackowiak, R. S., et al. (1995). Other minds in the brain: A functional imaging study of "theory of mind" in story comprehension. *Cognition, 57*(2), 109–128.

Frith, C. D. & Frith, U. (1999). Interacting minds—A biological basis. *Science, 286*, 1692–1695.

Gabrieli, J. D. E., Fleischman, D. A., Keane, M. M., Reminger, S. L., & Morrell, F. (1995). Double dissociation between memory systems underlying explicit and implicit memory in the human brain. *Psychological Science, 6*, 76–82.

Gallagher, H. L. & Frith, C. D. (2003). Functional imaging of "theory of mind." *Trends in Cognitive Science, 7*(2), 77–83.

Gallagher, H. L., Happé, F., Brunswick, N., Fletcher, P. C., Frith, U., & Frith, C. D. (2000). Reading the mind in cartoons and stories: An fMRI study of "theory of mind" in verbal and nonverbal tasks. *Neuropsychologia, 38,* 11–21.

Gallagher, H. L., Jack, A. I., Roepstorff, A., & Frith, C. D. (2002). Imaging the intentional stance in a competitive game. *Neuroimage, 16*(3 Pt 1), 814–821.

Gallup, G. G., Jr. (1985). Do minds exist in species other than our own? *Neuroscience and Biobehavioral Reviews, 9*(4), 631–641.

Gilbert, D. T. & Malone, P. S. (1995). The correspondence bias. *Psychological Bulletin, 117,* 21–38.

Goel, V., Grafman, J., Sadato, N., & Hallett, M. (1995). Modeling other minds. *Neuroreport, 6*(13), 1741–1746.

Gregory, C., Lough, S., Stone, V., Erzinclioglu, S., Martin, L., Baron-Cohen, S., et al. (2002). Theory of mind in patients with frontal variant frontotemporal dementia and Alzheimer's disease: Theoretical and practical implications. *Brain, 125*(Pt 4), 752–764.

Gusnard, D. A., Akbudak, E., Shulman, G. L., & Raichle, M. E. (2001). Medial prefrontal cortex and self-referential mental activity: Relation to a default mode of brain function. *Proceedings of the National Academy of Sciences, 98,* 4259–4264.

Hamilton, D. L., Driscoll, D. M., & Worth, L. T. (1989). Cognitive organization of impressions: Effects of incongruency in complex representations. *Journal of Personality and Social Psychology, 57,* 925–939.

Hamilton, D. L., Katz, L. B., & Leirer, V. O. (1980). Cognitive representation of personality impressions: Organizational processes in first impression formation. *Journal of Personality and Social Psychology, 39*(1 suppl 6), 1050–1063.

Hartwick, J. (1979). Memory for trait information: A signal detection analysis. *Journal of Experimental Social Psychology, 15*(6), 533–552.

Hastie, R. & Kumar, P. A. (1979). Person memory: Personality traits as organizing principles in memory for behaviors. *Journal of Personality and Social Psychology, 37*(1), 25–38.

Hauk, O., Johnsrude, I., & Pulvermuller, F. (2004). Somatotopic representation of action words in human motor and premotor cortex. *Neuron, 41*(2), 301–307.

Johnson, S. C., Baxter, L. C., Wilder, L. S., Pipe, J. G., Heiserman, J. E., & Prigatano, G. P. (2002). Neural correlates of self-reflection. *Brain, 125*(Pt 8), 1808–1814.

Kamil, A. C. (2004). Sociality and the evolution of intelligence. *Trends in Cognitive Science, 8*(5), 195–197.

Kelley, W. M., Macrae, C. N., Wyland, C. L., Caglar, S., Inati, S., & Heatherton, T. F. (2002). Finding the self? An event-related fMRI study. *Journal of Cognitive Neuroscience, 14,* 785–794.

Kosslyn, S. M. (1994). *Image and Brain: The Resolution of the Mental Imagery Debate.* Cambridge: MIT Press.

Macrae, C. N., Moran, J. M., Heatherton, T. F., Banfield, J. F., & Kelley, W. M. (2004). Medial prefrontal activity predicts memory for self. *Cerebral Cortex, 14*(6), 647–654.

Martin, A. (2001). Functional neuroimaging of semantic memory. In R. Cabeza & A. Kingstone (Eds.), *Handbook of Functional Neuroimaging of Cognition* (pp. 153–186). Cambridge: MIT Press.

Mason, M. F., Banfield, J. F., & Macrae, C. N. (2004). Thinking about actions: The neural substrates of person knowledge. *Cerebral Cortex, 14*(2), 209–214.

Mitchell, J. P., Banaji, M. R., & Macrae, C. N. (in press-a). General and specific contributions of the medial prefrontal cortex to knowledge about mental states. *NeuroImage.*

Mitchell, J. P., Banaji, M. R., & Macrae, C. N. (in press-b). The link between social cognition and self-referential thought in the medial prefrontal cortex. *Journal of Cognitive Neuroscience.*

Mitchell, J. P., Heatherton, T. F., & Macrae, C. N. (2002). Distinct neural systems subserve person and object knowledge. *Proceedings of the National Academy of Sciences, 99*, 15238–15243.

Mitchell, J. P., Macrae, C. N., & Banaji, M. R. (2004). Encoding specific effects of social cognition on the neural correlates of subsequent memory. *Journal of Neuroscience, 24*(21), 4912–4917.

Mitchell, J. P., Macrae, C. N., & Banaji. M. R. (in press). Forming impressions of people versus inanimate objects. Social-cognitive processing in the medial prefrontal cortex. *NeuroImage, 26*, 251–257.

Montague, P. R., Berns, G. S., Cohen, J. D., McClure, S. M., Pagnoni, G., Dhamala, M., et al. (2002). Hyperscanning: Simultaneous fMRI during linked social interactions. *NeuroImage, 16*(4), 1159–1164.

Ostrom, T. M. (1984). The sovereignty of social cognition. In R. S. Wyer & T. K. Srull (Eds.), *Handbook of Social Cognition* (vol. 1, pp. 1–37). Hillsdale, NJ: Erlbaum.

Parker, S. T. & Gibson, K. R. (1990). *"Language" and Intelligence in Monkeys and Apes: Comparative Developmental Perspectives.* Cambridge: Cambridge University Press.

Povinelli, D. J., Parks, K. A., & Novak, M. A. (1991). Do rhesus monkeys (*Macaca mulatta*) attribute knowledge and ignorance to others? *Journal of Comparative Psychology, 105*(4), 318–325.

Pylyshyn, Z. (1981). The imagery debate: Analogue media versus tacit knowledge. *Psychological Review, 88*(1), 16–45.

Pylyshyn, Z. (2003). Return of the mental image: Are there really pictures in the brain? *Trends in Cognitive Science, 7*(3), 113–118.

Ristau, C. A. (1991). *Cognitive Ethology: The Minds of Other Animals: Essays in honor of Donald R. Griffin.* Mahwah, NJ: Erlbaum.

Rogers, T. B., Kuiper, N. A., & Kirker, W. S. (1977). Self-reference and the encoding of personal information. *Journal of Personality and Social Psychology, 35*(9), 677–688.

Salmon, E., Garraux, G., Delbeuck, X., Collette, F., Kalbe, E., Zuendorf, G., et al. (2003). Predominant ventromedial frontopolar metabolic impairment in frontotemporal dementia. *Neuroimage, 20*(1), 435–440.

Schacter, D. L. & Buckner, R. L. (1998). Priming and the brain. *Neuron, 20,* 185–195.

Schmitz, T. W., Kawahara-Baccus, T. N., & Johnson, S. C. (2004). Metacognitive evaluation, self-relevance, and the right prefrontal cortex. *Neuroimage, 22*(2), 941–947.

Srull, T. K. (1981). Person memory: Some tests of associative storage and retrieval models. *Journal of Experimental Psychology: Human Learning and Memory, 7*(6), 440–463.

Srull, T. K. & Wyer, R. S. (1989). Person memory and judgment. *Psychological Review, 96,* 58–83.

Tager-Flusberg, H., Boshart, J., & Baron-Cohen, S. (1998). Reading the windows to the soul: Evidence of domain-specific sparing in Williams syndrome. *Journal of Cognitive Neuroscience, 10*(5), 631–639.

Tomasello, M. (1999). *The Cultural Origins of Human Cognition.* Cambridge: Harvard University Press.

Tomasello, M. & Call, J. (1997). *Primate Cognition.* Oxford: Oxford University Press.

Warrington, E. K. (1975). The selective impairment of semantic memory. *Quarterly Journal of Experimental Psychology, 27*(4), 635–657.

Wyer, R. S., Bodenhausen, G. V., & Srull, T. K. (1984). The cognitive representation of persons and groups and its effect on recall and recognition memory. *Journal of Experimental Social Psychology, 20,* 445–469.

Zysset, S., Huber, O., Ferstl, E., & von Cramon, D. Y. (2002). The anterior frontomedian cortex and evaluative judgment: An fMRI study. *Neuroimage, 15*(4), 983–991.

5 Four Brain Regions for One Theory of Mind?

Rebecca Saxe

In developmental psychology, the paradigmatic task for assessing a child's ability to reason about the mental states of others (theory of mind) is the false belief task (Wimmer & Perner, 1983; for reviews of this literature see Flavell, 1999; Wellman & Lagattuta, 2000; Wellman, Cross, & Watson, 2001). In the standard version of this task (the object transfer problem), the child is told a story in which a character's belief about the location of a target object becomes false when the object is moved without the character's knowledge. In the original version, Maxi puts his chocolate in the green cupboard. Then, while Maxi is outside playing, his mother moves the chocolate from the green to the blue cupboard. The child is then variously asked to report the content of the character's belief (Where does Maxi *think* the chocolate is?), to predict the character's action (Where will Maxi *look* for the chocolate?), or sometimes to explain the completed action (Why did Maxi look for the chocolate in the *green* cupboard?). The critical feature of a false belief task is that the correct answers to all three of these questions, even ones that do not specifically query a belief content, require the child to pay attention to Maxi's belief, and not the actual location of the chocolate (Dennett 1978; Premack & Woodruff, 1978). Dozens of versions of the false belief problem have been used, and although the precise age of success varies among children and among task versions (Wellman, Cross, & Watson, 2001), in general children younger than three or four years old do not correctly solve the problems, but older children do.

Many neuroimaging studies have followed developmental psychology using false belief problems as definitive theory of mind tasks (Fletcher et al., 1995; Goel et al., 1995; Gallagher et al., 2000; Brunet et al., 2000; Vogeley et al., 2001; Saxe & Kanwisher, 2003; Ruby & Decety, 2003; Grezes, Frith, & Passingham, 2004). These studies have revealed an impressively consistent pattern of brain regions involved when subjects are required to reason about someone's false belief, including the medial prefrontal cortex (MPFC), posterior cingulate (PC), and bilateral temporoparietal

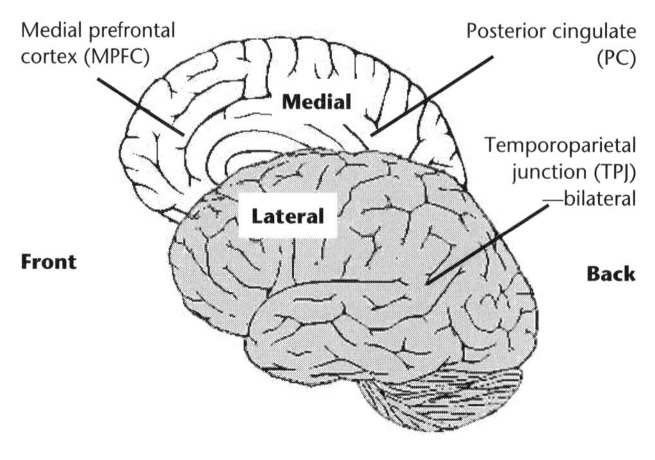

Figure 5.1
Approximate locations of four brain regions involved in reasoning about other minds.

junction extending into the posterior superior temporal sulcus (left: LTPJ, right: RTPJ; figure 5.1).

These results raise important questions. First, could any or all of these brain regions be a specialized neural substrate for theory of mind? In other words, is reasoning about other minds a necessary condition for the activity of each of these regions? Is it a sufficient condition? Second, what is the distinct contribution of each of these regions to the subject's reasoning about other people? That is, the coactivation of at least four brain regions might reflect deployment of at least four different computational ingredients of theory of mind. Can we identify and distinguish those ingredients?

Certainly the RTPJ and the MPFC, and to a slightly lesser extent the LTPJ and PC, are good candidates for the neural substrate of theory of mind. However, questions about the distinct contributions of these regions remain open. Contrary to some claims in the recent social neuroscience literature, I contend that the specific computational ingredients contributed by the RTPJ, LTPJ, MPFC, and PC are not yet known.

Caveats: Claims I'm Not Making

The question I begin with is whether any of the brain regions recruited when subjects read false belief problems could be a specialized neural substrate for a theory of mind. To illustrate how neuroimaging data can help to answer this question, a relatively in-depth characterization of the functional profile of the RTPJ is in order. It establishes both the specificity of RTPJ involvement in reasoning about other minds, and the generality of this response beyond false belief attribution. That is, reasoning about another mind appears to be both a necessary and a sufficient condition for a robust BOLD response in the RTPJ.

However, a few caveats are in order to preclude some common confusions. From what follows, it is important not to conclude

1. That the RTPJ is the sole (or most important) brain region responsible for reasoning about other minds;
2. That a functional region of interest is a single thing, and/or a functionally homogeneous unit; or
3. That theory of mind is a simple, homogeneous process.

First, in providing a functional profile of the RTPJ, I do not intend to claim that this region of the brain is solely, or even primarily, responsible for carrying out inferences about another mind. I take it for granted that many different brain regions will always be recruited for such reasoning, including those that are specifically involved in reasoning about other minds (as established empirically), and lots of others required for domain-general aspects of making inferences, such as parsing the sentences, attending to problems, executing responses, and so on. Furthermore, the LTPJ, MPFC, and PC share many of the features of the RTPJ's response profile. Data from the RTPJ are astonishingly clean and reliable, and so serve as a useful illustration.

Second, the language of regions of interest (talking about *the* RTPJ or *the* MPFC) can encourage misinterpretation, especially reification. The error of reification is to assume that regions of interest refer to chunks of cortex in which all of the neurons (hundreds of millions of them) are functionally homogeneous. Undeniably, there is a finer grain of structure, some of which will be detectable with stronger-field functional magnetic resonance imaging (fMRI), and some of which will be susceptible only to techniques with much higher temporal and spatial resolution, such as single-cell recording. Thus in future work the RTPJ is bound to be broken up into smaller units, with different, perhaps more specific, functional roles. In the meantime, the utility of the coarser-grain region of interest is determined by the consistency of data that emerge from it and the richness of theoretical progress those data support.

Finally, theory of mind itself is not a simple computation, and must not be identified with successful performance on a single task. For extended discussions of the components of reasoning about other minds, combining both developmental psychology and neuroimaging data, see Saxe, Carey, and Kanwisher (2004) and Frith and Frith (2003).

The RTPJ

In previous studies, as described, increased activation in the RTPJ was reported when subjects reasoned about the false belief of a character in a story (Fletcher et al., 1995; Gallagher et al., 2000; Vogeley et al., 2001), a cartoon (Gallagher et al., 2000; Brunet et al., 2000), an imaginary ill-informed protagonist such as Christopher Columbus (Goel et al., 1995) or a medical lay person (Ruby & Decety, 2003), relative to when no false belief attribution was required. Control conditions were scrambled texts, scrambled pictures, texts describing logical relations between events (Fletcher et al., 1995; Gallagher et al., 2000; Vogeley et al., 2001), judgments about the true function of an object (Goel et al., 1995) or medical condition (Ruby & Decety, 2003).

Based on these preliminary hints, what might be the role of the RTPJ? One possibility is that the RTPJ is genuinely engaged in the core responsibility of a theory of mind: achieving a representation of other people's mental states. However, there are at least three important alternative accounts of the RTPJ response. Relative to control conditions as described, false belief problems may have recruited the involvement of a region involved in:

1. Making causal inferences, specifically, inferences about invisible causal processes;
2. Maintaining two competing representations of reality (actual state of affairs and reality represented in the character's head) and/or inhibiting prepotent responses based on the actual state of the world;
3. Representing the presence of a person or intentional action in the story, prerequisites for the operation of a theory of mind, but not its fundamental core.

Causal Inference

Could the RTPJ response observed in previous studies reflect a domain-general process for making causal inferences? The data suggest not. First, even in earliest studies implicating the RTPJ in solving false belief problems, some control stories included elements of physical causality, such as breaking a laser beam to set off an alarm (Fletcher et al., 1995; Gallagher et al., 2000). However, whereas all of the stories in previous studies required inferences about invisible (in this case, mental) causal processes, only some control stories did so. We therefore devised a new version of false belief stories task to address directly the role of the RTPJ in reasoning about causes, mental and physical (Saxe & Kanwisher, 2003).

We had subjects read theory of mind stories that described a character's action caused by his or her false belief (mental cause), and mechanical inference (MI) stories that described the consequences of invisible mechanical causal processes, such as melting and rusting. The results clearly undermined any account of activation in the RTPJ based on domain-general causal inference. Both random effects analyses of the group and individual subject analyses revealed higher activity in the RTPJ while subjects read theory of mind stories compared with MI stories. In fact, the response to MI stories was barely higher than the response of the RTPJ during passive fixation. Our results were therefore consistent with those of earlier studies in suggesting that the RTPJ's role in causal inference, if any, is truly specific to reasoning about mental causes.

Meta-Representation

A second domain-general account of the response of the RTPJ to false belief problems is available. False belief stories have a specific and particularly demanding logical structure. Subjects must not only keep track of how the world actually is, but also of how the character believes the world to be. Meta-representation, as this is called, requires the ability to juggle simultaneous and competing pictures of the world, and to inhibit one of these representations in order to facilitate a response based on the other. Any part of this complex process might explain activation of the RTPJ during false belief stories, relative to previous controls.

To test this hypothesis, we used a control condition that would force subjects to represent literal pictures of the world, like photographs and maps (Saxe & Kanwisher, 2003; Saxe, unpublished data), that are incompatible with the true state of affairs. In these false photograph stories (Zaitchik, 1990) a physical representation was formed (e.g., a photograph of an apple tree was taken) and the actual state of the world changed (the apple fell to the ground). As in false belief stories, false photograph stories require subjects to juggle two simultaneous and incompatible representations of reality. Adults, like children (Zaitchik, 1990), found false photograph problems slightly more difficult than corresponding false belief problems. What differs between the conditions is the kind of false representation that subjects must reason about: mental representations in the false belief task versus physical representations in the false photograph task.

Once again, data supported a specific role for the RTPJ in reasoning about other minds. The RTPJ responded robustly when subjects read stories about false beliefs, but the response during false photograph stories was not higher than during passive fixation (figure 5.2). Thus, neither the logical demands of meta-representation nor the requirement to make inferences about invisible causes can account for the response of the RTPJ to false belief stories.

(a)

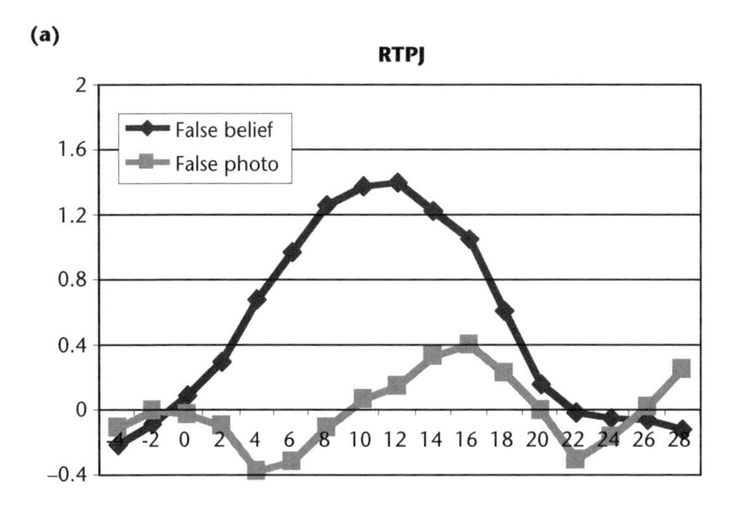

(b)

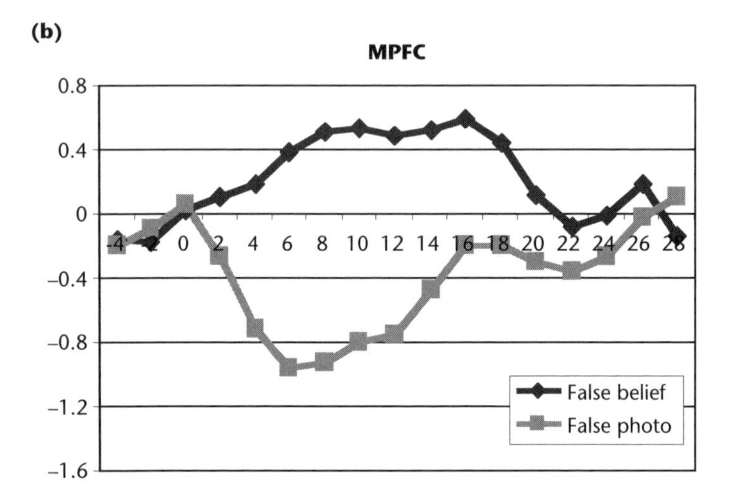

Figure 5.2

The percentage signal change with respect to passive fixation in four brain regions when subjects read false belief and false photograph stories. Time (seconds) on the x axis (Saxe, unpublished data). These data are typical, but they were not extracted from an independent regions of interest analysis in this case, and so are merely illustrative. (a) Right temporoparietal junction (RTPJ). The average response to the false photograph condition was no different from passive fixation. (b) Medial prefrontal cortex (MPFC). This region significantly deactivated to the control condition, but the response to false belief stories was not significantly above fixation.

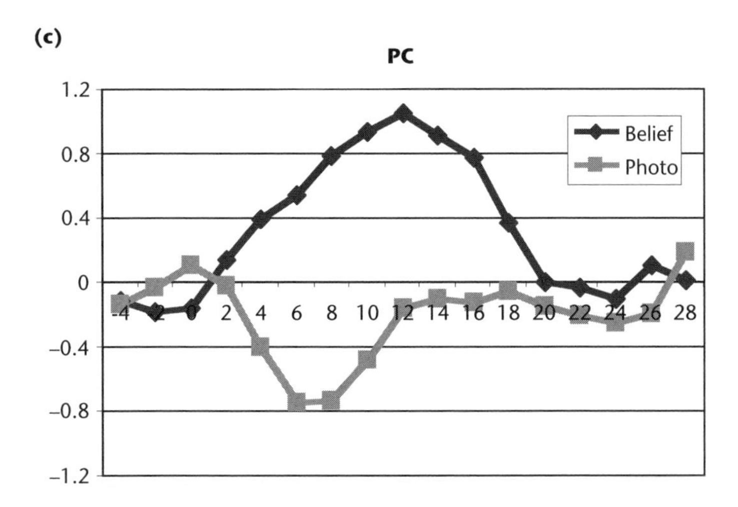

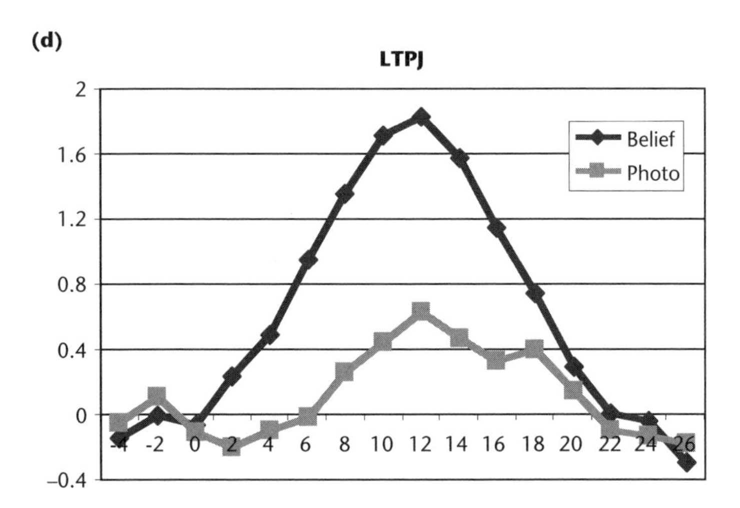

Figure 5.2 (continued)
(c) Posterior cingulate (PC). (d) Left temporoparietal junction (LTPJ).

Agents and Their Behavior

Previous authors concurred that the RTPJ seems to play a role social cognition, and yet denied to the RTPJ responsibility for theory of mind proper; that is, for attributing mental states. Rather, they suggest that the RTPJ reflects a precursor for genuine theory of mind: identifying and representing people, and/or people's physical behavior (Gallagher & Frith, 2003; Frith & Frith, 2003).

Consistent with a precursor role for the RTPJ, many studies reported activation in the nearby right posterior superior temporal sulcus in response to the mere presence of another person (i.e., a human face—Haxby, Hoffman, & Gobbini, 2000, 2002), movements of a person (body motion—Vaina et al., 2001; Grossman & Blake, 2002; Pelphrey et al., 2003a; eye motion—Hooker et al., 2003; Pelphrey et al., 2003b; other animate motion—Castelli et al., 2000; Allison, Puce, & McCarthy, 2000), the intentional action of a person (Zacks et al., 2001; Saxe et al., 2004; Grezes et al., 2004), or in the case of reciprocal imitation (Decety et al., 2002). Also, the RTPJ responds more to control conditions requiring physical or logical reasoning about a character in a cartoon or story (but no false belief attribution) than to scrambled sentence or picture controls (Fletcher et al., 1995; Gallagher et al., 2000; Brunet et al., 2000). Authors have sought a unifying explanation of all of these patterns and proposed a homogeneous role for the RTPJ and the neighboring posterior superior temporal sulcus for understanding other people. However, our results suggest a different resolution.

The mere presence of a person in the stimulus is not sufficient to elicit a strong response in the RTPJ. In two experiments, we directly tested the contribution of the presence of a person in the stimulus to the RTPJ response (Saxe & Kanwisher, 2003). The BOLD response in the RTPJ was not significantly higher for photographs of people than for photographs of other familiar objects (experiment 1), or for verbal descriptions of the physical appearance of a person relative to descriptions of the appearance of nonhuman objects (experiment 2).

To confirm that the RTPJ is recruited specifically during the attribution of mental states, we created new stories in which a character was introduced immediately, but information about that character's mental state was delayed (Saxe & Wexler, in press). The first sentences described the character's background (financial, geographical). Six seconds later, sentences describing what the character wanted or believed were added to the screen. This manipulation allowed us directly to compare the BOLD response in the RTPJ when mental states attribution could begin immediately (original false belief stories), and when the background of a character was described in the story but mental state attribution was delayed. The results of this comparison are illustrated in figures 5.3 and 5.4. The RTPJ response to the beginning of the stories was significantly

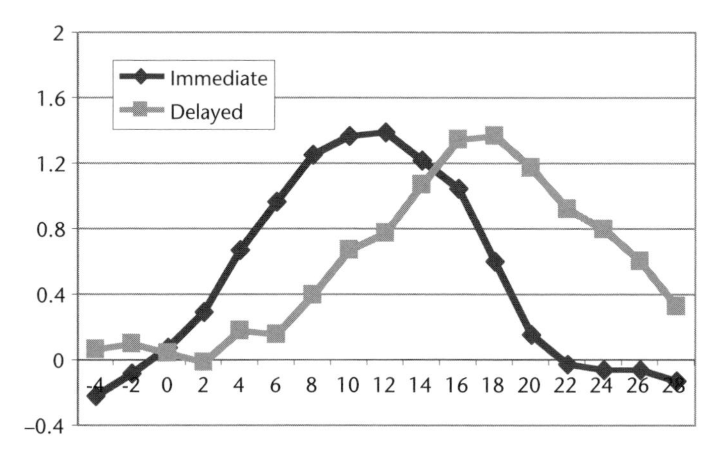

Figure 5.3
The effect of delaying the onset of mental state information in the RTPJ. In the immediate condition, stories began with a character's mental state. In the delay condition, the first section of the story (corresponding to time points 4–8) provided information about a character's social or geographical background, but not about her mental states. Mental state information became available only in the second section of the story (time points 10–14). These data were collected from 12 naïve subjects in a 3.0T scanner at Massachusetts General Hospital in the Charlestown Navy Yard (Saxe & Wexler, in press).

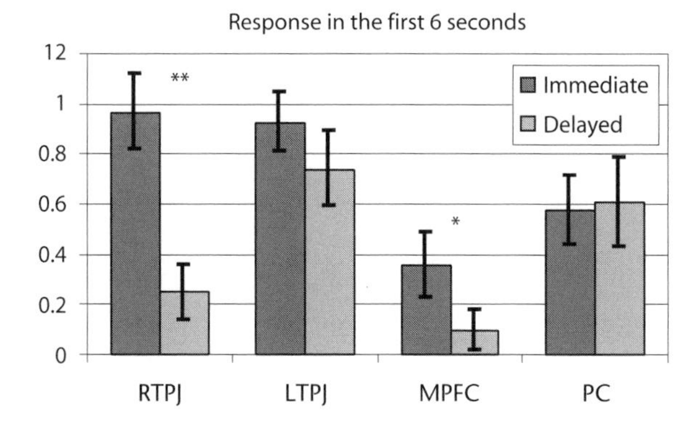

Figure 5.4
Immediate and delayed onset of mental state information in four brain regions. Error bars show the standard error of the mean. * $p < 0.01$, ** $p < 0.001$. Patterns in the left TPJ, MPFC, and PC were each significantly different from the response in the right TPJ. All interactions $p < 0.01$ (Saxe & Wexler, in press).

higher when mental state information was available (average percentage signal change 0.98), relative to when the mental state information was delayed (PSC 0.17, $p < 0.001$, paired-samples *t* test). Clearly, the response of the RTPJ reflects not just the onset of the story or the detected presence of a person, but the beginning of mental state attribution.

These data rule out the possibility that the RTPJ is involved in detecting, perceiving, or reasoning about the mere presence of a person in the stimulus. The same results argue against a role for the RTPJ restricted to the representation of intentional action. The middle section of the new stories described only what a character wanted or believed, and not an action (Saxe & Wexler, in press). However, the response of the RTPJ was robust; the peak response was as high as the response to the original false belief stories, which included both mental states and resulting actions.

Perceiving another person's intentional action is therefore not necessary to induce a robust response in the RTPJ. Other data suggest that it is not sufficient either. If the RTPJ response reflected subjects' imagining the intentional actions described in false belief stories, then having subjects actually watch a human actor engaged in intentional action should produce an even larger response. On the contrary, when subjects watched a video of an intentional action, the response of the RTPJ was significantly lower than while subjects read false belief stories (Saxe, Perrett, & Kanwisher, unpublished data).

The evidence reviewed so far strongly suggests that the RTPJ plays a role in the attribution of mental states—the core of a theory of mind—and not merely in a precursor process. Neither perception of a person nor of an intentional action is sufficient to elicit a robust response in the RTPJ. Mental state attribution is necessary.

If so, how can we explain the consistent finding of a selective response to people and/or intentional actions, clearly precursor processes, in the nearby right posterior superior temporal sulcus? One possibility is that there are at least two distinct regions near the right temporoparietal junction, with different roles in social cognition. Indeed, our data suggest that a region of the right posterior superior temporal sulcus is involved specifically in the representation of intentional action, which is both anatomically and functionally distinct from its neighboring RTPJ (Saxe, Perrett, & Kanwisher, unpublished data). These results highlight the importance of regions of interest analyses: the distinct functional profiles of neighboring regions are not always apparent from group data, and not always amenable to homogenizing explanations.

Not Just False Beliefs

There is one more criterion for a candidate neural substrate of theory of mind: its role must not be limited to the attribution of false beliefs. Observers attribute false beliefs to explain the rare deviance between an action and its surrounding context, but this is not the general case. Reasoning about other minds would not be possible if we did not attribute to others mostly true beliefs, reasonable desires, and rational actions (Dennett, 1996; Bloom & German, 2000). Also, the attribution of false beliefs per se is particularly dependent on the subject's ability to juggle simultaneous competing representations of the world and inhibit the dominant one—the actual state of affairs. A neural substrate for theory of mind should be recruited whenever the subject has to reason about the beliefs and desires of another person, not only in the unusual circumstance of a false belief. The RTPJ meets this requirement; the response to true beliefs and reasonable desires is robust (Saxe & Wexler, in press).

Therefore, the RTPJ met our initial criteria for a neural substrate of theory of mind. Its response is specific to reasoning about mental (rather than physical) causes and mental (rather than physical) representations. It cannot be assimilated into the neighboring posterior superior temporal sulcus as a precursor of genuine theory of mind. Finally, its response generalizes beyond the attribution of false beliefs.

The Medial Prefrontal Cortex, Posterior Cingulate, and Left TPJ

So far, we have focused exclusively on the RTPJ, but our introduction named four brain regions that reliably show increased BOLD response while subjects solve false belief problems, relative to scrambled sentence controls. The other three regions are the LTPJ, PC, and MPFC. Why are all four of these brain regions activated together?

It seems likely that each of these regions is responsible for different components (or computational ingredients) of solving the false belief task. Thus, for instance, some authors suggested that the MPFC is the only region specifically involved in reasoning about other minds, whereas the TPJ (bilaterally) was assigned a precursor function (Frith & Frith 2003; Gallagher & Frith, 2003). Another group hypothesized that the MPFC is required for "integration of emotion into decision making and planning," whereas the PC is involved in "integration of emotion . . . and memory, especially for coherent social narratives" (Greene & Haidt, 2002).

How could we establish whether indeed the right and left TPJs, MPFC, and PC make distinct contributions to solving the false belief problems, and what those contributions are? One common method is to tabulate, in a meta-analysis, all of the other contrasts in which activation was reported for each brain region, and from the union of these contrasts to infer the underlying function of the region. Such a meta-analysis has the advantage of a much broader scope than any single study, canvassing the results of dozens of different paradigms, and generalizing across a much larger population of subjects.

However, meta-analyses have at least two important limitations. The first is statistical. A meta-analysis can see only that the results of a particular experiment were significant in one brain region and not significant in another. But a difference of significances is not necessarily a significant difference. To conclude that two brain regions make significantly different contributions to a task, it is necessary to compare the two response profiles directly. The second drawback is that a meta-analysis has only limited anatomical precision. Thus, activations that may not overlap at all within each individual, as we saw for the RTPJ and right posterior superior temporal sulcus, could appear to overlap when the same data are averaged across subjects. This problem is exacerbated when comparing activations across subject groups or across studies.

Therefore, to evaluate what, if anything, is each brain region's distinct contribution to theory of mind, the ideal is to compare directly (1) the response of each region across a range of conditions, as we did with the RTPJ, and (2) the response of different regions to one another. Differences among these four brain regions emerge under such scrutiny. The overall pattern so far, however, is one of striking functional similarity. Distinguishing the contributions of each region to reasoning about other minds will be a critical problem for future research.

MPFC

The functional profile of the MPFC has been repeatedly dissociated from the RTPJ. For instance, in one of the first studies to investigate the neural correlates of the false belief task (Gallagher et al., 2000), unlike bilateral TPJ, it was not activated above threshold when subjects read stories about human actions in the absence of false beliefs, relative to scrambled sentence controls. To illustrate the difficulty of identifying the distinct functions of these two regions, two interpretations of this dissociation are possible, neither of which is right: first, that the MPFC is involved in representing the contents of other minds and the TPJ is involved in some precursor process, and second that the TPJ is involved in representing the contents of other minds while the MPFC is involved in executive control or inhibition.

The critical question is: what is the status, with respect to reasoning about other minds, of the human action control stories used by Gallagher et al. (2000). One possibility, the authors' own assumption, is that subjects did not reason about mental states when reading these control stories that did not involve false beliefs. This leads to the conclusion that the activity of the RTPJ, unlike the MPFC, is not specific to the need to reason about mental states, because the RTPJ was recruited for the human action control stories, but the MPFC was not. However, this conclusion is incompatible with all of the data described above, showing that the response of the RTPJ is actually highly specific to reasoning about mental states. So another interpretation of the observed dissociation between the RTPJ and MPFC is desirable.

The alternative view is that subjects *did* represent the mental states of the characters in the human action stories, even in the absence of false beliefs. I find this alternative very plausible. In one of the human action stories, for example, a burglar in the course of robbing a bank carefully climbs around the security laser beam across the door. But once inside he accidentally disturbs a small animal that runs out the bank door, breaking the beam and setting off the alarm. To me, this story strongly invites mental interpretation of what the burglar wants, believes, and feels about the outcome.

The assumption that subjects did reason about the mental states of the characters in human action stories is compatible with a role for the RTPJ (which did respond to these stories relative to the scrambled baseline) in reasoning about other minds. However, what are we to make of the fact that the MPFC did not respond to the stories? Since the MPFC was recruited selectively for reasoning about false beliefs, we might conclude that it is involved in reasoning about false representations per se, including inhibitory processing (Decety & Sommerville, 2003). That is, the MPFC, unlike the RTPJ, could aid in juggling two simultaneous and incompatible representations of reality, one of which is false, independent of whether the story involved any mental states. Unfortunately, this conclusion was undermined by further investigation.

This latter hypothesis predicts that responses of the MPFC and the RTPJ should differ in two ways in our own experiments. First, we would predict that the MPFC should be recruited for reasoning about false photographs, the control condition designed to have demands on inhibition and executive control to those of false belief stories, but without mental content. Second, the MPFC should not show a robust response when subjects attribute true beliefs and reasonable desires to a character. In our data, neither of these predictions was confirmed.

First, the magnitude and latency of the selectivity for false beliefs, relative to false photographs, were strikingly similar in the MPFC and the RTPJ (and indeed in the LTPJ and PC as well; figure 5.5). Second, PSC relative to fixation in the MPFC for attributing true beliefs and reasonable desires (average PSC 0.61) was no lower than when the same subjects attributed a false belief (0.50).

Also like the RTPJ, in our experiments the MPFC was selective for reasoning about mental states. The response of the MPFC was no higher during stories that described the physical appearance of a person than during descriptions of the appearance of inanimate objects (Saxe & Kanwisher, 2003). The response during the first six seconds

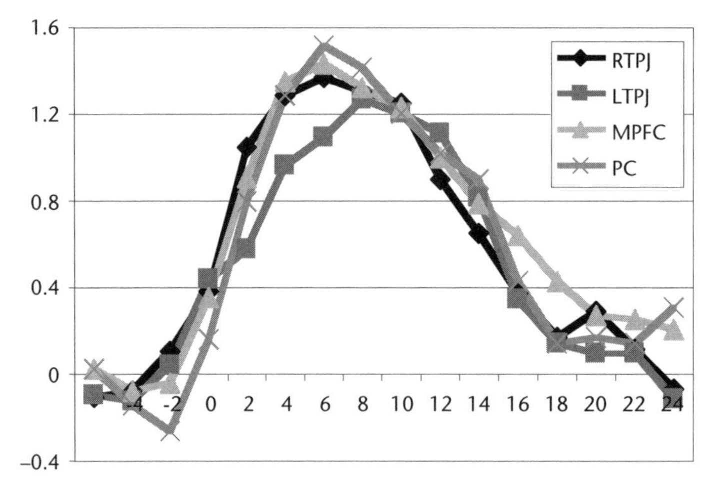

Figure 5.5
The latency and magnitude of selectivity for false belief over false photograph stories in four brain regions. Although the response profiles of these regions differ with respect to fixation, the results of this direct subtraction reveal striking similarity among the regions (Saxe, unpublished data).

of the delayed-attribution condition (average PSC 0.13) was significantly lower than during the same interval of stories that allowed immediate mental state attribution (PSC 0.36, $p < 0.01$, paired-samples *t* test; see figure 5.4; Saxe & Wexler, in press).

These data suggest that both the MPFC and the RTPJ are specifically involved in the core of theory of mind, representing and reasoning about mental states. Moreover, we did not replicate the earlier finding (Gallagher et al., 2000) that the RTPJ but not the MPFC responded more to human action stories than to control stories. Using similar human action stories to those in the original study, we found an intermediate response in both the RTPJ and MPFC. We speculated that the relatively unconstrained human action stories may have elicited mental state attribution on some trials but not on others. The response in both regions was half-way between false belief stories that require mental state attribution all the time and mechanical inference stories, which never do.

There is one gross functional difference between the RTPJ and the MPFC in all of our data: the magnitude relative to fixation of the response to control conditions such as false photograph stories. In the RTPJ, the response to false photograph stories is, on average, no higher than the response during passive fixation (see figure 5.2a). In the MPFC, in contrast, the BOLD response to the false photograph stories is actually significantly below the fixation baseline ($p < 0.01$, figure 5.2b). The deactivation in MPFC is consistent with reports of the MPFC's generally high resting metabolic rate (Raichle et al., 2001). However, the functional significance of this difference is not clear. As described, the latency and magnitude of selectivity in the MPFC is overall remarkably similar to those observed in the RTPJ, and so the difference seems to lie entirely in the relative BOLD signal in these regions during passive fixation, the condition over which this paradigm gives the experimenter least control.

In all, direct tests in individual subject regions of interest analyses reveal no clear functional difference between the RTPJ and the MPFC. Both regions fulfill all of the initial criteria for a region genuinely involved in theory of mind. Of course, these few tests do not show—or even suggest—that there are no functional differences at all between the structures. The data described above can only rule out the simplest alternative hypotheses—that the MPFC is involved in domain general meta-representation, for instance, or that the RTPJ is only involved in precursors of theory of mind.

The question for the future remains: within the domain of theory of mind, what is the specific contribution of each of these regions? What kind of representation of others' mental states, and/or what process or computation defined over those representations, does the activation of each region reflect? From the current literature, only one hint is available. Two studies had subjects play a simple game, such as "rock, paper, scissors," in two conditions: in some trials, subjects believed that they were playing against a computer opponent, and in others subjects believed that they were playing against a live human (McCabe et al., 2001; Gallagher et al., 2002). Both studies reported greater activation in the MPFC, but not the RTPJ, when subjects thought they were playing against a human (although unfortunately, neither region of interest analyses nor region by condition interactions were available). These results are suggestive, but their interpretation is unclear. Selective MPFC activation could reflect arousal and emotion associated with playing a human opponent, or choosing one's own actions in a social interaction, or predicting a human opponent's reply. Green and Haidt (2002), for instance, suggested that the medial prefrontal cortex is required for "integration of emotion into decision-making and planning." The RTPJ, on the other hand, might be more involved in inferring the content of another person's beliefs and desires (which are simple and unchanging in a game such as "rock, paper, scissors"). These are just post hoc speculations however; more work is necessary.

PC and LTPJ

Determining the distinct contribution of the other two regions, the PC and left TPJ, is as yet equally difficult. Both regions are robustly activated by false belief relative to false photograph or mechanical inference stories (Saxe & Kanwisher, 2003). Neither region's response was higher to pictures of people than pictures of objects, or to verbal descriptions of people relative to verbal descriptions of objects (Saxe & Kanwisher, 2003). And both regions do respond robustly when reasoning about mental states that are not false. That is, in these experiments the attribution of mental states seems to be sufficient and (mostly) necessary for involvement of the LTPJ and PC, just like the RTPJ and MPFC.

However, data from our most recent experiment do contain one small hint about the distinctive contribution of these two regions: responses of the PC and LTPJ are not perfectly selective for the availability of mental state information (Saxe & Wexler, in press). In both regions, the response to stories about people was robust, even when mental state information was not yet available. That is, the LTPJ and the PC both did show a significant increase in BOLD when only geographical-social background about the protagonist was available. These data were significantly different from the pattern in the RTPJ (both interactions $p < 0.01$), but we can only speculate about what this dissociation means. The LTPJ and the PC may be activated in anticipation of needing to reason about the character's mental states, in forming an impression of the personality of the protagonist, or in retrieving plausible mental state candidates for this character from long-term memory. Greene and Haidt's (2002) proposal that the PC is recruited for the creation of coherent social narratives is also consistent with these data. The task for the future will be to move beyond post hoc speculations by testing these hypotheses directly.

Conclusion

Developmental psychology has so far proved to be a very rich vein for neuroimaging miners interested in the neural substrate of theory of mind. The canonical false belief problem revealed at least four brain region candidates for that neural substrate. Control conditions adopted from developmental psychology, such as false photograph stories, help to establish the selectivity of these regions. Just like developmental psychology, however, social neuroscience may have to move on. Important questions for the future will not be whether given brain regions are involved in theory of mind reasoning, but rather how those regions are involved. The particular computational ingredients of theory of mind indexed by activation in the RTPJ, MPFC, PC, and LTPJ are as yet unknown.

Acknowledgments

Supported by grant NIHM 66696. None of this work would have been possible without Nancy Kanwisher. Also, thanks to Anna Wexler, Lindsey Powell, Laura Schulz, Yuhong Jiang, Johannes Haushoffer, Tania Lombrozo, and Jonah Steinberg for comments on previous drafts.

References

Allison, T., Puce, A., & McCarthy, G. (2000). Social perception from visual cues: Role of the STS region. *Trends in Cognitive Science, 4*, 267–278.

Bloom, P. & German, T. P. (2000). Two reasons to abandon the false belief task as a test of theory of mind. *Cognition, 77*, B25–31.

Brunet, E., Sarfati, Y., Hardy-Bayle, M. C., & Decety, J. (2000). A PET investigation of the attribution of intentions with a nonverbal task. *Neuroimage, 11*, 157–166.

Castelli, F., Happe, F., Frith, U., & Frith, C. (2000). Movement and mind: A functional imaging study of perception and interpretation of complex intentional movement patterns. *Neuroimage, 12*, 314–325.

Decety, J., Chaminade, T., Grezes, J., & Meltzoff, A. N. (2002). A PET exploration of the neural mechanisms involved in reciprocal imitation. *Neuroimage, 15*, 265–272.

Decety, J. & Sommerville, J. A. (2003). Shared representations between self and other: A social cognitive neuroscience view. *Trends in Cognitive Sciences, 7*(12), 527–533.

Dennet, D. (1996). *Kinds of Minds: Towards an Understanding of Consciousness*. New York: HarperCollins.

Dennett, D. (1978). Beliefs about beliefs. *Behavioural and Brain Sciences, 1*, 568–570.

Flavell, J. H. (1999). Cognitive development: Children's knowledge about the mind. *Annual Review of Psychology, 50*, 21–45.

Fletcher, P. C., Happe, F., Frith, U., Baker, S. C., Dolan, R. J., Frakowiak, R. S., et al. (1995). Other minds in the brain: A functional imaging study of "theory of mind" in story comprehension. *Cognition, 57*, 109–128.

Frith, U. & Frith, C. D. (2003). Development and neurophysiology of mentalizing. *Philosophical Transactions of the Royal Society of London, Series B, Biological Sciences, 358*, 459–473.

Gallagher, H., Jack, A., Roepstorff, A., & Frith, C. (2002). Imaging the intentional stance in a competitive game. *Neuroimage, 16*, 814.

Gallagher, H. L. & Frith, C. D. (2003). Functional imaging of "theory of mind." *Trends in Cognitive Sciences, 7*, 77–83.

Gallagher, H. L., Happe, F., Brunswick, N., Fletcher, P. C., Frith, U., & Frith, C. D. (2000). Reading the mind in cartoons and stories: An fMRI study of "theory of mind" in verbal and nonverbal tasks. *Neuropsychologia, 38*, 11–21.

Goel, V., Grafman, J., Sadato, N., & Hallett, M. (1995). Modeling other minds. *Neuroreport, 6*, 1741–1746.

Greene, J. & Haidt, J. (2002). How (and where) does moral judgment work? *Trends in Cognitive Sciences, 6*(12), 517–523.

Grezes, J., Frith, C. D., & Passingham, R. E. (2004). Inferring false beliefs from the actions of oneself and others: An fMRI study. *Neuroimage, 21*, 744–750.

Grossman, E. D. & Blake, R. (2002). Brain areas active during visual perception of biological motion. *Neuron, 35*, 1167–1175.

Haxby, J. V., Hoffman, E. A., & Gobbini, M. I. (2000). The distributed human neural system for face perception. *Trends in Cognitive Sciences, 4*, 223–233.

Haxby, J. V., Hoffman, E. A., & Gobbini, M. I. (2002). Human neural systems for face recognition and social communication. *Biological Psychiatry, 51*, 59–67.

Hooker, C. I., Paller, K. A., Gitelman, D. R., Parrish. T. B., Mesulam, M. M., & Reber, P. J. (2003). Brain networks for analyzing eye gaze. *Cognitive Brain Research, 17*(2), 406–418.

McCabe, K., Houser, D., Ryan, L., Smith, V., & Trouard, T. (2001). A functional imaging study of cooperation in two-person reciprocal exchange. *Proceidings of the National Academy of Sciences of the United States, 98*, 11832–11835.

Pelphrey, K. A., Mitchell, T. V., McKeown, M. J., Goldstein, J., Allison, T., & McCarthy, G. (2003a). Brain activity evoked by the perception of human walking: Controlling for meaningful coherent motion. *Journal of Neuroscience, 23*(17), 6819–6825.

Pelphrey, K. A., Singerman, J. D., Allison, T., & McCarthy, G. (2003b). Brain activation evoked by perception of gaze shifts: The influence of context. *Neuropsychologia, 41*, 156–170.

Premack, D. & Woodruff, G. (1978). Does the chimpanzee have a theory of mind? *Behavioural and Brain Sciences, 1*, 515–526.

Raichle, M. E., MacLeod, A. M., Snyder, A. Z., et al. (2001). A default mode of brain function. *Proceidings National Academy Sciences of the United States, 98*(2), 676–682.

Ruby, P. & Decety, J. (2003). Effect of subjective perspective taking during simulation of action: A PET investigation of agency. *Nature Neuroscience, 4*(5), 546–550.

Saxe, R., Carey, S., & Kanwisher, N. (2004). Understanding other minds: Linking developmental psychology and functional neuroimaging. *Annual Review of Psychology, 55*, 87–124.

Saxe, R. & Kanwisher, N. (2003). People thinking about thinking people: fMRI investigations of theory of mind. *Neuroimage, 9*(4), 1835–1842.

Saxe, R. & Wexler, A. (in press). Making sense of another mind: The role of the right temporo-parietal junction. *Neuropsychologia.*

Saxe, R., Xiao, D.-K., Kovacs, G., Perrett, D. I., & Kanwisher, N. (2004). A region of right posterior superior temporal sulcus responds to observed intentional actions. *Neuropsychologia, 42,* 1435–1446.

Vaina, L. M., Solomon, J., Chowdhury, S., Sinha, P., & Belliveau, J. W. (2001). Functional neuroanatomy of biological motion perception in humans. *Proceidings of the National Academy of Sciences of the United States, 98*(20), 11656–11661.

Vogeley, K., Bussfeld, P., Newen, A., Herrmann, S., Happe, F., et al. (2001). Mind reading: Neural mechanisms of theory of mind and self-perspective. *Neuroimage, 14,* 170–181.

Wellman, H. M., Cross, D., & Watson, J. (2001). Meta-analysis of theory-of-mind development: The truth about false belief. *Child Development, 72,* 655–684.

Wellman, H. M. & Lagattuta, K. H. (2000). Developing understandings of mind. In S. Baron-Cohen, H. Tager-Flusberg, & D. J. Cohen (Eds.), *Understanding Other Minds,* 2nd, ed. (pp. 21–49). New York: Oxford University Press.

Wimmer, H. & Perner, J. (1983). Beliefs about beliefs: Representation and constraining function of wrong beliefs in young children's understanding of deception. *Cognition, 13,* 103–128.

Zacks, J. M., Braver, T. S., Sheridan, M. A., Donaldson, D. I., Snyder, A. Z., Ollinger, J. M., et al. (2001). Human brain activity time-locked to perceptual event boundaries. *Nature Neuroscience, 4,* 651–655.

Zaitchik, D. (1990). When representations conflict with reality: The preschooler's problem with false beliefs and "false" photographs. *Cognition, 35,* 41–68.

6 Theory of Mind and the Evolution of Social Intelligence

Valerie E. Stone

I would like to place social cognition, and thus the social brain, in an evolutionary context. Humans are social animals, adapted to living in groups. Group living probably goes back at least 54 million years in our family tree, to our common ancestor with other primates (Yoder et al., 1996; Foley, 1997). Many of our social behaviors are shared with our primate cousins, so it is likely that many of our social cognitive abilities are as well. On the other hand, each species has its uniqueness. Humans, unlike other primates, use language, plan for the future, and make complex inferences, and all of these abilities affect our social behavior and cognition. These abilities, however, may depend on a relatively small number of uniquely human capacities, with most of our social cognition being in common with our primate cousins. Recent studies in evolutionary biology and primatology have revealed surprising cognitive capacities in great apes—orangutans, with whom we share a common ancestor 14 million years ago (mya), and chimpanzees and bonobos, with whom we share a common ancestor 5 to 7 mya (Foley, 1997; Gibbons, 2002; figure 6.1). African and Asian monkeys (common ancestor 26 mya) for the most part do not seem to share these abilities, but apes engage in tool use, cultural learning, and insightful problem solving (Suddendorf, 1999; McGrew, 2001). These discoveries have forced psychologists and neuroscientists to define more narrowly what is unique about human minds, and give us a basis for understanding how our complex cognition can be continuous with that of our closest relatives and our hominid ancestors.

Social behavior can be defined as any interaction with members of one's own species. Social cognition, then, is the information-processing architecture that enables us to engage in social behavior. Social neuroscience is the study of how the brain implements the information-processing architecture for sociality. A key question for social neuroscience is to what extent brain systems subserving social behavior are socially specific, consisting of neural processes that operate only on social information, and to what extent these systems are more general, consisting of neural processes

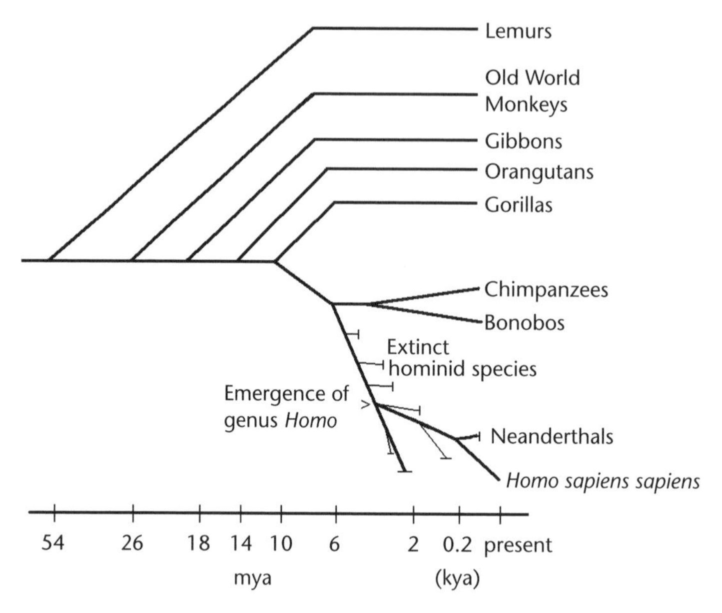

Figure 6.1
Our primate family tree, showing branching points leading to monkeys, great apes, hominids, and humans. Time is measured in mya (millions of years ago), except where marked kya (thousands of years ago). Information on when certain branchings occurred is based on genetic analyses and was compiled from Yoder et al. (1996), Foley (1997), Gibbons (2000), and Wildman et al. (2003).

that subserve multiple areas of cognition. Research in patients with neurological damage can give us insight into these questions, as we find evidence from dissociations either that social cognition can be impaired independently of more general cognitive processes, or that deficits in social cognition are always accompanied by more general cognitive impairments. Neuroimaging can reveal whether brain systems involved in different kinds of social tasks are in the same or different locations in the brain, and how much those systems overlap with areas involved in more general cognitive functions.

I contend that key aspects of our social cognition depend not on specifically social processes, but on some very general and powerful cognitive abilities that are unique to our species, and are used in many other contexts besides social cognition. Our capacity for executive control over cognition, for metarepresentation, and for recursion enable not only our complex social cognition, but many of our other uniquely human abilities as well[1]: symbolic language, syntax, future planning, and episodic memory, to name a few (Suddendorf, 1999; Corballis, 2003). The remarkable changes that have taken place in hominid brain evolution have been due to expansion of our general cognitive capacities, particularly those subserved by the frontal lobes. One could posit that these general cognitive capacities were selected for in primate and hominid evolution because they were so useful for social behavior; indeed Brothers (1997) made such a case for the general perceptual computations that subserve face recognition. Such arguments are historical speculations rather than scientific theories; it is difficult to know what selection pressures operated in the past. One can make some inferences from the adaptive design of a cognitive system, proposing that a cognitive system will perform the function it was selected for most efficiently, and that it will have design features specific to that function (Tooby & Cosmides, 1992). However, executive function and recursion clearly fail that test, as they are equally useful in memory, language, social cognition, and tool manufacture, and have no design features specific to the social domain (Corballis, 2003).

Much of our social behavior, however, is similar to that of other primates and other mammals. It is parsimonious to assume that many of the brain systems that subserve social behavior in humans and in other social mammals with whom we share common descent are inherited from a common social mammal ancestor, and will have common features. The use of animal models in social neuroscience depends on such assumptions. In comparative biology, this is known as homology, to indicate that two species share similar structures because of common descent and common genetics, not because they evolved independently in two different lines. If forming a mother-infant bond looks similar in many branches of our family tree, we can reasonably expect that the brain systems that implement that behavior will be similar in those branches. Furthermore, even our higher-level social cognitive abilities, those that depend on language and complex self-representation, should be seen as continuous with those of our primate ancestors, in whom the building blocks of our complex sociality and cognition first emerged.

Although some domain-specific brain systems for sociality may have been preserved for tens of millions of years, social neuroscientists should be wary of assuming too much social specificity in social cognition. With each domain of social cognition—attachment, hierarchy negotiation, cooperation, in-group–out-group categorization—one must carefully define what the components of an ability are, consider whether each component is an aspect of cognition that we share with other social primates, and then do careful work to define whether each component is specific to social behavior or is used in nonsocial contexts, and whether it depends on more general cognitive processes.

Social Cognition in Monkeys, Apes, and Humans: Components of Theory of Mind

As an example of how to approach a topic in social neuroscience in this way, I would like to review research on our ability to understand other people's mental states, a cognitive capacity known as theory of mind (ToM). Humans make inferences about and interpret others' behavior in terms of their mental states, meaning their emotions, desires, goals, intentions, attention, knowledge, and beliefs. Thus ToM encompasses a variety of cognitive processes and takes several years to unfold in human development (Baron-Cohen, 1995; Wellman & Liu, 2004). By breaking ToM down into components, we can ask which of those components are shared with our ape relatives, which are uniquely human, and which seem to be socially specific.

First, however, we must clarify some issues of terminology, as different scholars and different fields sometimes use the term theory of mind differently. The developmental and primate literature on theory of mind makes distinctions among several different types of mental state inferences. Before age four years, children can make inferences about others' intentions, goals, desires, wants, and feelings. Somewhere between three and four they can infer what are called epistemic mental states: knowledge, belief, perception. Sometimes ToM is used to refer specifically to the ability to represent the contents of one's own and others' mental states, something that younger children cannot do. Theory of mind is often seen as equivalent to metarepresentation, the ability to represent representations, as in, "He thinks that [his car is in the garage]," or "She saw that [the lion had escaped from its cage]" (Leslie, 1987; Perner, 1991; Baron-Cohen, 1995; Suddendorf, 1999). Not all of the mental states that we routinely infer require metarepresentational inferences. For example, inferring another person's emotional state does not require representing someone else's representations, but only his or her external appearance; "She looks angry." For this reason, Leslie and Frith (1990) proposed that inferring emotional states should not be considered ToM. Inferring others' intentions, goals, and desires is another gray area for theory of mind terminology, as such inferences also do not necessarily require metarepresentation. Children can make these inferences much earlier than they can do metarepresentation (Suddendorf & Whiten, 2001). In social neuroscience as in developmental psychology, ToM is used broadly to mean inferring a variety of mental states, not limited to metarepresentation. It is important to be clear which type of mental state is meant when using the term.

Developmental psychology can reveal the building blocks and components of ToM, and comparative ethology can tell us whether or not our mammal and primate relatives have each component (Saxe, Carey, & Kanwisher, 2004; Suddendorf, 1999). The groundwork is then laid for neuroscience to investigate the brain systems involved in each component. Many building blocks of ToM are present in our ape relatives (Suddendorf, 1999; Suddendorf & Whiten, 2001; Hare, Call, & Tomasello, 2001; Leavens, Hopkins, & Thomas, 2004). Thus, those building blocks at least are more ancient than the hominid line of the past 5 to 7 million years.

Below, I discuss developmental studies on how humans develop ToM and relevant primate studies, before reviewing the neuroscience data on the brain systems involved. In reviewing the developmental literature, I remain agnostic on theories of how ToM develops, whether through simulation, theory building, or modular maturation. Rather, I focus on when certain abilities emerge and which abilities might share common processes. From such a review, it is clear that we have gaps in research on the neural basis of ToM. For example, although we have ample research on systems underlying our ability to detect eye gaze direction, there is almost no research on what we do with eye gaze; that is, following gaze or establishing joint attention. Much research on theory of mind in the brain does not include control conditions with equivalent task demands (I include some of my own research in this), and thus leaves open the question of whether certain areas are involved specifically in ToM. As a corrective to this, I suggest taking a careful look at components of ToM to see how future research might take account of these issues.

Building Blocks of ToM: Inferring Goals and Intentions

Infants from very early on begin to distinguish actions that are intentional, and to discern an actor's goal. Between five and nine months of age they can differentiate accidental from intentional behavior (Woodward, 1999), and by fifteen months they classify actions according to goals of actions (Csibra et al., 2003). These results show implicit understanding of intentions and goals. Chimpanzees (*Pan troglodytes*) and orangutans (*Pongo pygmaeus*) also distinguish visually between accidental and intentional actions (Call & Tomasello, 1998). Monkeys do not seem to make this distinction. Assuming homology, this would put the date for this ability in a common ancestor at about 14 mya.

Jellema et al. (2000) investigated the neural networks involved in detection of goals. They recorded cells in the anterior superior temporal sulcus (STS) of macaque monkeys that responded to the sight of an agent reaching for something, but only when the agent was looking at the point reached for, with "looking at" indicated by eye gaze, head orientation, and upper body orientation. They proposed that these cells integrate input from cells that respond to gaze direction with cells that respond to limb movement direction. Such integration is a necessary part of the cognitive architecture of detecting goals, which is the first step in understanding intentional action. After apes branched off from monkeys, further elaborations of this architecture may have occurred, involving areas beyond STS.

Building Blocks of ToM: Joint Attention

Between the first and second years, children treat others' gaze direction as a source of information, indicative of that person's focus of attention. In joint attention, emerging between 18 and 24 months of age, the child takes an active step beyond gaze monitoring, to call adults' attention to particular objects by pointing or holding up something for them to see. Establishing empirically that a child is using joint attention usually depends on clear evidence that the child has either moved an object deliberately into another person's line of view, or that he or she is using "protodeclarative pointing," pointing to something and alternating gaze between another person and the object (Baron-Cohen, 1995; Franco & Butterworth, 1996).

It is worth noting that children at this stage do not always successfully bring things to adults' attention. They make systematic errors about what others can and cannot see (Mossler, Marvin, & Greenberg, 1976; Liben, 1978; Flavell et al., 1981), and call adults' attention to things the adult cannot see (such as holding up a watch and saying to a parent on the telephone, "Mommy, look at my new Mickey Mouse watch!"). Thus, there is still no evidence that children at this stage can correctly understand the contents of others' mental states. The child's representation may only integrate information about whether or not an adult is paying attention (binary yes-no), the rough location of objects in space, and a repertoire of actions that generally succeed in engaging adult's attention (holding things up, pointing). Nevertheless, children at this age become more active in trying to affect others' attention.

What evidence is there for joint attention in primates? Kumashiro et al. (2002) presented suggestive evidence that a Japanese macaque monkey (*Macaca fuscata*) engaged in gaze monitoring and learned to use protodeclarative pointing. In a later study, the same group noted that monkeys that had been trained to learn this kind of joint attention could copy complex motions by human experimenters, whereas monkeys that did not engage in joint attention did not copy a human (Kumashiro et al., 2003). Chimpanzees and orangutans have clearly been observed to engage in gaze monitoring as a kind of visual joint attention. They respond differently to situations in which another animal or human experimenter can see an object clearly, and situations in which the other's gaze is occluded (Hare et al., 2000). Chimpanzees seem to use gestures differently when a human experimenter is or is not looking at a food item they desire (Povinelli et al., 2003). Apes, like monkeys, do not seem to use referential pointing spontaneously in their natural environment. Even apes in captivity usually use pointing to request an object (Povinelli & O'Neill, 2000). Given that in their natural environment monkeys walk on all fours and apes knuckle-walk, and that they all rely on their hands for climbing, it is perhaps not surprising that they do not use their hands for such gestures, and instead rely on a cue that is easily available. Since the empirical standard for joint attention is pointing with a finger, there may be a somewhat human-centric methodological bias against finding evidence of joint attention in primates.

No patient or neuroimaging research to date has focused specifically on joint attention. It is an important stage in ToM development, however, and thus this is a significant omission.

Building Blocks of ToM: Pretend Play

At the same age, children begin to engage in pretend play (18–24 months). Pretending involves decoupling the pretend reality ("this is my baby") from perceptual reality ("this is an inanimate doll"). Considerable debate in developmental psychology surrounds what children understand about pretense as a mental state. Leslie (1987) strongly posited that pretense involves representing one's own and others' mental states; that is, that children have a representation such as "Mommy is pretending that → [the doll is a baby]." However, children at this age still fail perspective-taking tasks and make systematic errors about what others can and cannot see (Mossler, Marvin, & Greenberg, 1976; Liben, 1978; Flavell et al., 1981). Thus, it is difficult to make a convincing case that they can represent the contents of a playmate's mental states. Furthermore, younger children do not always understand the role of mental states in pretense (Lillard et al., 2000). In debates over whether children at this age truly understand the representational nature of pretend play, the more parsimonious alternative hypothesis is that they treat pretense as a special kind of action (Wellman & Lagattuta, 2000). Neuroscience research into the brain substrates for pretense might help resolve this debate. If understanding pretense is just a special kind of action, areas involved in pretense might overlap with areas involved in representing actions not currently being done, such as the supplementary motor area. If pretense does involve representing the contents of others' mental states, the same areas active for ToM should be active for pretense. No neuroscience studies of pretense have yet been reported in the literature, so the question remains open.

ToM and Implicit Mentalistic Understanding: Acting Based on Others' Mental States

Something genuinely new emerges between two and three years of age. Children begin to show understanding of some properties of mental things as opposed to physical things. They seem to understand that mental states such as desire and knowledge are private, cannot be observed directly, and can change or not change independent of reality. I refer to this understanding as mentalism, to denote understanding of the properties of mental things. Wellman (1990) discussed this as "belief-desire psychology." Children at about age three also begin to demonstrate implicit knowledge of the contents of others' mental states, although not explicit knowledge. This aspect of ToM, mentalism, may be what is socially specific.

Desire Beginning at around age two, children readily use language about desire, such as "she likes," and seem to understand that people's attitudes and emotions toward various objects can be used to predict what they will do (Wellman & Lagattuta, 2000; Wellman, Cross, & Watson, 2001). Thus, understanding desire may bootstrap off of an understanding of intentions and goals in development. At this age, children can also understand that different people's desires are distinct, that, for example, they don't want to eat their vegetables, but grownups seem to like this yucky-tasting stuff. Experimental evidence indicates that the ability to understand diverse desires emerges between eighteen and twenty-four months (Wellman & Woolley, 1990; Repacholi & Gopnik, 1997; Wellman & Liu, 2004). Likes, wants, and desires are private mental states that can change independent of external reality changing. This developmental step represents the first stage at which children display mentalism, an understanding of something that is specifically mental, private and, decoupled from the external world.

Belief and Knowledge Adult human theory of mind involves understanding epistemic mental states, knowledge, and beliefs. Children and primates may or may not fully understand these mental states. Knowledge is cumulative compared with desire, but like desire, it is changeable and decoupled from the external world. Children do show an implicit understanding of the changeable nature of knowledge and belief before they can talk about it or understand it explicitly.

To test whether children or primates know about someone else's knowledge state, one has to distinguish their representation of someone else's mental state from their representation of the state of reality. If one probes what a subject thinks someone else knows, and what that person knows is true, it is always possible that the subject is just responding with what he or she knows himself or herself. Thus, testing whether a subject can understand that someone else holds a false belief has long been held to be the key test of ToM. But what does it mean to understand false belief? Does someone understand it if he or she can act based on someone else's false belief, but cannot talk about it, or does someone have to be able to talk or answer verbal questions about it explicitly? Understanding false belief, it appears, can be either implicit or explicit.

Two basic kinds of false belief tasks have been used with children, location-change and unexpected contents tasks. In a location-change task, the subject is told a short story (and shown pictures to go with the story, or the story is acted out with toy figures) in which character A puts an object in location 1 and then turns away or goes out of the room. Character B moves the object to location 2 while character A cannot see, and then the subject is asked where character A will look for the object, location 1 or location 2. Children generally can pass this test between $3^1/_2$ to 4 years of age, but it is rare that three-year-olds can pass it (Wellman et al., 2001).

However, three-year-olds can pass an implicit version of the task. Perner and Garnham (2001) used an ingenious test to demonstrate this. The child was told that another person was going to slide down one of two slides, and the child was supposed to place a mat so the person could land safely at the bottom. The situation was manipulated so that sometimes the person who was supposed to slide down had a true belief about which slide he or she was supposed to come down, and sometimes a false belief, and would therefore come down the wrong slide. The task was set up so that the children had to act quickly, without time for deliberation. Children age 36 months were likely to place the mat under the correct slide, showing that they had an implicit ability to track the other person's belief state, true or false. These same children failed a standard false belief task that asked for an explicit choice of which slide the person would come down.

Children age forty months passed a seeing-leads-to-knowing test at an age when children generally cannot pass false belief tasks (Pratt & Bryant, 1990). Children were shown two pictures, one of a girl looking into a box and one of a girl touching the box but looking away, and then asked which girl knew what was in the box. This task does not require explicitly reporting the contents of the girl's mental state, and thus is a more implicit task than false belief tasks. Thus, implicit understanding of the fact that knowledge can change independent of reality, and that such changes are linked to perception, seems to emerge around age three years.

Chimpanzees (*Pan troglodytes*) can do a task that may also reflect implicit understanding of knowledge and ignorance. In chimpanzee society, if two animals see the same piece of food, the dominant animal will almost always get it, and trouble follows if he does not. Chimpanzees were given a choice to head toward one of two food items, one a dominant animal could see and one he could not see, or one location in which he had seen food being hidden but had not seen it being moved (Hare et al., 2001). The subject, the nondominant chimp, could see what the dominant animal had or had not seen. In each case, he preferentially chose to head toward food of which the dominant animal was ignorant or had a false belief about its location. Thus, chimpanzees seem to be able to act on an implicit understanding of other animals' knowledge and ignorance when testing conditions are ecologically valid (competition for food), and their behavior must be guided by tracking other animals' knowledge (Suddendorf & Whiten, 2003).

Like understanding desire, this stage of implicit belief understanding requires understanding that others' mental states are private, internal, and can change independent of reality. The mentalism that emerges with the understanding of desire is thus further extended into implicit understanding of belief (Wellman & Lagattuta, 2000). Even with evidence for implicit belief understanding, however, we still have no water-tight evidence that either three-year-old children or apes can explicitly represent the contents of others' mental states. Indeed, the fact that children that age fail perspective-taking tasks (even with controls for nonmentalizing task demands) is evidence that they cannot explicitly represent the content of others' mental states (Mossler et al., 1976; Liben, 1978; Flavell et al., 1981). To represent others' mental contents explicitly, another cognitive ability must emerge first.

Theory of Mind Proper: Metarepresentation

Although three-year-olds and chimpanzees show implicit tracking of others' belief states, this does not mean that they understand the representational nature of beliefs. Knowledge and belief are referred to as epistemic mental states, as they are about knowledge representations and referents: agent—represents → [proposition]. A statement about belief can be true whether or not the proposition that the belief represents is true. Understanding this representational nature of knowledge and belief means understanding the way that epistemic mental states refer to propositions about the world. Mentalism does not suffice for understanding representation. Rather, a new step in ToM development must occur, metarepresentation, the ability explicitly to represent representations as representations (Perner, 1991; Leslie, 1994; Baron-Cohen, 1995). It is metarepresentation that enables children to pass explicit false belief tasks, and it is metarepresentation that apes lack (Suddendorf, 1999). The child can now understand that beliefs refer to propositions about the world, can explicitly represent the contents of those beliefs, and thus can represent explicitly that beliefs may be mistaken. Passing an explicit false belief task is certain evidence of theory of mind capacities (Dennett, 1987).

However, the converse is not true. Many other cognitive abilities also contribute to being able to pass an explicit false belief task. Thus, if a person fails a false belief task, it does not necessarily mean that she lacks metarepresentation. It might be that he or she lacks one of the other cognitive abilities on which successful false belief task performance depends. In particular, solving such tasks depends on executive control, being able to inhibit the inappropriate response—what the subject knows to be the true state of reality—in order to answer with the perhaps less salient correct response—what the other person's mental state is (Carlson & Moses, 2001; Flynn, O'Malley, & Wood, 2004; Carlson, Moses, & Claxton, 2004). In fact, children can pass false belief tasks slightly earlier if the task demands are changed in such a way that not so much inhibitory control is required, for example, by making the current state of reality less salient (Wellman & Lagattuta, 2000). False belief tasks also depend on working memory and sequencing, as the subject has to keep in mind all the elements of the story as it unfolds in order, and how those elements are changing with respect to each other (Keenan, 1998; Stone, Baron-Cohen, & Knight, 1998). Thus, someone who has deficits in inhibiting a prepotent response or in working memory could easily fail a false belief task while having intact metarepresentational abilities.

Furthermore, metarepresentation may be just one example of a more general cognitive ability, embedding/recursion. Explicitly to represent X represents → [proposition] requires the ability to embed one proposition in another. If metarepresentation is simply one type of recursion rather than a separate ability, difficulties with recursion could cause failures on the false belief task (Corballis, 2003).

Many other cognitive tasks use metarepresentation and recursion: complex syntax, self-representation, creativity, episodic memory and future planning (a.k.a. mental time travel), metamemory, and counterfactual reasoning (Shimamura, Janowsky, & Squire, 1990; Knight & Grabowecky, 1995; Suddendorf, 1999; Suddendorf & Fletcher-Flinn, 1999; De Villiers, 2000; Shimamura, 2000). Thus, recursion and metarepresentation may be general cognitive abilities, not limited to social cognition, that interact with mentalism to produce what we call explicit theory of mind. Indeed, evidence from neuroimaging and patient studies reveals that understanding beliefs can be dissociated from metarepresentation and counterfactual reasoning (Saxe & Kanwisher, 2003; Samson et al., 2004). If metarepresentation is found to be used by many other cognitive abilities besides ToM, it would not be socially specific.

As an example of metarepresentation and recursion in another domain, development in syntactical abilities enables and precedes development in explicit belief representation (De Villiers, 2000). De Villiers proposes that the ability to form embedded sentence complements, those of the form "agent—says → subordinate clause" (e.g. "He said that he finished his peas," or "She says that she saw the movie)" provides the representational structure necessary for explicitly representing belief and knowledge. Sentences such as "Agent says that X," however, are about observable things, utterances, rather than about private and changeable things such as mental states. Thus, the metarepresentational ability that is necessary to use and understand sentence complements is distinct from mentalism, from understanding the relationship between mental states and reality. In development, the ability to use and understand embedded sentences, both sentence complements and embedded relative clauses, precedes the ability to pass false belief tests (De Villiers & Pyers, 2002; Smith, Apperly, & White, 2003). Although not a strict test of cause and effect, this suggests that a general metarepresentational capacity could be necessary before children can perform successfully on explicit false belief tasks.

The idea that explicit ToM is dependent on the metarepresentational competence necessary for such complex grammatical structures is consistent with results on the cognitive abilities of chimps. Chimps failed an explicit false belief task, indicating that they lack explicit metarepresentation (Call & Tomasello, 1999). They also do not show recursive abilities: chimpanzees who have been taught to use signs and symbols to refer to things have never been observed to use complex syntax at all, much less any kind of syntactical embedding (Snowdon, 2001). Apes also do not show any evidence of either episodic memory or future planning, also abilities that depend on metarepresentation (Suddendorf & Busby, 2003). Thus, metarepresentation and recursion seem to be uniquely human abilities.

The union of two abilities—an implicit understanding of the changeable nature of mental states and the ability to do metarepresentation of the explicit contents of those mental states—results in having an explicit ToM in humans. Below, I discuss neuroscience research on ToM, and interpret the findings in terms of implicit mental state understanding (mentalism) and metarepresentation.

Neuroscience Research on ToM: Metarepresentation ≠ ToM

Social neuroscience has been studying ToM for less than a decade, and thus research in this area is still very much in its infancy. We are only now beginning to learn from the methodological issues that developmental and comparative psychologists have had to work out over the past thirty years. Much ToM research in neuroscience has not been done with proper controls for working memory, inhibitory demands of tasks, or other executive functions, nor has it been done with a clear definition of which types of mental states (intention, belief, desire) are tapped by various tasks. The body of research in this area claims variously that ToM might be processed in superior temporal areas, temporal pole, the amygdala, temporoparietal junction (TPJ), medial frontal cortex, orbitofrontal cortex (OFC), and/or frontal pole (Goel et al., 1995; Stone et al., 1998, 2003; Gallagher et al., 2000, 2002; Fine, Lumsden & Blair, 2001; Happé, Mahli, & Checkley, 2001; Stuss, Gallup, & Alexander, 2001; Gregory et al., 2002; Frith & Frith, 2003; Gallagher & Frith, 2003; Snowden et al., 2003; Grèzes, Frith, & Passingham, 2004; Samson et al., 2004; Saxe et al., 2004). Having just painted a picture of the complexity of ToM and the many cognitive abilities that may contribute to successful performance of ToM tasks, not to mention ToM developments after age four, I believe it is not surprising that the brain basis of ToM has not been narrowed down more. One reason that such a variety of brain areas have emerged as important for ToM in different studies could be that these areas may be subserving different aspects of ToM.

Given the review of ToM's components, I believe the following four questions have to be answered before we will have a clear answer about ToM in the brain.

1. Do patients fail ToM tasks because of non-ToM task demands? Do patients show deficits on a ToM task but not a control task that has the same executive function (EF) demands or verbal comprehension demands? Does changing the task to lessen the executive/comprehension demands improve patients' performance?
2. Are ToM and EF independent? Do patients' deficits in ToM correlate with deficits on EF measures tapping into relevant areas of EF: inhibitory control and working memory? Can some patients perform highly on relevant EF measures while being impaired in ToM? Are there patients with EF deficits who can perform well on ToM tasks that minimize executive demands?
3. Does inferring belief require different brain systems as inferring other mental states (desire, intention), or are the same brain areas involved? Can some patients perform poorly on measures tapping explicit metarepresentation of belief while still being able to perform well on tasks measuring understanding of desire or intention, and vice versa?
4. Is metarepresentation/recursion separable from ToM? Can some patients perform poorly on ToM measures while still being able to perform well on other tasks requiring metarepresentation and recursion, such as comprehension of embedded sentences or passing a false photograph test?

Below, I discuss how social neuroscience has or has not provided answers to these questions in more detail. Each of the four questions above can also be addressed using neuroimaging, looking for commonalities and differences in areas activated by different kinds of tasks. I will focus primarily on patient research, as neuroimaging research is reviewed elsewhere in this volume (Saxe, chapter 5; see also Saxe, Carey & Kanwisher, 2004), and as only patient research can answer questions about whether an area is crucial for a particular ability.

Questions 1 and 2 Do patients fail ToM tasks because of non-ToM task demands? Are ToM and EF independent? Science consists of finding evidence that is consistent with one hypothesis and inconsistent with alternative hypotheses. If a patient with neurological damage is impaired on a ToM task, the obvious alternative hypothesis is that the patient failed because of task demands that have nothing to do with ToM, such as inhibitory control or working memory. The correct way to test ToM in patients is to use control conditions and comparison tasks to rule out such an alternative hypothesis. These careful controls have sometimes been done, but not always.

One way to control for task demands is to vary non-ToM task demands to see if this makes a difference in ToM performance. When patients who had lesions in left dorsolateral frontal cortex (DFC) had to hold the elements of the story in working memory (which is the standard false belief task format), they often failed the false belief task (66% correct). However, they performed almost at ceiling on these same tasks when we removed the working memory load (98% correct), showing that their ToM metarepresentational capacities were intact (Stone et al., 1998). Clearly, patients can fail a false belief task because of non-ToM task demands.

In patients with frontal damage, it is particularly important to do these types of controls. If no such controls have been done, then direct correlations between EF measures and ToM performance should be reported, as these can be informative. Patients with right orbitomedial or bifrontal lesions were impaired in tasks measuring perspective taking and deception, but these tasks had strong working memory and inhibitory demands, as they had to track a sequence of actions involving hiding an object (Stuss et al., 2001). There were no control tasks with the same demands but no ToM component. Because these patients were impaired on some EF measures as well, it is difficult to interpret their deficits as truly reflecting deficits in ToM, particularly since no direct correlation between ToM performance and EF measures was reported. Patients with frontotemporal dementia (OFC damage) were impaired on making ToM inferences from eye gaze (Snowden et al., 2003). On a test measuring whether patients would infer what someone else wanted from that person's eye gaze direction (Which one does X want?), they would answer with the item representing what they wanted rather than what the stimulus person was looking at. They could have responded this way because of impulsivity and lack of inhibitory control rather than because of failure in ToM per se. These patients were also impaired on EF measures, although again, no direct correlations between EF and ToM were reported.

Researchers working with patients with frontal damage can learn from the example of Samson et al. (2004), who used an elegant control condition in their video false belief task in patients with TPJ lesions. In the control tasks, memory and inhibitory demands were matched, but false belief attribution was not required. Thus, the patients' poor false belief task performance is more clearly attributable to deficits in ToM rather than non-ToM demands.

One can also control for non-ToM task demands by using different kinds of ToM tasks. Happé et al. (2001), Stone et al. (1998), and Gregory et al. (2002) got around some of the demands in ToM tasks by assessing ToM in neurological patients without using false belief tasks, giving patients cartoons to interpret, or stories that required understanding, for example, desire, emotion, belief, deception, white lies, and social faux pas (Happé's "Strange Stories Task," and Stone and Baron-Cohen's "Faux Pas Recognition Task"). Deficits on ToM questions were found on these tasks in a patient with resection of medial frontal cortex (Happé et al., 2001). Deficits on the Faux Pas task were seen in patients with damage to OFC (Stone et al., 1998; Gregory et al. 2002), as well as in two patients with bilateral amygdala lesions (Stone et al., 2003).

However, even in using ToM tasks that are not false belief tasks, it is still important to look at the relationship to EF. Happé et al. (2001) report that the patient with medial frontal damage had severe EF deficits, particularly in inhibition and working memory tasks, so there remains some question as to why this patient failed ToM tasks. Generally no correlation was noted between Faux Pas task performance and some EF measures, but a relationship was seen between perseverative errors on the Wisconsin and Faux Pas task performance (Gregory et al., 2002). This correlation may have been driven by a couple of patients whose errors on the Faux Pas task were perseverative, in which they kept giving the same answer in the same words. One patient with bilateral but primarily left amygdala damage was impaired in making belief and intention judgments, but also had EF deficits (Stone et al., 2003). In these studies, it is thus difficult to conclude that ToM deficits are independent of EF deficits.

There are examples, however, of patients who have impaired ToM without impaired EF. One patient from Gregory et al. (2002) showed a striking dissociation between his ToM performance, which was poor, and his EF performance, which was close to ceiling (Lough, Gregory, & Hodges, 2002). Stone et al. (2002) reported a patient from the Stone et al. (1998) study with OFC, temporal pole, and amygdala damage, who was impaired on a variety of ToM tasks and had intact executive function. The patient was further impaired in the ability to tell whether someone might be cheating another person, which could possibly tap into ToM, but performed normally on a control task matched exactly for executive and nonexecutive task demands (Stone et al., 2002). A patient with left amygdala damage acquired in childhood had high scores on EF tasks, particularly inhibition tasks, but was severely impaired on a variety of ToM tests, including false belief tasks (Fine, Lumsden, & Blair, 2001). Thus, his poor performance on false belief tasks clearly cannot be accounted for by difficulties with inhibition. A patient with bilateral amygdala damage acquired in adulthood, primarily on the right, was impaired in the Faux Pas Recognition task and in Reading the Mind in the Eyes, and unimpaired on EF (Stone et al., 2003). In looking for ToM deficits independent of EF deficits, patient research points to the amygdala, OFC, and TPJ as possible key areas. Medial frontal cortex may also be involved, but results on independence from EF are inconclusive.

Question 3 Does inferring belief require different brain systems as inferring other mental states (desire or intention), or are the same brain areas involved? Many ToM researchers in neuroscience have used tasks that measure multiple types of inferences, including epistemic inferences about belief and inferences about desire or intention ("Strange Stories Task," and "Faux Pas Recognition Task", Saxe & Kanwisher, 2003). Attributing intentions is an early building block of ToM, something that great apes and very young children can do. Attributing desires taps into mentalism, but does not require metarepresentation. Thus, it is important to try to tease apart whether all of these abilities are using the same neural substrates, or whether different kinds of mentalistic inferences depend on different brain areas.

We can answer these questions if we use more fine-grained methods. Researchers can give patients or participants in a scanner multiple tasks, some that require only belief and others that require desire or intention, and report the results for belief, desire, and intention separately. Patients with OFC damage were impaired on the Recognition of Faux Pas task, but performed at ceiling on false belief tasks (Stone et al., 1998). Many of these patients' errors on the faux pas task (Stone et al., 1998; Gregory et al., 2002) were reflected in statements such as, "Well, he meant to put him down," or "He wanted to make her feel bad so he could feel like the big man." They seemed to make errors in judging whether or not the faux pas was committed accidentally or intentionally. Only the most severely affected patients with frontotemporal dementia, whose damage may have spread beyond OFC, had difficulty with false belief tasks (Gregory et al., 2002).

Further support for the idea that OFC is not involved in any kind of metarepresentational ToM inference comes from results with the same OFC-damaged patients who were tested in Stone et al. (1998) on the Soap Opera Task, which measures ability to make zero-, first-, second-, and third-order mental state inferences. Participants read fairly complex stories about topics such as spies, embezzling, or extramarital affairs, and then were asked to make true-false judgments about statements involving mental state inferences and control statements about details of the stories. Statements about mental states requiring no metarepresentation (zero order) were all about character's likes or desires,[2] such as, "Tim fancies Maria," or "The children like Easter candy." Third-order belief statements required the highest level of recursion, such as, "John thought that Sue believed that Mary thought that X." Control statements were matched for grammatical complexity with ToM questions, and had equal levels of embedded clauses. Both mental state and control statements were constructed so that all levels of grammatical embedding had to be parsed to get the correct answer; the participant could not simply make the correct choice based on one clause. Patients with OFC damage and controls performed equally on this task, scoring high on all questions. Both groups made more errors on second- and third-order statements, but made no more errors on ToM than on non-ToM statements. Thus, with a more difficult task tapping metarepresentation in both ToM and non-ToM linguistic statements, these patients showed no deficits. These results make it unlikely that OFC is involved in any metarepresentational aspects of ToM.

The picture of OFC's role in ToM is complicated by neuroimaging results concerning belief. Most neuroimaging does not find OFC activation specific to belief tasks, although it is difficult to obtain a good signal from OFC in fMRI. However, a recent nonverbal belief task did find OFC activation specific to watching someone perform an action with a false expectation versus a true expectation (Grèzes, Frith, & Passingham, 2004). Since the task looked at expectations, rather than belief statements with content, it is possible that OFC involvement could reflect judgments of intended versus unintended actions. Liu et al. (2004) used the false belief task, with a few true belief catch trials, and looked at event-related potential (ERP) components that are closely time locked to the point at which participants make a belief judgment. New statistical techniques allow some rough localization of where the signal generator for a component is. They report that the ERP component specific to belief questions and not to control questions is statistically inconsistent with a signal generator in dorsal or medial frontal cortex, but is consistent with a generator in left OFC. However, it might be difficult to rule out a generator close to OFC, such as temporal pole. It would also be important to see the results for true and false belief items separately, to rule out the possibility that the signal results from inhibition required when answering about false belief. It may take research with a technology such as magnetoencephalography (MEG), with better spatial and temporal resolution, and using both verbal and nonverbal ToM tasks, to clarify the meaning of these studies. I believe the most parsimonious interpretation of all results on ToM and OFC may be that OFC is mediating judgments about intentional actions, but not belief. The OFC is considered an evolutionarily older part of the frontal lobes, and thus it makes some sense that it would handle judgments of intentional behavior rather than computing metarepresentational mental state inferences.

Unfortunately, it has been difficult to separate belief and desire in both patient and neuroimaging results. Happé et al. (2001) did not report separate results for the stories in Strange Stories task or cartoons that assess belief attribution versus understanding of desire, so it is unclear if medial frontal cortex is involved in belief, desire, or both. If it were specific to representing belief, then it should be active for both true and false beliefs, yet it is not as active during true belief as false belief attribution (Fletcher et al., 1995; Saxe et al., 2004). Fine et al. (2001) used false belief tasks and the same tasks used by Happé et al. (2001) with their patient with amygdala damage. He was impaired on all these tasks, but again, on those that asked about both belief and desire, separate results for belief inferences and other mental state inferences are not reported. With both the medial frontal and the amygdala patients, it is possible that their understanding of desire was completely unimpaired while their understanding of belief was impaired. Samson et al. (2004) report only results for location change false belief tasks, thus we have no information about TPJ patients' understanding of desire as a mental state. According to Saxe and Kanwisher (2003), TPJ is more active during stories requiring desire inferences than physical inferences, as is true for belief compared with physical inferences. Saxe et al. (2004) note that the TPJ is more active for belief than desire inferences, but also that the desire stories may also have elicited belief attributions. In future research in patients, I believe it is important to report results separately for desire and belief, and in neuroimaging to use tasks that cleanly assess the two separately. There are also a range of mental states related to desire (adoration, disgust), and a pure test of belief versus desire would ideally sample all of them. If areas active for belief tasks always turn out to also be active for understanding desire, this would point to these areas being involved in mentalism more generally, rather than just in belief.

Question 4 Is metarepresentation/recursion separable from ToM? Developmental and evolutionary considerations point to mentalism; that is, an understanding of the nature of mental things, being separable from metarepresentation/recursion required to do false belief tasks. If so, these abilities might be dissociable in patients or neuroimaging research. No published patient research has addressed the question of whether areas involved in ToM might be involved in other non-ToM tasks that require metarepresentation and recursion. Saxe and Kanwisher (2003) did evaluate separability of ToM and metarepresentation in fMRI. The false photograph task requires an inference about whether or not a photograph, a nonmental representation, has changed if the state of reality changes after the picture is taken. Thus it requires metarepresentation, representing the representational nature of the photograph, but not mentalism. Participants were scanned while doing false belief tasks and false photograph tasks. Activation in the TPJ, superior temporal pole, and medial portions of frontal pole was significantly greater during belief tasks. However, the researchers did not report whether these areas were more active during the false photograph stories than during stories about physical descriptions of people and objects, which did not involve metarepresentation. If these areas were differentially active during the false photograph test, that would be some evidence for their involvement in metarepresentation. Clearly, TPJ and medial portions of frontal pole and temporal pole seem to be involved in mentalism, and possibly, this ability can be separated from metarepresentation.

With respect to TPJ in particular, further research distinguishing it from areas involved in recursion would be helpful. Some regions very close to the parts of TPJ that were reported for ToM (Saxe & Kanwisher, 2003) have also been found to be active during a specific kind of grammatical task that requires recursion, the processing of embedded relative clauses (frontal areas were also active; Caplan et al., 2002; Cooke et al., 2001). These areas do not completely overlap with the TPJ areas for ToM, but some direct tests would be useful. Using a test such as the Soap Opera task that I used in patients with OFC impairment would directly compare embedding/recursion in both ToM and non-ToM control questions. Using this test with patients with medial frontal damage, TPJ damage (provided they were not aphasic), or frontal pole damage, all areas thought to be involved in belief representation, could help uncover dissociations between recursion and mentalism in ToM inferences. A study directly comparing the processing of embedded sentences and ToM in fMRI would solidify the conclusion that TPJ is involved in mentalism rather than recursion and metarepresentation.

Neuroimaging research with implicit belief tasks would also be important. The first study ever to test ToM with fMRI used a task that required only implicit inferences about beliefs, which represents a true advantage over many imaging studies (Goel et al., 1995). Most imaging studies have evaluated explicit, deliberate reasoning about ToM, often including instructions to think about character's motivations or mental states (Liu et al., 2004; Saxe et al., 2004). As ToM inferences in everyday life are made on the fly, implicitly, tasks requiring deliberate inferences may not tap into these processes in the same way. Implicit belief attribution tasks, perhaps styled after this first one used, or after Perner and Garnham's (2001) implicit false belief task (without the running around with mattress pads), would be valuable additions to imaging research in this area, as they might help identify areas that are involved in mentalism but not in metarepresentation of belief.

The Maturation of ToM Research in Neuroscience

Decades of research on ToM in developmental psychology and primatology have given us a detailed picture of its precursors and components. Developmental research into why children have difficulty with false belief tasks, in particular, provides insight into how people can seem to be impaired on certain ToM tasks because of limitations in non-ToM cognitive abilities. Neuroscience research on ToM is just beginning to take these methodological lessons into account. We are also just starting to make distinctions between ToM and other related abilities such as recursion.

I believe the mentalism evident in children's and primates's understanding of desire and their implicit understanding of belief should be carefully distinguished from metarepresentation of belief and recursion, because they emerge at different points in development and evolution. Although further research will solidify this conclusion, medial portions of frontal pole, TPJ, and superior temporal pole are probably activated in ToM tasks because they subserve mentalism, rather than metarepresentation and recursion in general. I suggest that these areas mediate what is specifically social about ToM, the mentalism required to understand that belief states and desires can change independent of reality, and form the neural basis of implicit mental state understanding. As such, the computations carried out by these areas would be maturing between twenty-four and forty months of age in humans and could be those shared with other great apes. To the extent that brain regions are involved in specifically social computations, they may not be involved in uniquely human computations.

Metarepresentation, recursion, and executive control are not at all limited to ToM. They enable language, complex tool manufacture, future planning, episodic memory, and explicit cultural transmission of knowledge, all things that are hallmarks of the cognitive uniqueness of *Homo sapiens sapiens* (Suddendorf, 1999; Corballis, 2003). Uniquely human abilities are likely to be frontal and temporal, as these areas are clearly disproportionately larger in humans compared with other recent hominids and great apes, more so than other cortical regions (Semendeferi et al., 2001; Lieberman et al., 2002; Stone, in press). The degree of executive control over cognition in humans is unique among primates. This at least we know is mediated by the frontal lobes. In contrast, brain areas involved in metarepresentation and recursion remain unknown. Uniquely human aspects of language depend on recursion, and such complex syntactical abilities seem to involve areas in left temporoparietal and inferior frontal regions (Caplan et al., 2002; Cooke et al., 2001).

Overall, areas involved in the social aspects of ToM seem well positioned anatomically to interact with areas involved in executive function and recursion. Further research using both patient data and neuroimaging, testing a variety of both ToM and non-ToM tasks that require executive control, metarepresentation, and recursion, can clarify these issues. For now, we can understand the processing of ToM in the brain as the interaction of several regions, some specifically social, some not. The operation of human social intelligence in the brain involves areas and functions that are shared with other social mammals, such as gaze direction-detection in the STS; some shared only with primates, such as gaze following; some shared only with great apes, such as joint attention or detecting intentionality; and some that are only human. As neuroscience research on ToM matures, and takes these complexities into account, we will have a clearer picture of how brain areas involved in ToM might have evolved and how different components of ToM interact with each other.

Notes

1. In an unexpected contents task, the subject is shown a container that is clearly labeled as if it contains one kind of thing; for example, a candy box clearly indicates that it contains candy. The subject is shown that it really contains a quite different thing, such as pencils. Then the subject is asked what another person, who has not seen what is inside the box, will think is in there. Control questions usually ask about what was true originally, what the subject thought originally, and what is true right now.

2. Some would describe desire statements as involving first-order intentionality (Dennett, 1987). When I use zero-order, first-order, second-order, and third-order here, I am referring to the level of grammatical embedding.

References

Baron-Cohen, S. (1995). *Mindblindness: An Essay on Autism and Theory of Mind.* Cambridge: MIT Press.

Brothers, L. (1997). *Friday's Footprint: How Society Shapes the Human Mind.* Oxford: Oxford University Press.

Call, J. & Tomasello, M. (1998). Distinguishing intentional from accidental actions in orangutans (*Pongo pygmaeus*), chimpanzees (*Pan troglodytes*) and human children (*Homo sapiens*). *Journal of Comparative Psychology, 112*(2), 192–206.

Call, J. & Tomasello, M. (1999). A nonverbal false belief task: The performance of children and great apes. *Child Development, 70*(2), 381–395.

Caplan, D., Vijayan, S., Kuperberg, G., West, C., Waters, G., Greve, D. D., et al. (2002). Vascular responses to syntactic processing: Event-related fMRI study of relative clauses. *Human Brain Mapping, 15*(1), 26–38.

Carlson, S. & Moses, L. (2001). Individual differences in inhibitory control and children's theory of mind. *Child Development, 72*(4), 1032–1053.

Carlson, S., Moses, L., & Claxton, L. (2004). Individual differences in executive functioning and theory of mind: An investigation of inhibitory control and planning ability. *Journal of Experimental Child Psychology, 87*(4), 299–319.

Cooke, A., Zurif, E. B., DeVita, C., Alsop, D., Koenig, P., Detre, J., et al. (2001). Neural basis for sentence comprehension: Grammatical and short-term memory components. *Human Brain Mapping, 15*, 80–94.

Corballis, M. (2003). Recursion as the key to the human mind. In K. Sterelny & J. Fitness (Eds.), *From Mating to Mentality: Evaluating Evolutionary Psychology* (pp. 155–171). New York: Psychology Press.

Csibra, G., Biro, S., Koos, O., & Gergely, G. (2003). One-year-old infants use teleological representations of actions productively. *Cognitive Science, 27*(1), 111–133.

Dennett, D. (1987). *The Intentional Stance.* Cambridge: MIT Press.

De Villiers, J. (2000). Language and theory of mind: What are the developmental relationships? In S. Baron-Cohen, H. Tager-Flusberg, & D. Cohen, (Eds.), *Understanding Other Minds: Perspectives from Developmental Cognitive Neuroscience*, 2nd ed. (pp. 83–123). Oxford: Oxford University Press.

De Villiers, J. & Pyers, J. (2002). Complements to cognition: A longitudinal study of the relationship between complex syntax and false-belief understanding. *Cognitive Development, 17*, 1037–1060.

Fine, C., Lumsden, J., & Blair, J. (2001). Dissociation between "theory of mind" and executive functions in a patient with early left amygdala damage. *Brain, 124*, 287–298.

Flavell, J., Everett, B. A., Croft, K., & Flavell, E. (1981). Young children's knowledge about visual perception—Further evidence for the level 1–level 2 distinction. *Developmental Psychology, 17*, 99–103.

Fletcher, P., Happé, F., Frith, U., Baker, S., Dolan, R. J., Frackowiak, R., et al. (1995). Other minds in the brain: A functional imaging study of "theory of mind" in story comprehension. *Cognition, 57*(2), 109–128.

Flynn, E., O'Malley, C., & Wood, D. (2004). A longitudinal, microgenetic study of the emergence of false belief understanding and inhibition skills. *Developmental Science, 7*(1), 103–115.

Foley, R. (1997). *Humans before Humanity.* London: Blackwell.

Franco, F. & Butterworth, G. (1996). Pointing and social awareness: Declaring and requesting in the second year. *Journal of Child Language, 23*(2), 307–336.

Frith, U. & Frith, C. D. (2003). Development and neurophysiology of mentalizing. *Philosophical Transactions of the Royal Society of London, Series B, Biological Sciences, 358,* 459–473.

Gallagher, H. L. & Frith, C. D. (2003). Functional imaging of "theory of mind." *Trends in Cognitive Science, 7*(2), 77–83.

Gallagher, H. L., Happe, F., Brunswick, N., Fletcher, P., Frith, U., & Frith, C. D. (2000). Reading the mind in cartoons and stories: An fMRI study of "theory of the mind" in verbal and nonverbal tasks. *Neuropsychologia, 38*(1), 11–21.

Gallagher, H. L., Jack, A. I., Roepstorff, A., & Frith C. D. (2002). Imaging the intentional stance in a competitive game. *Neuroimage, 16,* 814–821.

Gibbons, A. (2002). In search of the first hominids. *Science, 295,* 1214–1219.

Goel, V., Grafman, J., Sadato, N., & Hallett, M. (1995). Modeling other minds. *Neuroreport,* 6(13), 1741–1746.

Gregory, C., Lough, S., Stone, V. E., Erzinclioglu, S., Martin, L., Baron-Cohen, S., et al. (2002). Theory of mind in frontotemporal dementia and Alzheimer's disease: Theoretical and practical implications. *Brain, 125,* 752–764.

Grezes, J., Frith, C. D., & Passingham, R. E. (2004). Inferring false beliefs from the actions of oneself and others: An fMRI study. *Neuroimage, 21,* 744–750.

Happé, F., Mahli, G. S., & Checkley, S. (2001). Acquired mind-blindness following frontal lobe surgery. A single case study of impaired "theory of mind" in a patient treated with stereotactic anterior capsulotomy. *Neuropsychologia, 39,* 83–90.

Hare, B., Call, J., Agnetta, B., & Tomasello, M. (2000). Chimpanzees know what conspecifics do and do not see. *Animal Behaviour, 59,* 771–785.

Hare, B., Call, J., & Tomasello, M. (2001). Do chimpanzees know what conspecifics know? *Animal Behaviour, 61,* 139–151.

Jellema, T., Baker, C. I., Wicker, B., & Perrett, D. I. (2000). Neural representation for the perception of the intentionality of actions. *Brain and Cognition.* Special issue: *Cognitive Neuroscience of Actions, 44*(2), 280–302 .

Keenan, T. (1998). Memory span as a predictor of false belief understanding. *New Zealand Journal of Psychology, 27*(2), 36–43.

Knight, R. T. & Grabowecky, M. (1995). Escape from linear time: Prefrontal cortex and conscious experience. In M. S. Gazzaniga (Ed.), *The Cognitive Neurosciences*. Cambridge: MIT Press.

Kumashiro, M., Ishibashi, H., Itakura, S., & Iriki, A. (2002). Bidirectional communication between a Japanese monkey and a human through eye gaze and pointing. *Current Psychology of Cognition, 21*(1), 3–32.

Kumashiro, M., Ishibashi, H., Uchiyama, Y., Itakura, S., Murata, A., & Iriki, A. (2003). Natural imitation induced by joint attention in Japanese monkeys. *International Journal of Psychophysiology, 50*, 81–99.

Leavens, D. A., Hopkins, W. D., & Thomas, R. K. (2004). Referential communication by chimpanzees (*Pan troglodytes*). *Journal of Comparative Psychology, 118*(1), 48–57.

Leslie, A. M. (1987). Pretence and representation: The origins of "theory of mind." *Psychological Review, 94*, 412–426.

Leslie, A. M. (1994). Pretending and believing: Issues in the theory of ToMM. *Cognition, 50*, 211–238.

Leslie, A. M. & Frith, U. (1990). Prospects for a cognitive neuropsychology of autism: Hobson's choice. *Psychological Review, 97*(1), 122–131.

Liben, L. S. (1978). Perspective-taking skills in young children: Seeing the world through rose-colored glasses. *Developmental Psychology, 14*(1), 87–92.

Lieberman, D. E., McBratney, B. M., & Krovitz, G. (2002). The evolution and development of cranial form in *Homo sapiens*. *Proceedings of the National Academy of Sciences of the United States of America, 99*(3), 1134–1139.

Lillard, A. S., Zeljo, A., Curenton, S., & Kaugars, A. S. (2000). Children's understanding of the animacy constraint on pretense. *Merrill–Palmer Quarterly, 46*(1), 21–44.

Liu, D., Sabbagh, M. A., Gehring, W. J., & Wellman, H. (2004). Decoupling beliefs from reality in the brain: An ERP study of theory of mind. *Neuroreport, 15*(6), 991–995.

Lough, S., Gregory, C., & Hodges, J. (2002). Dissociation of social cognition and executive function in frontal variant frontotemporal dementia. *Neurocase*. Special issue: *Frontotemporal Dementia*, Part II, *7*(2), 123–130.

McGrew, W. C. (2001). The nature of culture. In F. deWaal (Ed.), *Tree of Origin: What Primate Behavior Can Tell Us about Human Social Evolution* (pp. 231–254). Cambridge: Harvard University Press.

Mossler, D. G., Marvin, R. S., & Greenberg, M. T. (1976). Conceptual perspective taking in 2- to 6-year-old children. *Developmental Psychology, 12*(1), 85–86.

Perner, J. (1991). *Understanding the Representational Mind.* Cambridge: MIT Press.

Perner, J. & Garnham, W. A. (2001). Actions really do speak louder than words—But only implicitly: Young children's understanding of false belief in action. *British Journal of Developmental Psychology, 19*(3), 413–432.

Povinelli, D. J. & O'Neill, D. K. (2000). Do chimpanzees use their gestures to instruct each other? In S. Baron-Cohen, H. Tager-Flusberg, & D. Cohen (Eds.), *Understanding Other Minds: Perspectives from Developmental Cognitive Neuroscience*, 2nd ed. (pp. 459–487). New York: Oxford University Press.

Povinelli, D. J., Theall, L. A., Reaux, J. E., & Dunphy-Lelii S. (2003). Chimpanzees spontaneously alter the location of their gestures to match the attentional orientation of others. *Animal Behaviour, 66*(1), 71–79.

Pratt, C. & Bryant, P. (1990). Young children understand that looking leads to knowing (so long as they are looking into a single barrel). *Child Development, 61*(4), 973–982.

Repacholi, B. M. & Gopnik, A. (1997). Early reasoning about desires: Evidence from 14 and 18-month-olds. *Developmental Psychology, 33*(1), 12–21.

Samson, D., Apperly, I. A., Chiavarino, C., & Humphreys, G. W. (2004). The left temporoparietal junction is necessary for representing someone else's belief. *Nature Neuroscience, 7*(5), 499–500.

Saxe, R. & Kanwisher, N. (2003). People thinking about people: The role of the temporo-parietal junction in "theory of mind." *Neuroimage, 19*, 1835–1842.

Saxe, R., Carey, S., & Kanwisher, N. (2004). Understanding other minds: Linking developmental psychology and functional neuroimaging. *Annual Review of Psychology, 55*, 87–124.

Semendeferi, K., Armstrong, E., Schleicher, A., Zilles, K., & Van Hoesen, G. W. (2001). Prefrontal cortex in humans and apes: A comparative study of area 10. *American Journal of Physical Anthropology, 114*(3), 224–241.

Shimamura, A. P. (2000). Toward a cognitive neuroscience of metacognition. *Consciousness and Cognition, 9*, 313–323.

Shimamura, A. P., Janowsky, J. S., & Squire, L. R. (1990). Memory for the temporal order of events in patients with frontal lobe lesions and amnesic patients. *Neuropsychologia, 28*(8), 803–813.

Smith, M., Apperly, I., & White, V. (2003). False belief reasoning and the acquisition of relative clause sentences. *Child Development, 74*(6), 1709–1719.

Snowden, J. S., Gibbons, Z., Blackshaw, A., Doubleday, E., Thompson, J., Craufurd, D., et al. (2003). Social cognition in frontotemporal dementia and Huntington's disease. *Neuropsychologia, 41*, 688–701.

Snowdon, C. T. (2001). From primate communication to human language. In F. deWaal (Ed.), *Tree of Origin: What Primate Behavior Can Tell Us about Human Social Evolution* (pp. 195–227). Cambridge: Harvard University Press.

Stone, V. E. (in press). The evolution of ontogeny and human cognitive uniqueness: Selection for extended brain development in the hominid line. In S. Platek, J. P. Keenan, & T. Shackelford (Eds.), *Evolutionary Cognitive Neuroscience*. Cambridge, Massachusetts: MIT Press.

Stone, V. E., Baron-Cohen, S., & Knight, R. T. (1998). Frontal lobe contributions to theory of mind. *Journal of Cognitive Neuroscience, 10*, 640–656.

Stone, V. E., Baron-Cohen, S., Calder, A. C., Keane, J., & Young, A. W. (2003). Acquired theory of mind impairments in individuals with bilateral amygdala lesions. *Neuropsychologia, 41*, 209–220.

Stone, V. E., Cosmides, L., Tooby, J., Kroll, N., & Knight, R. T. (2002). Selective impairment of reasoning about social exchange in a patient with bilateral limbic system damage. *Proceedings of the National Academy of Sciences of the United States of America, 99*(17), 11531–11536.

Stuss, D. T., Gallup, G., & Alexander, M. (2001). The frontal lobes are necessary for "theory of mind." *Brain, 124*(2), 279–286.

Suddendorf, T. (1999). The rise of the metamind. In M. C. Corballis & S. Lea (Eds.), *The Descent of Mind: Psychological Perspectives on Hominid Evolution* (pp. 218–260). London: Oxford University Press.

Suddendorf, T. (2004). How primatology can inform us about the evolution of the human mind. *Australian Psychologist, 39*(3), 180–187.

Suddendorf, T. & Busby, J. (2003). Mental time travel in animals? *Trends in Cognitive Sciences, 7*, 391–396.

Suddendorf, T. & Fletcher-Flinn, C. M. (1999). Children's divergent thinking improves when they understand false beliefs. *Creativity Research Journal*, Special issue*: Longitudinal Studies of Creativity, 12*, 115–128.

Suddendorf, T. & Whiten, A. (2001). Mental evolution and development: Evidence for secondary representation in children, great apes and other animals. *Psychological Bulletin, 127*(5), 629–650.

Suddendorf, T. & Whiten, A. (2003). Reinterpreting the mentality of apes. In K. Sterelny & J. Fitness (Eds.), *From Mating to Mentality: Evaluating Evolutionary Psychology* (pp. 173–196). New York: Psychology Press.

Tooby, J. & Cosmides, L. (1992). The psychological foundations of culture. In J. Barkow, L. Cosmides, & J. Tooby (Eds.), *The Adapted Mind: Evolutionary Psychology and the Generation of Culture* (pp. 19–136). New York: Oxford University Press.

Wellman, H. (1990). *The Child's Theory of Mind.* Cambridge: MIT Press.

Wellman, H., Cross, D., & Watson, J. (2001). Meta-analysis of theory-of-mind development: The truth about false belief. *Child Development, 72*(3), 655–684.

Wellman, H. & Lagattuta, K. H. (2000). Developing understandings of mind. In S. Baron-Cohen, H. Tager-Flusberg, & D. Cohen (Eds.), *Understanding Other Minds: Perspectives from Developmental Cognitive Neuroscience*, 2nd ed. (pp. 21–49). Oxford: Oxford University Press.

Wellman, H., Cross, D., & Watson, J. (2001). Meta-analysis of theory of mind development: The truth about false belief. *Child Development, 72*(3), 655–684.

Wellman, H. & Liu, D. (2004). Scaling theory of mind tasks. *Child Development, 75*(2), 523–541.

Wellman, H. & Wooley, J. (1990). From simple desires to ordinary beliefs: The early development of everyday psychology. *Cognition, 35*(3), 245–275.

Wildman, D. E., Uddin, M., Liu, G., Grossman, L. I., & Goodman, M. (2003). Implications of natural selection in shaping 99.4% nonsynonymous DNA identity between humans and chimpanzees: Enlarging genus *Homo*. *Proceedings of the National Academy of Sciences of the United States of America, 100*(12), 7181–7188.

Woodward, A. (1999). Infants' ability to distinguish between purposeful and non-purposeful behaviors. *Infant Behavior and Development, 22*(2), 145–160.

Yoder, A. D., Cartmill, M., Ruvolo, M., Smith, K., & Vilgalys, R. (1996). Ancient single origin for Malagasy primates. *Proceedings of the National Academy of Sciences of the United States of America, 93*(10), 5122–5126.

7 Investigating Cortical Mechanisms of Language Processing in Social Context

Howard C. Nusbaum and Steven L. Small

Language is a social enterprise determined by society and culture, and shaped by interpersonal interaction. Consequently it provides the foundation for society, culture, and interpersonal interaction. Almost all of our social institutions (government, school, church) and almost all of our daily interactions with other individuals depend on language. To researchers interested in understanding language processing, this is, of course, all very basic and assumed. Yet, little of the research focused on mechanisms that mediate language processing takes this into account explicitly. Such research seldom starts with the recognition of the social reality of language as a core definition of language processing, largely because of a set of assumptions about the computational architecture of the language system.

Language Is Treated as Cold and Not Cognitive

A number of years ago, Zajonc (1980) raised the issue of whether affective and cognitive processes should be viewed as independent, thereby segregating social and cognitive psychology from each other. However, interactions between social psychology and cognitive psychology have turned out to be extremely productive over a number of years. For example, research in social cognition is informed by work on mechanisms of attention and automaticity (Bargh & Chartrand, 1999; Wegner & Bargh, 1998) and categorization (Hugenberg & Bodenhausen, 2004; Wittenbrink, Hilton, & Gist, 1998). However, basic research on language processing has been affected only to a small degree by work in social psychology, despite the entirely social nature of language use. As brain mechanisms underlying social and cognitive processes are increasingly elaborated, it becomes increasingly difficult to ignore the role of neural substrates for social functions arising in the context of purely cognitive functions (Siegal & Varley, 2002).

For over half a century, the scientific study of language was disproportionately influenced by the perspective of linguistics (Chomsky, 1957). Linguistic investigation represents just one way to study language, and psychology, sociology, computer science, anthropology, neurobiology, and philosophy all play important roles in understanding language. During this time, two basic assumptions shaped a broad range of language research, particularly in psychology and neurobiology. The first assumption is that we can separate out the study of language from the study of language use. The competence-performance distinction is rooted in the notion that a Platonic idealization of language can be studied independent of the moment-by-moment limitations, distractions, and disturbances to which all human performance is subject. In other words, language can be studied independent of other aspects of psychology including cognitive processing, social interaction, and affective processing, and language areas of the brain can be understood apart from other complementary regions and networks, such as the motor system or the limbic system.

The second assumption derives from the notion that we can analytically separate putative language functions from each other and, having done so, empirically study each as if truly dissociated from other language functions. The prototypical example is syntax. Chomsky (1957) initially viewed the explanation of syntactic processing as a component of the language system that was scientifically tractable and isolable from other aspects of language. It is somewhat amusing to consider (in hindsight) that he was so optimistic about the ease of explaining syntax, once isolated from the more complicated aspects of language, that he expressed the concern that that there might be several equally explanatory theoretical accounts of syntax. He was convinced that this would raise a serious problem of finding other means (other than linguistic intuition) to distinguish among equally explanatory syntactic theories. Needless to say, this particular concern has not yet become a problem, given the lack of any sufficient theory of syntax. But one could draw the inference that the entire enterprise of separating out a psychological process that appears to be tractable on its own terms may be misleading, taking scientific inquiry down a garden path.

Consider attempts to understand airflow over a wing of an airplane (see Gleick, 1987 for a discussion). Under some conditions, airflow over a wing is smooth and laminar whereas under other conditions it is noisy and turbulent. For a long time, these two physical behaviors were viewed as having entirely different theoretical explanations, one being simple and linear and the other complex and nonlinear. However, this separation into systems based on apparently different modes of behavior led researchers away from a single, simpler explanation in terms of a chaotic dynamical system.

The assumption that, because we can analytically introspect about syntactic structure or any particular psychological process, it has an independent psychological and biological reality from other psychological processes may be deeply problematic. As described by Fodor (1983), the notion of an autonomous, biologically fixed, evolutionarily expert, mandatory, and automatic syntactic processor has given rise to a great deal of research and controversy (Appelbaum, 1998; Garfield, 1987; Sperber, 2001). As in the case of all language-processing modules (e.g., lexical module), controversy generally focused on whether processing falling within the domain of expertise of the putative module interacts with other kinds of processing outside the module (beyond the operation of discrete input and output stages). Proponents of modularity deny such interaction whereas opponents support it; however, this debate has had several important consequences. Language processing is often studied independent of context and separately for each putative unit of analysis—syllables, words, or sentences. Language processing is typically studied without much consideration for the relationship of interlocutors to each other in a conversation, in those relatively rare instances when actual face-to-face conversation is studied (Pickering & Garrod, 2004). It is often studied as if it were independent of other cognitive, affective, and social mechanisms, and without consideration for either the speaker's or the listener's intentions or goals. In theoretical terms, it is generally accepted that messages can be understood independent of situation (pragmatic context) or even of the contents of the message. In other words, it is assumed that the impact or possible importance of a message's meaning to a listener has little or no bearing on the processing of the message as language. Understanding the message is typically viewed as independent of what we understand about the person who is talking (see Holtgraves, 1994). On one hand, the assumption of modular language mechanisms is often viewed as a null hypothesis to be rejected by specific experiments and therefore only a testable theoretical proposition. On the other hand, this is a theoretical foundation that shifts research away from basic evolutionary forces that actually shaped the biological development of linguistic communication—the need for social interaction between people.

Dewey (1896) pointed out that although researchers can analytically decompose a mental process into separate components, the reality of such components is not substantiated by that analysis. Furthermore, just starting with the assumption of separability or isolability leads to research questions and methods that would not otherwise be employed but that could distort, in various ways, our understanding of a psychological system such as language. Dewey recognized that in many cases goals and motives may actually determine (that is, restructure or change) the nature of the entire processing system rather than simply affect the outcome of processing. This notion goes well beyond the standard view of interactions in which one process provides inputs to another process. Instead this is closer to the idea emerging from some neuroscience research (Barrie, Freeman, & Lenhart, 1996; Freeman, 2003; Freeman &

Skarda, 1990) that neural processes and representations are context sensitive and can change dynamically with goals and motivations and experience.

Although the original view of modularity was spelled out in cognitive terms and explicitly distinguished from phrenology (Fodor, 1983), Gazzaniga (1985) proposed a more specific cortical modularity. Whereas Fodor's (and most linguists' and psychologists') view of modularity is agnostic about the relationship between cognitive mechanisms and cortical mechanisms, based on evidence such as double dissociation of damage to cortical systems, Gazzaniga (1985) argued for separable cortical modules for specific cognitive functions. Indeed, on the face of it, some examples seem clear. For intance, in achromatopsia, damage to V4 eliminates consciousness of colors even though color processing still exists in V1 (Kohler & Moscovitch, 1997). Or consider blindsight in which dorsal-stream visual information is preserved compared with cases in which ventral-stream information is preserved (Weiskrantz, 1986). In these cases, as reported and investigated, it seems that a single psychological function is impaired by focal cortical damage. It is perhaps not surprising that such examples typically derive from basic sensory systems such as vision. When it comes to language function, however, which is often treated as the prototype for cortical modularity (Gazzaniga, 1985; Geschwind, 1965), structure-function relationships become much more complicated.

The landmark ninetieth-century work of Broca (1861) and Wernicke (1874) shaped much of our understanding of the way language and brain are related. The association between anatomical locations of brain injury and disruption of particular language behaviors (e.g., production and comprehension) provided an important functional definition of language processing (Benson, 1979; Geschwind, 1971). For example, the theoretical division between expressive and receptive language processing derives in part from gross deficits seen in patients with damage located in more anterior or posterior cortical regions, and research questions emerging from this division focus on characterizing the processing of those regions (such as agrammatism versus working memory deficits for Broca's area). However, the clear divisions between Broca's aphasia as expressive aphasia and Wernicke's aphasia as receptive aphasia are no longer as clear cut as they once seemed (Blumstein & Milberg, 2000; Saygin et al., 2003).

With increasing use of neuroimaging measures, methods of lesion analysis and psycholinguistic experimentation seem to have formed the conceptual foundation for the methodological toolbox of functional brain imaging. An assumption underlying both of these approaches is the componential reduction of language processing, with a focus on language competence—basic linguistic knowledge—rather than on language performance (Chomsky, 1965; de Saussure, 1959). Over the past fifty years we have learned a great deal about many levels of language processing, from phonology to discourse, using this approach. However, this approach may be limited when it comes to neuroimaging studies, imposing a different set of distortions on the results we obtain.

Studying linguistic competence by definition abstracts language processing away from its grounding in behavior. By shifting to studying language use rather than linguistic competence, however, we may gain, rather than lose, in our ability to understand language processing (see Clark, 1996, for a discussion) when using neuroimaging measures.

There can be no doubt that language evolved for communication between people, or for multimodal, face-to-face communication, and its use occurs in a rich environmental context that can ground communication for cognitive purposes. Rather than start from the position of looking for evidence of specific types of language processing "in" the brain or for evidence of language processing by "the brain," we suggest that it may be useful to examine cortical activity during language behavior that most closely matches conditions of evolution: language use by people at a time and place, aiming to understand and to be understood, fulfilling a purpose. The utility of this approach is that it considers how language processing, in service of specific goals and uses, interacts with a broad set of neural circuits that are involved in more general cognitive, affective, and social processing.

By examining the distribution of such network activity during language use we can begin to investigate the richness of the neural interactions that occur in real time, integrating linguistic knowledge with putatively nonlinguistic processes such as motor activity, working memory, and attention. A tendency in neuroimaging research has been to try to isolate language processing from these other processes using a variety of analytic and design methods. However, it is important to remember that language use in the real world interacts fundamentally with motor behavior—all language expression *is* motor behavior—and the systems for language use and motor behavior are functionally intertwined, affecting our ability to investigate and ultimately to understand the neurobiology of language. Furthermore, real language use entails cognitive, sensorimotor, and affective operations in addition to linguistic ones. To study the biology of language use, understanding the relationships among these interrelated neural processes will be a central aspect of the basic scientific problem. Psychological and computational theories typically isolate language processing from both cognitive and affective mechanisms, so that research on the biology of language processing tends to take neither into account.

Components of Processing

A common feature of both lesion analysis and much psychological research is the emphasis on functional decomposition, which views the brain as organized into anatomically segregated parts (Gall, 1825) and complex behavior as being mediated by a collection of functionally independent units (Fodor, 1983). Work in dynamic systems theory (Freeman & Barrie, 1994) suggests an alternative approach: rather than viewing different patterns of behavior as the result of the operation of different and independent subsystems each responsible for a different pattern, such patterns of behavior can arise from a single complex system operating in different modes at different parameter values. This has produced significant scientific breakthroughs, including in psychology (Smith & Thelen, 1993).

Our argument against strict functional decomposition is not an argument in favor of the older holographic view of the brain as a mass of equipotential tissue (Lashley, 1950). We do not assume that all parts of the brain participate equally in all behaviors. Nor do we assume that each part of the brain provides an identifiably unique and functionally separate process. Rather, we postulate that the neural circuits that operate within and across different anatomical regions, are both interdigitated and interactive, and operate differently depending on their dynamic patterns of activity. This intrinsic neural context (McIntosh, 2000) complements the extrinsic environmental context, producing different modes of processing in different circumstances, leading to specific patterns of behavior. In some cases, the apparent specializations of different anatomical regions may not have clear psychological interpretations, which has been an underlying assumption of much neuroimaging work.

The information processing era of the cognitive revolution led to a plethora of serial componential "boxological" models of behavior (Neisser, 1976). For example, language comprehension was studied as a series of processing stages that match the propositional encoding of a sentence against a propositional encoding of a picture (Clark, Carpenter, & Just, 1973). This decomposition provided the basis for important experimental manipulations to investigate subprocesses of sentence comprehension. These information-processing models assumed, however, that each processing stage was independent of the others and was necessarily completed before starting the next (Sternberg, 1969).

This approach to cognitive research has continued through recent times. Just as Fodor (1983) viewed the mind as composed of modules, neurosciences viewed the brain as modular, consisting of functionally specialized and independent locations (e.g., Shallice, 1988). In the study of language, the frontal operculum (Broca, 1861) and posterior superior temporal region (Wernicke, 1874) played special roles in this localizational view, representing sites for language production (early view) or syntax (later view) and language comprehension (early view) or semantics (later view), respectively. In part, these componential views are rooted in other studies of biological specialization. Just as the heart and the lungs are anatomically and mechanically specialized for specific distinct physiological functions but operate together as integrated systems, anterior and posterior cortices have been viewed as specialized for motor and sensory functions, replicating the notion of structure-function relationships found elsewhere in biology.

Many systems are not decomposable into independent functional parts (Runeson, 1977), even though the standard operating assumption in psychology is to reduce systems to putative functional components. In psychological research, this componential view is critical to the interpretation of response-time experiments: in broad terms, these experiments generally assume that (1) the duration of any particular cognitive process is composed of the sum of a set of constituent subprocesses (Donders, 1868/1969), and (2) these putative subprocesses provide the basis for the manipulation of experimental variables from which to infer the processing characteristics of component subsystems (Sternberg, 1969).

Neurology has also taken a componential (anatomical decomposition) approach to understanding the neural mechanisms that mediate complex behaviors. The inferential logic of "double dissociation" (Shallice, 1988) depends on the notion that there are component mechanisms that have independent functions. Damage to one component should produce patterns of behavior change that are different from and complementary to the change produced by damage to a different component.

Ultimately, this conceptual framework is the basis for many studies in functional brain imaging with position emission tomography (PET) and functional magnetic resonance imaging (fMRI). In research on language and the brain, some studies focused on validating certain models derived from information-processing psychology, which themselves often were derived from the analytic considerations of theoretical linguistics. Consider the example of lexical access, in which the process of recognizing a spoken word is viewed as isolable from the rest of the language processing system by comparing neural activity produced by (1) repeatedly speaking a word with (2) hearing reverse speech and uttering a standard word (Howard et al., 1992). This elucidates brain regions for lexical access, based on the assumption that the two tasks contain all the same components except one (access component), in the same order and with the same feedback (Sergent et al., 1992).

Neuroimaging studies often assume a one-to-one correspondence between neural (brain locations) components and psychological (behaviorally isolable) components. Typical tasks used to study language in the brain include, at different levels of language processing, rhyme judgment and phoneme discrimination (phonological level), lexical decision (lexical level), or grammaticality judgment (sentence level). To carry out any of these tasks, responses depend on the use of a specific kind of linguistic competence. For example, to judge that two words rhyme, the listener must compare the phonological patterns of the words, thereby exercising phonological processing. (Of course this assumes that the nature of phonological processing used in a metalinguistic rhyme-judgment task depends on the same phonological competence used in fluent language use.) By designing tasks based on well-defined (in theoretical terms) specific areas of linguistic competence, it is assumed that the operation of a component mechanism that mediates that competence will be selectively illuminated. The success of this approach depends on the assumption that the explicit judgment of a linguistic property of an utterance exercises the same kind of processing (same mechanism used the same way) as the implicit routine use of this processing in daily language use.

A study conducted in our laboratory illustrated this concern and the nature of the problem. Phoneme discrimination was compared with nonspeech tone discrimination in a context in which the former required phonological segmentation and another in which it did not (Burton, Small, & Blumstein, 2000). By contrasting two discrimination tasks (one phonological, one auditory), both calling for stimulus comparison and planned motor behavior, we intended to isolate those neural processing components that mediate phonological segmentation. We concluded that "it is the process of segmentation of the initial consonant from the following vowel, probably requiring articulatory recoding, that appears to involve left . . . inferior and middle frontal [gyri]" (Burton, et al., 2000).

Of course, the contrasts that are carried out in these kinds of studies assume that we understand a priori the componential structure of the tasks we use. Do listeners actually segment the speech stream into phonemes before recognizing the phonemes, or do they just recognize linguistic units without segmentation? Are phonemes truly the basic units of speech perceptual analysis, or are syllables or diphones or onset-rime structures the basic unit of perception? Although these are standard assumptions in much speech research, and may reflect consistency in information conveyed in speech (Studdert-Kennedy, 1981), this does not necessarily license a neural reality for these assumptions. If tone discrimination and phoneme discrimination are carried out by complex neural networks that are simply modulated differently across conditions, isolable anatomical components may have little or no relationship to behavioral components, if there really are any (Runeson, 1977).

Indeed, it turns out that the conclusions of our first study depended critically on the specific nature of the task comparison, as we later learned. A follow-up study, using a different nonspeech tone-discrimination control task (requiring pattern segmentation, similar to the meta-phonological judgment made with syllables) found no frontal activation (Burton & Small, 2001) because this component was "subtracted off" when more comparable speech–nonspeech comparisons were carried out.

Holding aside for the moment that listeners never have to make explicit phonological discriminations during real conversations (thus making discrimination a very unnatural task), the presence or absence of apparent frontal activity in this study depends on the comparison task that is used for subtraction, as should be the case. However, this leaves us with a very real question: which result is more indicative of real phonological perception, involvement or noninvolvement of the frontal lobe? If one nonspeech-control task emphasizes working memory and the motor system more than another, this will moderate the appearance of neural activity in the frontal region during the phonological discrimination task. Since we can modulate this involvement easily with the control task, how can we ascertain the correct degree of match between control and target experimental tasks? The only possible way to make this decision is by an a priori theoretical assumption, which may be of questionable validity. This particular issue is broader than the study of language in the brain. Tasks that involve judgments of any kind—self-relevance, affective valence—involve working memory, attention, and motor planning specific to task-defined goals. Thus any neuroimaging study that involves explicit judgments or decisions that are not the behavior of interest must be considered in the same light.

Language-Motor Interaction in Cortical Activity

Studies such as the phonological segmentation experiment are intended to investigate independent components of a complex behavior as if the parts can be inserted or removed without changing *ceteris paribus* the functioning of other components (Donders, 1868/1969). Since most neuroimaging experiments are designed with explicit decision-making components and overt motor responses, and these aspects of processing are not the focus of the scientific investigation, the contribution of these components to dependent measures of brain activity must be eliminated. This requires that decision making and button pressing must be treated as (or at least assumed to be) independent and isolable from the cognitive and linguistic processes of interest in both behavioral terms and in the brain. In general, this has been a productive strategy for understanding some basic aspects of linguistic competence and cognitive functioning. To understand language use, rather than competence, however, it is important to understand interactions that occur between language processes and cognitive, affective, and motor systems. With this research goal, it is likely that assumptions regarding component isolability may be problematic, and that matched-task subtractions could mask or eliminate activity from brain regions of interest. Thus, applying the common experimental method for functional brain imaging to the study of language use may involve the inadvertent study of language-motor integration in task-dependent (as suggested with the example of phonological segmentation) rather than language use-dependent ways.

To understand the difference between a behavioral psychology experiment and its transplanted form in a brain imaging experiment, two things must be considered. First, how do dependent measures differ in brain imaging and behavioral experiments? Second, what is the role of decision making and motor output in producing the imaging result?

The dependent measure in fMRI brain imaging—hemodynamic response—is fundamentally and critically different from the dependent measures in typical psychology studies—response time and accuracy. Behavioral measures, such as response time and accuracy, typically give us a relatively univariate view of language processing, providing only a measure at the outcome of the overall process. In essence, this compresses a complicated network of neural computation into a single behavioral output. By contrast, neuroimaging gives us a multivariate data set reflecting all of the activity in this network over time (Nyberg & McIntosh, 2001). Every subprocess can manifest itself relatively simultaneously (depending on temporal sensitivity) and in parallel across the brain. In the behavioral measure, experimental manipulations of specific variables can modulate the mean difference across conditions such that the contribution of some subprocesses is swamped by the variance due to "manipulated" subprocesses of interest. However, in a neuroimaging study, manipulated target subprocesses and ancillary subprocesses are all manifest distributed across the dependent measure. We call this difference between behavioral and neurophysiological measurements the dependent measure problem.

Brain imaging offers the opportunity to observe all the components operating in parallel, overlapping, and distributed in time. However, unlike response time (Sternberg, 1969) or error rate, the dependent measure reflects aggregate system behavior in a very different way. It is important to note that in neuroimaging, dependent measures are themselves directly linked to system components of interest—anatomy.

Variation in one dependent measure is no longer a reflection of the entire chain of processing in a task; rather the dependent measure can reflect the contribution of any one anatomical component to the task, as well as modulation of that component by linked components.

A corollary issue then is that the decompositional or subtractive approach to imaging can lead to the inadvertent study of motor interactions with cognitive, affective, or social processes. Since much of the cortical activity of interest in social neuroscience involves frontal and prefrontal brain areas that are closely associated with motor behavior as well as other psychological processes, inadvertent motor interactions can obscure activity due to psychological processing under investigation. We call this the motor output problem. It is obvious that virtually all measurable behavior involves the motor system. A central feature of most neuroimaging studies has been to use measurable behavioral outputs (rhyme decision, button presses) to establish that the brain activity being measured corresponds to the intended processing. In other words, if listeners are making accurate rhyme decisions, they must be using phonological processing. Rhyme-based button-pressing behavior itself is not the processing of interest in these studies. However, due to the dependent measure problem, without appropriate treatment, cortical activity underlying the behavior will show up in the dependent measures of putative phonological processing.

This has meant that for the results of imaging studies to be interpretable, it necessary to assume that motor planning and control are independent of cognitive, social, or affective processes under investigation. This assumption would allow the motor activity to be subtracted off using appropriately matched control conditions. Yet this assumption seems questionable—we know that complex motor circuits interact with many other networks throughout the brain. In fact, areas of the brain that have been associated with language (Broca, 1861; Burton et al., 2000; Zatorre et al., 1996), emotional experience (Lane et al., 1997), attentional control (Banich et al., 2000), and working memory (Cohen et al., 1997; Smith et al., 1998) are also closely identified with motor processing. For example, it is known that much of the anterior cingulate gyrus, an area frequently implicated in attention mechanisms (Smith & Jonides, 1999) and in emotional experience (Lane et al., 1997), plays an integral role in motor processes (Grafton, Hazeltine, & Ivry, 1998; Morecraft & van Hoesen, 1998; Picard & Strick, 1996). If a motor task is imposed on a neuroimaging experiment to guarantee that the brain activity reflects the intended psychological processing, a significant degree of the dependent measure will reflect the motor system activity produced by aspects of the task that may be irrelevant to the psychological process under investigation. This activity may not be easily (if at all) dissociable from that psychological process.

Understanding Mechanisms of Language Use

In studying language use rather than component linguistic competencies, it may be possible to avoid or at least moderate both the dependent measure problem and the motor output problem. Rather than impose artificial metalinguistic probe tasks on participants, it is possible to use more ecologically plausible language tasks, such as conversation. This immediately sets the study of language use into a social context in which social goals, affective reactions, relationships, and status are directly part of the experiment. Whereas such aspects of language interaction may be controlled in a study, perhaps to avoid their investigation, this approach affords a natural opportunity to study the interaction of social, affective, and cognitive mechanisms.

Brain-activation patterns during language use might be particularly revealing, since these social interactions are likely to have played a role in the ontogeny and phylogeny of brain development and language development. Ecologically valid language use, in contrast with metalinguistic judgment tasks, may be more closely suited to the nature of the dependent measure of brain imaging and to the fundamental questions of social neuroscience.

Brain imaging studies of ecological language processing in multimodal naturalistic context might be a valuable way to avoid problems associated with componential modeling assumptions and decision-making tasks. This is not new to psychology or neurology. In fact, the Chicago school of psychology emphasized the study of cognitive processing in context, interactivity of component parts, and investigation of naturalistic phenomena (Dewey, 1896; James, 1904). Furthermore, Brunswik (1955) argued that psychological research should contrast conditions that display the full range of natural variation observed in behavior. Whereas true ecologically valid language behavior is difficult under the conditions of neuroimaging, particularly with fMRI, it is possible to move studies more in that direction, both by changing the nature of the tasks and by changing the kind of information provided to participants.

For example, language evolved in the context of face-to-face communication, not in the context of telephone conversation. Although most theories of speech perception emphasize the role of the acoustic signal in spoken language understanding (Diehl, Lotto, & Holt, 2004; Fant, 1967), faces are considered one of the basic aspects of social stimuli, and it should not be surprising that information about faces is important to understanding speech. Simply seeing a picture of a face, identified to listeners as the talker, can change the perception of speech. For example, for Caucasians, seeing an Asian face identified as a talker reduces the intelligibility and comprehensibility of a speech signal, compared with seeing a Caucasian face (Rubin, 1992) paired with the same speech. However, it is important to begin to understand the mechanism by which visual information can change auditory speech perception.

Visual information showing movements of the mouth and lips during talking enhances speech comprehension in noise, even though we often think of speech perception as being defined by the acoustic signal alone (Sumby & Pollack, 1954; Summerfield, 1992). In the McGurk effect (McGurk & Macdonald, 1976). When presented with an acoustic signal specifying one phoneme (e.g., /k/) and visual information about mouth movements specifying a different phoneme (e.g., /p/), listeners may report hearing an emergent phoneme (e.g., /t/). Visual and auditory information jointly specify a percept that is present in neither modality but combines information from both. Furthermore, other visual information about motor movements produced by an interlocutor while speaking is important to communication, such as the manual gestures that accompany speech, which clearly affect our understanding of that speech (McNeill, 1992). In addition, manual gesturing during speech improves cognitive efficiency as measured by memory capacity (Goldin-Meadow et al., 2001), suggesting an interaction among language system, visual system, and motor system for cognitive functions. Face-to-face communication clearly provides important information about an interlocutor that is not present in the acoustic signal alone and that changes our understanding of that signal.

What is the mechanism by which this visual information affects spoken language understanding? Behavioral measures alone can indicate only the nature of the interaction among sources of information but not how this interaction takes place. However, neuroimaging reveals more directly the neural systems that are active when different sources of information are available to the comprehender. Subjects were imaged with fMRI while listening to interesting stories (audio only), listening to stories while seeing the storyteller (audiovisual), or just seeing the storyteller (visual). We found far more activation in the inferior frontal cortex (BA 44/45) in the audiovisual condition than in either other condition (Skipper, Nusbaum, & Small, 2002, 2005). Moreover, the presence of the visuomotor information changed the laterality of the activity in superior temporal cortex, demonstrating the interaction in processing between face information and acoustic speech in more traditional speech perception areas. These data suggest that visual information about mouth movements must be decoded (in some fashion) by the motor system (inferior frontal gyrus, supplemental motor area, etc.) that interacts with the process of interpreting speech using superior temporal cortex.

It is important to note that listeners were required only to understand the spoken stories in this study, and not perform an additional judgment task. If we had designed a specific task to measure comprehension, the motor behavior in responding and the working memory used during judgment could have masked Broca's area activity observed during comprehension. However, the limitation of this approach is that without specific behavioral measures of comprehension processing, we cannot directly relate patterns of cortical activity to details of behaviors. Although posttask questioning can establish gross aspects of processing, such as whether listeners understood the stories and some of what they remember, these measures are not sufficiently sensitive to diagnose more specific hypotheses. One challenge is to develop new methods that allow us to assess more directly the relationships between brain activity and behavior without changing either. We can think of this as a kind of Heisenberg uncertainty principle in cognitive neuroimaging research.

Mechanisms of Social Interaction

On one hand, our results could be taken as a simple extension of a bottom-up view of speech perception. Visual information from mouth movements is combined with auditory information to determine a new interpretation of an utterance. This by itself would not suggest much of a paradigmatic change to language research and would only open the door to social and affective information in terms of additional streams of input.

On the other hand, these results may suggest a different way to think about language processing more broadly. Rather than think about language as a signal (multimodal rather than just acoustic, of course) that is a means of transmitting information, we can think about the communicative interaction itself as a psychologically significant act that was part of the basic force shaping the evolution of the brain. By this construal, the listener's goal may not be to interpret the linguistic message but to interact with the interlocutor in a way that satisfies specific social goals and motives. This would suggest that communicative behavior, broadly construed, should be affected by a conversational partner's behavior, even beyond the simple process of interpretation.

Indeed, substantial evidence supports this. Giles (1973; Giles & Smith, 1979) showed speech accommodation in conversations, in which one interlocutor (or both) converges on the speech of the other, in terms of speaking patterns. This vocal accommodation or indexical mimicry is increased between members of the same social group and decreased between groups (Giles & Coupland, 1991). It is also increased, when one interlocutor is trying to persuade the other of something (Giles & Coupland, 1991). Moreover, this behavioral convergence in a conversation is not restricted to speech patterns. Chartrand and Bargh (1999) reported that other motor behaviors that are not speech related show similar accommodation between conversational partners and depend on social goals. One conversational partner tapping her foot can start the other partner tapping as well, even though this is not a linguistically relevant behavior. And this can serve to link interlocutors socially, increasing the sense of interpersonal affiliation (Lakin & Chartrand, 2003) as well as shifting attention from one's self to the broader environment (van Baaren, 2004).

Clearly, in face-to-face communication the motor system plays an important role in moderating social cohesion that goes well beyond comprehension and production of language. Whereas this motor mimicry may be mediated in part by the observation-execution matching system (Rizzolatti, Craighero, & Fadiga, 2002) in terms of the operation of mirror neurons, this cannot account for the attentional, affective, and social consequences of such interpersonal interaction. Such motor mimicry changes the way we attend to stimuli and the way we feel about conversational partners. This should not be surprising, given the close anatomical relationships among cortical areas involved in the motor system, attention, working memory, and affect.

Roland (1993) proposed that frontal cortex can generate sensory expectations that can be used to "tune" the sensitivity of more posterior areas. Bar (2003) concluded that the dorsal visual stream operates to tune the ventral stream by using prefrontal cortex in just this way, specifically because of the relationship between dorsolateral prefrontal cortex and orbitofrontal cortex and connections to the limbic system and memory systems. His theory is that affective goals and evaluations are important in shifting attention for perception. This process of shifting attention involves frontal and prefrontal systems changing the sensitivity of posterior sensory processes, presumably by modifying receptive fields (Moran & Desimone, 1985). This is quite similar to the way in which visual information about mouth movements during speech may change the sensitivity of superior temporal cortex during comprehension of the speech.

This clearly goes beyond the use of visual information about motor movements in phonetic perception. The race of a talker's face (shown as a static image) can change the perception of the person's speech (Rubin, 1992) and the gender of a talker's face can change interpretation of speech (Johnson, Strand, & D'Imperio, 1999). Even without dynamic real-time perceptual input about a talker's mouth movements, expectations can change the interpretation of speech. Furthermore, language processing is shaped by expectations even when they do not involve information about a talker's face. For example, when presented with sinusoidal replicas of a sentence, listeners generally reported that these signals sound like nonspeech bird chirps, but when told that the signals are language, they correctly understood the speech (Remez et al., 1981).

We measured cortical activity of listeners presented with sine wave speech before and after language instructions. The results suggested that the effect of linguistic expectations on language comprehension may be mediated by activity in an attentional-motor network involving inferior frontal gyrus and superior parietal cortex (Wymbs, Nusbaum, & Small, 2004). Similarly, changing listeners' expectations that the same prosodic information conveys affective or syntactic information changes the ear advantage suggesting a change in cortical processing (Luks, Nusbaum, & Levy, 1998). Changing a listener's expectations about speech fundamentally changes the pattern of cortical processing that mediates perception of speech.

Summary and Conclusion

Every utterance is inherently ambiguous. Comprehension of an utterance depends on understanding the speaker's intended meaning. Given the inherently nondeterministic nature of the comprehension process (Nusbaum & Magnuson, 1997), listeners must use more information than is present in the utterance itself, such as race, gender, and social status of the speaker, to interpret the message. However, none of these categories is itself assessed deterministically. Listeners must use a variety of weak constraints to determine an intended message.

One solution to this problem is to treat language comprehension as an active, inferential process (Nusbaum & Schwab, 1986; Small & Rieger, 1982). Listeners may form hypotheses about a message from expectations, context, and inference. The relationship among the hypotheses may provide information that would be useful in determining which hypothesized message is actually intended. Given this contrast, listeners may shift attention to give differential weight to this information in real time. This suggests that comprehension will depend on working memory (for holding communicative goals, different hypotheses formation, and testing among them), attention and attention shifts, affective systems (for evaluating sources of information and for setting goals), perceptual processes, and motor processes. This view of language processing is quite different from the cold and noncognitive view represented by a modular perspective.

By shifting the focus of research questions from meta-linguistic tasks to understanding language use in service of social interaction, brain imaging allows us investigate the breadth of neural mechanisms that interact during real language behavior. This approach will depend on the analysis of activation across network structures rather than in specific localized regions. This presents substantial new challenges for experimental design and image-processing methods. This combination of context-dependent naturalistic imaging with monitoring of natural behaviors, novel experimental design, and network-based analysis could lead to tremendous new insights into language and the brain.

Acknowledgments

Supported by National Institutes of Health grant DC-3378. Additional support from the Brain Research Foundation and the McCormick Tribune Foundation. We thank Ana Solodkin, John Cacioppo, and Jeremy Skipper for helpful discussions about these topics. Finally, we thank Elizabeth Bates for many conversations over the past ten years about the strengths and weaknesses of brain imaging for the study of human language.

References

Appelbaum, I. (1998). Fodor, modularity, and speech perception. *Philosophical Psychology, 11*, 317–330.

Banich, M. T., Milham, M. P., Atchley, R., Cohen, N. J., Webb, A., Wszalek, T., et al. (2000). fMRI studies of Stroop tasks reveal unique roles of anterior and posterior brain systems in attentional selection. *Journal of Cognitive Neuroscience, 12*, 988–1000.

Bar, M. (2003). A cortical mechanism for triggering top-down facilitation in visual object recognition. *Journal of Cognitive Neuroscience, 15*, 600–609.

Bargh, J. A. & Chartrand, T. L. (1999). The unbearable automaticity of being. *American Psychologist, 54*, 462–479.

Barrie, J. M., Freeman, W. J., & Lenhart, M. D. (1996). Spatiotemporal analysis of prepyriform, visual, auditory, and somesthetic surface EEGs in trained rabbits. *Journal of Neurophysiology, 76*, 520–539.

Benson, D. F. (1979). *Aphasia, Alexia, and Agraphia*. New York: Churchill Livingstone.

Blumstein, S. E., & Milberg, W. P. (2000). Language deficits in Broca's and Wernicke's aphasia: A singular impairment. In Y. Grodzinsky & L. P. Shapiro (Eds.), *Language and the Brain: Representation and Processing*. Foundations of Neuropsychology Series (pp. 167–183). San Diego: Academic Press.

Broca, P. P. (1861). Nouvelle observation d'aphémie produite par une lesion de la partie postérieure des deuxième et troisième circonvolutions frontales. *Bulletin de Societe Anatomique Paris, 6*, 398–407.

Brunswik, E. (1955). Representative design and probabilistic theory in a functional psychology. *Psychological Review, 62*, 193–217.

Burton, M. W. & Small, S. L. (2001). Functional neuroanatomy of segmentation of speech and nonspeech. *Neuroimage, 13*, S511–S511.

Burton, M. W., Small, S. L., & Blumstein, S. E. (2000). The role of segmentation in phonological processing: An fMRI investigation. *Journal of Cognitive Neuroscience, 12*, 679–690.

Chartrand, J. L. & Bargh, J. A. (1999). The chameleon effect: The perception-behavior link and social interaction. *Journal of Personality and Social Psychology, 76*, 893–910.

Chomsky, N. (1957). Syntactic *Structures*. The Hague: Mouton.

Chomsky, N. (1965). *Aspects of the Theory of Syntax*. Cambridge: MIT Press.

Clark, H. H. (1996). *Using Language*. Cambridge: Cambridge University Press.

Clark, H. H., Carpenter, P. A., & Just, M. A. (1973). On the meeting of semantics and perception. In W. G. Chase (Ed.), *Visual information processing*. Oxford: Academic Press.

Cohen, J. D., Perlstein, W. M., Braver, T. S., Nystrom, L. E., Noll, D. C., Jonides, J., et al. (1997). Temporal dynamics of brain activation during a working memory task. *Nature, 386*, 604–608.

de Saussure, F. (1959). *Course in General Linguistics* (W. Baskin, Trans.). New York: Philisophical Library.

Dewey, J. (1896). The reflex arc concept in psychology. *Psychological Review, 3*, 357–370.

Diehl, R. L., Lotto, A. J., & Holt, L. L. (2004). Speech perception. *Annual Review of Psychology, 55*, 149–179.

Donders, F. C. (1868/1969). On the speed of mental processes. *Acta Psychologica, 30*, 412–431.

Fant, G. (1967). Auditory patterns of speech. In W. Wathen-Dunn (Ed.), *Models for the Perception of Speech and Visual Form* (pp. 111–125). Cambridge: MIT Press.

Fodor, J. A. (1983). *The Modularity of Mind: An Essay on Faculty Psychology*. Cambridge: MIT Press.

Freeman, W. J. (2003). Neurodynamic models of brain in psychiatry. *Neuropsychopharmacology, 28*, S54–S63.

Freeman, W. J. & Barrie, J. M. (1994). Chaotic oscillations and the genesis of meaning in cerebral cortex. In G. Bizsaki (Ed.), *Temporal Coding in the Brain* (pp. 13–17). Berlin: Springer-Verlag.

Freeman, W. J. & Skarda, C. A. (1990). Representations: Who needs them? In J. L. McGaugh & N. M. Weinberger (Eds.), *Brain Organization and Memory: Cells, Systems, and Circuits* (pp. 375–380). New York: Oxford University Press.

Gall, F. J. (1825). *Sur l'origine des qualités morales et des facultés intellectuelles de l'homme: et sur les conditions de leur manifestation*. Paris: J. B. Baillière.

Gazzaniga, M. S. (1985). *The Social Brain: Discovering the Networks of the Mind*. New York: Basic Books.

Geschwind, N. (1965). The organization of language and the brain. *Science, 170,* 940–944.

Geschwind, N. (1971). Current concepts: Aphasia. *New England Journal of Medicine, 284,* 654–656.

Garfield, J. L. (Ed.). (1987). *Modularity in Knowledge Representation and Natural-Language Understanding.* Cambridge: MIT Press.

Giles, H. (1973). Accent mobility: A model and some data. *Anthropological Linguistics, 15,* 87–105.

Giles, H. & Coupland, N. (1991). *Language: Contexts and Consequences.* Pacific Grove, CA: Brooks–Cole.

Giles, H. & Smith, P. (1979). Accommodation theory: Optimal levels of convergence. In H. Giles & R. N. St. Clair (Eds.), *Language and Social Psychology* (pp. 45–65). Baltimore: University Park Press.

Gleick, J. (1987). *Chaos: Making a New Science.* New York: Viking.

Goldin-Meadow, S., Nusbaum, H., Kelly, S. D., & Wagner, S. (2001). Explaining math: Gesturing lightens the load. *Psychological Science, 12,* 516–522.

Grafton, S. T., Hazeltine, E., & Ivry, R. B. (1998). Abstract and effector-specific representations of motor sequences identified with PET. *Journal of Neuroscience, 18,* 9420–9428.

Holtgraves, T. (1994). Communication in context: Effects of speaker status on the comprehension of indirect requests. *Journal of Experimental Psychology: Learning, Memory, and Cognition, 20,* 1205–1218.

Howard, D., Patterson, K., Wise, R., Brown, W. D., Friston, K., Weiller, C., et al. (1992). The cortical localization of the lexicons. Positron emission tomography evidence. *Brain, 115,* 1769–1782.

Hugenberg, K. & Bodenhausen, G. V. (2004). Ambiguity in social categorization: The role of prejudice and facial affect in race categorization. *Psychological Science, 15,* 342–345.

James, W. (1904). The Chicago school. *Psychological Bulletin, 1,* 1–5.

Johnson, K., Strand, E. A., & D'Imperio, M. (1999). Auditory-visual integration of talker gender in vowel perception. *Journal of Phonetics, 27,* 359–384.

Jonides, J., Schumacher, E. H., Smith, E. E., Koeppe, R. A., Awh, E., Reuter-Lorenz, P. A., Marshuetz, C., & Willis, C. R. (1998). The role of parietal cortex in verbal working memory. *Journal of Neuroscience, 18,* 5026–5034.

Kohler, S. & Moscovitch, M. (1997). Unconscious visual processing in neuropsychological syndromes; a survey of the literature and evaluation of models of consciousness. In M. D. Rugg (Ed.), *Cognitive Neuroscience* (pp. 305–373). Cambridge: MIT Press.

Lakin, J. L. & Chartrand, T. L. (2003). Using nonconscious behavioral mimicry to create affiliation and rapport. *Psychological Science, 14,* 334–339.

Lane, R. D., Reiman, E. M., Ahern, G. L., Schwartz, G. E., & Davidson, R. J. (1997). Neuroanatomical correlates of happiness, sadness, and disgust. *American Journal of Psychiatry, 154,* 926–933.

Lashley, K. D. (1950). In search of the engram. *Symposia of the Society for Experimental Biology, 4,* 454–482.

Luks, T. L., Nusbaum, H. C., & Levy, J. (1998). Hemispheric involvement in the perception of syntactic prosody is dynamically dependent on task demands. *Brain and Language, 65,* 313–332.

McGurk, H. & MacDonald, J. (1976). Hearing lips and seeing voices. *Nature, 264,* 746–748.

McIntosh, A. R. (2000). Towards a network theory of cognition. *Neural Networks, 13,* 861–870.

McNeill, D. (1992). *Hand and Mind: What Gestures Reveal about Thought.* Chicago: University of Chicago Press.

Moran, J. & Desimone, R. (1985). Selective attention gates visual processing in the extrastriate cortex. *Science, 229,* 782–784.

Morecraft, R. J. & van Hoesen, G. W. (1998). Convergence of limbic input to the cingulate motor cortex in the rhesus monkey. *Brain Research Bulletin, 45,* 209–232.

Neisser, U. (1976). *Cognition and Reality.* San Francisco: W. H. Freeman.

Nyberg, L. & McIntosh, A. R. (2001). Functional neuroimaging: Network analyses. In R. Cabeza & A. Kingstone (Eds.), *Handbook of Functional Neuroimaging of Cognition* (pp. 49–72). Cambridge: MIT Press.

Nusbaum, H. C. & Magnuson, J. (1997). Talker normalization: Phonetic constancy as a cognitive process. In K. Johnson & J. W. Mullennix (Eds.), *Talker Variability in Speech Processing* (pp. 109–132). San Diego: Academic Press.

Nusbaum, H. C. & Schwab, E. C. (1986). The role of attention and active processing in speech perception. In E. C. Schwab & H. C. Nusbaum (Eds.). *Pattern Recognition by Humans and Machines,* vol. 1: *Speech Perception* (pp. 113–157). San Diego: Academic Press.

Picard, N. & Strick, P. L. (1996). Motor areas of the medial wall: A review of their location and functional activation. *Cerebral Cortex, 6,* 342–53.

Pickering, M. J. & Garrod, S. (2004). Toward a mechanistic psychology of dialogue. *Behavioral and Brain Sciences, 27,* 169–226.

Remez, R. E., Rubin, P. E., Pisoni, D. B., & Carrell, T. I. (1981). Speech perception without traditional speech cues. *Science, 212,* 947–950.

Rizzolatti, G., Craighero, L., & Fadiga, L. (2002). The mirror system in humans. In M. I. Stamenov & V. Gallese (Eds.), Mirror neurons and the evolution of brain and language. *Advances in Consciousness Research, 42,* 37–59.

Roland, P. E. (1993). *Brain Activation*. New York: Wiley–Liss.

Rubin, D. L. (1992). Nonlanguage factors affecting undergraduates' judgments of nonnative English-speaking teaching assistants. *Research in Higher Education, 33,* 511–531.

Runeson, S. (1977). On the possibility of smart perceptual mechanisms. *Scandinavian Journal of Psychology, 18,* 172–179.

Saygin, A. P., Dick, F., Wilson, S. W., Dronkers, N. F., & Bates, E. (2003). Neural resources for processing language and environmental sounds: Evidence from aphasia. *Brain, 126,* 928–945.

Sergent, J., Zuck, E., Lévesque, M., & MacDonald, B. (1992). Positron emission tomography study of letter and object processing: Empirical findings and methodological considerations. *Cerebral Cortex, 2,* 68–80.

Shallice, T. (1988). *From Neuropsychology to Mental Structure*. Cambridge: Cambridge University Press.

Siegal, M. & Varley, R. (2002). Neural systems involved in "theory of mind." *Nature Reviews Neuroscience, 3,* 463–467.

Skipper, J. I., Nusbaum, H. C., & Small, S. L. (2002). Speech perception and the inferior frontal neural system for motor imitation. *Journal of Cognitive Neuroscience, 14,* F103.

Skipper, J. I., Nusbaum, H. C., & Small, S. L. (2005). Listening to Talking Faces: Motor Cortical Activation During speech percestion. *NeuroImage, 25,* 76–89.

Small, S. L. & Rieger, C. J. (1982). Parsing and comprehending with word experts: A theory and its realization. In W. G. Lenhert & M. H. Ringle (Eds.), *Strategies for Natural Language Processing* (pp. 89–147). Hillsdale, NJ: Erlbaum.

Smith, E. E. & Jonides, J. (1999). Storage and executive processes in the frontal lobes. *Science, 283,* 1657–1661.

Smith, L. B. & Thelen, E. (Eds.) (1993). *A Dynamic Systems Approach to Development: Applications*. Cambridge, MA: MIT Press.

Sperber, D. (2001). In defense of massive modularity. In E. Dupoux (Ed.). *Language, Brain, and Cognitive Development: Essays in Honor of Jacques Mehler* (pp. 47–57). Cambridge: MIT Press.

Sternberg, S. (1969). The discovery of processing stages: Extensions of the Donders' method. *Acta Psychologica, 30,* 276–315.

Studdert-Kennedy, M. (1981). The emergence of phonetic structure. *Cognition, 10,* 301–306.

Sumby, W. H. & Pollack, I. (1954). Visual contribution to speech intelligibility in noise. *Journal of the Acoustical Society of America, 26,* 212–215.

Summerfield, Q. (1992). Lipreading and audio-visual speech perception. *Philosophical Transactions of the Royal Society of London Series B, Biological Sciences, 335,* 71–78.

van Baaren, R. B., Horgan, T. G., Chartrand, T. L., & Dijkmans, M. (2004). The forest, the trees, and the chameleon: Context dependence and mimicry. *Journal of Personality and Social Psychology, 86*, 453–459.

Wegner, D. M. & Bargh, J. A. (1998). Control and automaticity in social life. In D. T. Gilbert & S. T. Fiske (Eds.), *The Handbook of Social Psychology*, 4th ed., (vol. 1, pp. 446–496). New York: McGraw Hill.

Wernicke, C. (1874). *Der Aphasische Symptomenkomplex*. Breslau: Cohn & Weigert.

Weiskrantz, L. (1986). *Blindsight: A Case Study and Implications*. New York: Oxford University Press.

Wittenbrink, B., Hilton, J. L., & Gist, P. L. (1998). In search of similarity: Stereotypes as naive theories in social categorization. *Social Cognition, 16*, 31–55.

Wymbs, N. F., Nusbaum, H. C., & Small, S. L. (2004). The informed perceiver: Neural correlates of linguistic expectation and speech perception. Poster presented at the Cognitive Neuroscience Society, San Francisco.

Zajonc, R. B. (1980). Feeling and thinking: Preferences need no inferences. *American Psychologist, 35*, 151–175.

Zatorre, R. J., Meyer, E., Gjedde, A., & Evans, A. C. (1996). PET studies of phonetic processing of speech: Review, replication, and reanalysis. *Cerebral Cortex, 6*, 21–30.

8 Orbitofrontal Cortex and Social Regulation

Jennifer S. Beer

Human social interaction is a complex process that requires individuals to act not only on their instincts, but to modify their behavior in reference to their own expectations and the expectations of others (social regulation). Self-perception processes and perceptive-processes of other people have been theorized to be central to the regulation of social behavior. However, studying the role of self-perception and person-perception processes in the regulation of behavior has been challenging. Individuals are socialized from a very early age to regulate their behavior. Just as walking and talking begin as effortful processes and eventually become automatic, aspects of behavioral regulation become a well-learned process. As a well-learned process, regulation of social behavior is difficult to override in experimental or naturalistic studies. However, the fact still remains that variance in social regulation is needed to make hypotheses about the underlying role of self- and person-perception processes.

Neuroscientific methods suggest new avenues for examining the role of self- and person-perception processes in social regulation. Neuropsychological populations provide one new method for understanding the mechanisms underlying an overlearned ability such as social regulation. For example, variance in adaptive social regulation can be created through comparisons between patient populations with social regulatory dysfunction and healthy control individuals. Functional neuroimaging methods are useful complements to neuropsychological approaches. For example, if damage to a particular area is shown to selectively impair social regulation, neuroimaging can be useful for understanding which computations are made when brain tissue in this area is intact. Although not intended to replace more traditional methods used in social-personality psychology, neuroscience methods may provide additional avenues for studying phenomena of interest.

Social Regulation: A Brief Overview

What exactly is social regulation? The most basic explanation of behavior in psychology is a stimulus-response model in which particular stimuli in the environment elicit particular responses from the organism. For example, a basic response to the presence of food is to approach it. However, people do not simply react to their environments. They bring standards and expectations to the environment and these standards, in turn, influence their behavior (Baumeister & Heatherton, 1996; Carver & Scheier, 1990; Higgins, 1987). Therefore, descriptions of human social behavior must account for behavioral responses that are more complex than just a basic reaction to a stimulus.

More complex explanations of the relation between environmental stimuli and behavior take into account internal qualities of individuals. From this perspective, individuals actively control at least some aspects of their behavior; in other words, they may actively select behavioral responses. A number of factors within the organism contribute to perception of stimuli and to selection of behavioral responses. For example, even a hungry person may not immediately reach for the last cookie at lunch. Instead, people may offer the cookie to others because they are concerned about their weight or appearing gluttonous. In this case the person's response was mediated by a desire to maintain self-standards for physical appearance or politeness toward others. In summary, social regulation occurs when people exert control over their interpersonal behavior to match their internal standards and expectations, instead of merely reacting to properties of environmental stimuli.

Orbitofrontal Cortex: Social Executive Functioning

A first step toward answering questions about social regulation using neuroscience methods is to identify whether any neuropsychological population presents selective self-regulatory deficits. If so, comparisons between these people and healthy controls will be useful for understanding whether self- and person-perception processes are necessary for social-regulation. The complexity of social-regulatory processes suggests that no one area of the Brain will be wholly responsible for social regulation, but areas associated with cognitive or emotional control are a good place to begin. Generally, the prefrontal cortex is in a unique position to control both cognitive and social processes through its extensive reciprocal connections with cortical, limbic, and subcortical sites. More specifically, the orbitofrontal cortex (figure 8.1) is richly connected to areas associated with emotional and social processing, including the amygdala,

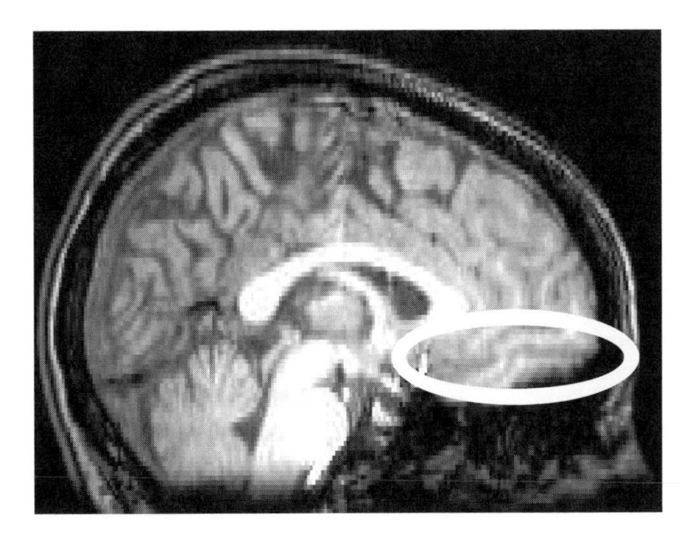

Figure 8.1
Orbitofrontal cortex indicated with a white circle on a sagittal brain slice.

anterior cingulate, and somatosensory areas I and II (Adolphs, 1999; Brothers, 1996). Furthermore, evidence suggests that damage to the orbitofrontal cortex might be selectively associated with poorly regulated social behavior.

Orbitofrontal Damage and Impaired Social Regulation

For a long time, the relation between orbitofrontal damage and social disinhibition was supported only by case studies and clinical anecdotes (table 8.1).

In fact, the most famous case study was first described in the 1800s. Phineas Gage was a railroad worker who accidently shot a tamping iron through his head. Harlow, the physician who treated him, noted that after his injury, Gage could walk, talk and remember things just fine, but he did have problems with his social behavior. For example, Harlow described Gage as "irreverent, indulging in the grossest profanity (which was not previously his custom), manifesting little deference to his fellows, impatient of restraint or advice when it conflicts with his desires" (MacMillan, 1986, p. 85). Other physicians who examined Gage noted that these changes were a problem to the extent that "persons of delicacy, especially women, found it impossible to endure his presence" and "his society was intolerable to decent people" (MacMillan, 1986, p. 99). Similarly, more recent descriptions of patients with orbitofrontal lesions associated damage to this area with impaired ability to prioritize solutions to inter-

Table 8.1
Examples of Social Disinhibition Associated with Orbitofrontal Damage

Social impairment	Study
Case Studies and Clinical Characterizations	
Profane, irreverent	Harlow, 1848, cited in MacMillan, 1986
Impaired problem-solving for interpersonal vignettes	Saver & Damasio, 1991
Riding on a hospital gurney	Blair & Cipolotti, 2000
Practicing karate in the waiting room, greeting nurses in an overly familiar fashion	Rolls et al., 1994
Unrestrained, disinhibited actions	Kretschmer, 1956, cited in Stuss & Benson, 1992
Puerile, jocular attitude, sexually disinhibited behavior, inappropriate self-indulgence	Blumer & Benson, 1975
Empirical Studies	
Digression from conversation topic	Kaczmarek, 1984
Teasing inappropriately, excessive self-disclosure	Beer, 2002
Excessive self-disclosure	Beer et al., 2003

personal problems (Saver & Damasio, 1991), a tendency to greet strangers in an overly familiar manner (Rolls et al., 1994), and disruptive behavior in a hospital setting (Blair & Cipolotti, 2000).

In addition to descriptive evidence, three empirical studies show that orbitofrontal damage does impair social regulation (table 8.1; Beer, 2002; Beer et al., 2003; Kaczmarek, 1984). They suggest that patients with orbitofrontal damage were unable to regulate their behavior, in many cases, behaving with strangers in ways that were more appropriate for interactions with close others. In one study, patients with orbitofrontal damage (figure 8.2) teased strangers in inappropriate ways and were more likely to include unnecessary personal information when answering questions (Beer et al., 2003). This task required regulation because teasing behavior is not common between strangers and requires apologetic, submissive behavior to be appropriate (Keltner et al., 2001). A second study compared the self-disclosure of patients with orbitofrontal and dorsolateral prefrontal lesions and healthy controls (Beer, 2002). Those with orbitofrontal damage were judged as much more intimate and inappropriate when speaking to a stranger. Finally, in the third study, orbitofrontal damage was

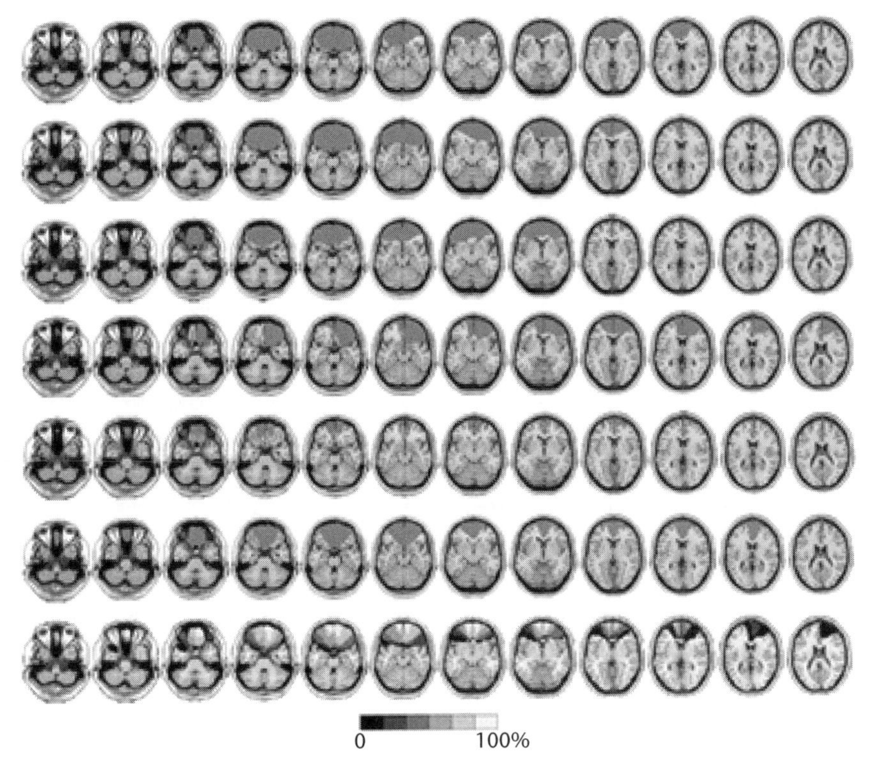

Figure 8.2
Brain slices of patients with orbitofrontal damage. Each row represents several slices of a single patient; gray areas indicate areas of damage. The bottom row represents the amount of overlap in lesion areas across the whole sample.

associated with failure to maintain conversational relevance (Kaczmarek, 1984). Together these results suggest that orbitofrontal cortex is critically involved in ensuring that behavior is appropriate given a particular social context. Therefore, understanding the exact mechanism by which orbitofrontal cortex supports social regulation will enhance understanding of critical psychological processes for social regulation.

The Role of Self-Perception and Person Perception in Self-Regulation

Two possible mechanisms by which orbitofrontal cortex may support appropriate social functioning are self- and person-perception processes. Both processes are given a central role in most theories of self-regulation (Baumeister & Heatherton, 1987;

Carver & Scheier, 1990; Gray & McNaughton, 2000; Higgins, 1982). Specifically, individuals must monitor their behavior in order to determine if it is appropriate for a given context. Therefore, individuals must have insight into what they are actually doing and determine if it is affecting other people in the intended manner.

Self-Perception Processes and Social Regulation

The fundamental role of self-perception processes in social regulation is exemplified by consensus on the importance of monitoring one's behavior to ensure appropriateness. Most, if not all, theories of regulation include some component of self-monitoring. This monitoring may occur implicitly or explicitly and is usually characterized by a comparison of actual behavior in reference to goals, social norms, and/or expectations of others (Carver & Scheier, 1990; Gray & McNaughton, 2000; Higgins, 1987).

The importance of self-monitoring for adaptive social regulation is reflected in research on self-focused attention. When people are placed in conditions of self-focused attention their behavior conforms closely to their intentions, presumably because the frequency of monitoring is increased (Carver & Scheier, 1990; Gibbons, 1990). In other words, regulation is more precise when monitoring is facilitated by focusing people's attention on themselves. Experimental inductions of self-focused attention include placing a mirror in front of people and showing them a videotape of themselves (Carver & Scheier, 1990; Gibbons, 1990; Robins & John, 1997).

Another suggested role between self-perceptive processes and social regulation comes from research on the accuracy of self-perception. Whereas the importance of self-perception for social regulation is somewhat agreed upon, the motivation underlying self-perceptions most beneficial for social regulation has been debated. Some studies suggest that people generally tend to inflate their assessments of themselves and that such inflation promotes adaptive social behaviors (Taylor & Brown, 1988), whereas others suggest that successful self-regulation relies on accurate self-perceptions (Heatherton & Ambady, 1993; Robins & Beer, 2001).

Self-Perception Processes and Orbitofrontal Cortex

A host of research supports the role of orbitofrontal cortex, as well as other frontal lobe areas, in self-perception processes identified as necessary for social regulation (monitoring, insight accuracy). Clinical characterizations and empirical research of damage to various areas of the frontal lobes suggest that the frontal lobes are intricately involved in self-monitoring (Beer, 2002; Lhermitte, 1986; Luria & Homskaya, 1970; Stuss & Benson, 1986). Orbitofrontal cortex may permit individuals to have accurate insight into their behavior and compare this behavior with an abstract standard of how they would like to be behaving.

Failure to monitor the appropriateness of behavior is exemplified in the typical stimulus-bound behavior of patients with frontal lobe impairment, including but not specific to orbitofrontal cortex (Luria & Homskaya, 1970; Lhermitte, 1986; Lhermitte et al., 1986). Stimulus-bound behavior is not under the control of the individual, but is determined by context. For example, objects placed in front of patients with prefrontal damage may be picked up and used (utilization behavior) without patients being asked to do so (Lhermitte, 1986). In addition, patients with prefrontal damage may imitate the behavior of the experimenter even when this behavior is bizarre and socially inappropriate. These studies suggest that prefrontal damage impairs the ability to choose behaviors as a function of whether or not they are appropriate for a given context.

As a complement to these classic case studies, one empirical study showed that patients with orbitofrontal damage tend to lack insight into the inappropriateness of their social behavior compared with those with dorsolateral prefrontal lesions and healthy controls (Beer, 2002). Participants took part in a task that required them to regulate how much personal information they revealed. Transcripts of the conversations were coded by trained judges for breadth, appropriateness, and intimacy of self-disclosure. A measure of accurate self-perception was computed by comparing self-reports with trained judges' codes. Patients with orbitofrontal damage tended to overestimate the appropriateness of the intimacy of their self-disclosure. No differences were found for those with dorsolateral prefrontal damage and controls.

Self-insight impairment is also suggested by failure of patients with orbitofrontal lesion to emotionally appraise their behavior in an appropriate way unless their ability to monitor their behavior accurately is facilitated. For example, patients with orbitofrontal damage became embarrassed by their inappropriate social behavior only when experimental methods increased their self-focused attention and specifically drew their attention to their behavior (feedback from others, video playback).

Not all self-processes are associated with the orbitofrontal cortex. Imaging research suggests that the medial prefrontal cortex is involved in particularly the effective encoding of information in reference to the self (the self-referent effect; Symons & Johnson, 1997). Studies using position emission tomography (Craik et al., 1999) and functional magnetic resonance imaging (fMRI; Kelly et al., 2002) have found increased activity in the medial prefrontal cortex (BA 9 and 10) when encoding information in relation to the self, compared with encoding information in relation to a famous political figure and syllabic structure. Similarly, fMRI studies consistently show that right BA 9 and 10 has increased activity when observing one's own face versus that of a close other or another person (Keenan et al., 2000).

Person-Perception Processes and Social Regulation

In addition to knowing one self, successful social regulation requires that people understand the mental and emotional states of others. Social regulation is inherently interpersonal, and often people strive to be accepted, or at least not rejected, by others. Therefore, behavior must be modified in response to reactions of others. The importance of understanding others for social regulation is exemplified by the notion of perceptiveness (Goffman, 1967). To ensure smooth social interactions, people must be skilled at perceptiveness, that is, inferring interpretations others place on one's acts (Goffman, 1967). Presumably, perceptiveness includes interpreting both evaluations and emotional responses of others.

Faulty understanding of others' mental and emotional states is associated with poor social regulation. For example, faulty information processing regarding the intentions of others has been shown to lead to aggressive behavior (Crick & Dodge, 1994, 1996). Another example of the relation between understanding others and appropriate social regulation comes from research on self-disclosure and liking. Self-disclosure is governed by strict norms prescribing gradual and reciprocal escalation of intimacy (Collins & Miller, 1994; Cozby, 1973). The reciprocal element makes it necessary to properly understand the perspective of a conversation partner. Understanding mental and emotional states is not just important for regulation of new relationships; it is also essential for well-established relationships. For example, the concept of minding includes making correct attributions for a partner's intentions and motivations and is considered essential for relationship satisfaction (Harvey & Omarzu, 1997).

Person-Perception Processes and Orbitofrontal Cortex

Both lesion and imaging researches suggest that prefrontal cortex, including orbitofrontal cortex, is involved in making inferences about others, such as understanding their mental and emotional states. For example, prefrontal areas may be involved in making mental inferences. Patients with dorsolateral prefrontal damage were impaired at giving directions to another person compared with healthy persons (Price et al., 1990). Although some studies suggest that orbitofrontal cortex, particularly the right side, is critically involved in theory of mind on a visual perspective task (Stuss, Gallup, & Alexander, 2001), another group found that patients with orbitofrontal and dorsolateral prefrontal lesions did not have trouble with basic level theory of mind tasks (Stone, Baron-Cohen, & Knight, 1998). However, in this study, patients with orbitofrontal damage were impaired at their ability to make inferences about the intentions of characters in vignettes about social faux pas compared with normal controls and those with dorsolateral prefrontal damage.

Lack of insight into others' minds is also reflected in the conversational style of patients with orbitofrontal damage. Damage to the left orbitofrontal cortex was associated with confabulation, misnamings, and digressions from the topic (Kaczmarek, 1984). These findings were interpreted to reflect a decreased appreciation for the necessity of a coherent description to ensure that the audience understood the speaker. It is important to note that it is also possible to interpret these data in support of impaired self-monitoring. Patients with orbitofrontal damage may have been tangential and incoherent in their responses because they failed to monitor the appropriateness of their responses to questions.

Imaging studies also suggest that both medial prefrontal and orbitofrontal cortex are implicated in making inferences about another's mental states. In two imaging studies, left medial prefrontal areas, BA 8 and 9, showed increased activation when participants were asked to make a mental inference versus a physical inference. In one study, participants had to decide whether a person living in the fifteenth century would know or could decide the function of an object. When these two conditions were compared, increased activity in left BA 9 was found (Goel et al., 1997). Activation was assessed in relation to story comprehension that required participants to make either a mental or physical inference about a character's actions. When the two conditions were compared, increased activity in left BA 8 was found (Fletcher et al., 1995). In contrast, another imaging study showed increased activation in right orbitofrontal regions compared with left polar frontal regions during a task that required participants to decide whether a word described a mind state or a body state (Baron-Cohen et al., 1994).

It is also possible that prefrontal areas are important for making inferences about the emotional states of others. Empirical support for the particular role of orbitofrontal cortex in this regard is mixed. Some studies found that patients with orbitofrontal damage were impaired on paper and pencil measures of empathy (Grattan et al., 1994; Shamay-Tsoory et al., 2003). Similarly, they were impaired at inferring emotional states from pictures of various emotional facial expressions (Hornak, Rolls, & Wade, 1996), although in one study this impairment held only for expressions of embarrassment and shame (Beer et al., 2003). However, orbitofrontal patients had no trouble inferring the feelings of story characters that had been on the receiving end of a social faux pas (Stone et al., 1998).

Conclusion

Research using neuroscience methods supports the importance of self-perception processes for social regulation. Damage to the orbitofrontal cortex selectively impairs the ability to modify interpersonal behavior in reference to social norms. This impairment occurs concurrently with impaired self-monitoring. This suggests that self-monitoring, or on-line comparison of immediate behavior with norms of appropriateness, may be fundamentally associated with social regulation. This is in contrast to more general self-reflection, which is associated with more superior portions of the prefrontal cortex. Evidence for the role of orbitofrontal cortex in understanding the intentions and emotions of other people is weak at best; however, it is not possible to conclude that person-evaluative processes are irrelevant for social regulation. Indeed, if selective damage to the medial prefrontal cortex were possible, is might also selectively impair social regulation. Therefore, the support for the role of self-monitoring in social regulation is not mutually exclusive with other theorized mechanisms such as person-evaluative processes or emotion-cognition synthesis.

References

Adolphs, R. (1999). Social cognition and the human brain. *Trends in Cognitive Sciences, 3*, 469–479.

Baron-Cohen, S., Ring, H., Moriarty, J., Schmitz, B., Costa, D., & Ell, P. (1994). Recognition of mental state terms. *British Journal of Psychiatry, 165*, 640–649.

Baumeister, R. F. & Heatherton, T. A. (1996). Self-regulation failure: An overview. *Psychological Inquiry, 7*, 1–15.

Beer, J. S. (2002). *Self-regulation of Social Behavior.* Dissertation, University of California, Berkeley.

Beer, J. S., Heerey, E. A., Keltner, D., Scabini, D., & Knight, R. T. (2003). The regulatory function of self-conscious emotion: Insights from patients with orbitofrontal damage. *Journal of Personality and Social Psychology, 85*, 594–604.

Blair, R. J. R. & Cipolotti, L. (2000). Impaired social response reversal: A case of "acquired sociopathy." *Brain, 123*, 1122–1141.

Blumer, D. & Benson, D. F. (1975). Personality changes with frontal and temporal lobe lesions. In D. F. Benson & D. Blumer (Eds.), *Psychiatric Aspects of Neurologic Disease* (pp. 151–169). New York: Grune & Stratton.

Brothers, L. (1996). Brain mechanisms of social cognition. *Journal of Psychopharmacology, 10*, 2–8.

Carver, C. S. & Scheier, M. (1990). Principles of self-regulation: Action and emotion. In E. T. Higgins & R. M. Sorrentino (Eds.), *Handbook of motivation and cognition: Foundations of social Behavior* (pp. 3–52). New York: The Guildford Press.

Collins, N. L. & Miller, L. C. (1994). Self-disclosure and liking: A meta-analytic review. *Psychological Bulletin, 116*, 457–475.

Cozby, P. C. (1973). Self-disclosure: A literature review. *Psychological Bulletin, 79*, 73–91.

Craik, F. I. M., Moroz, T. M., Moscovitch, M., Stuss, D. T., Wincour, G., Tulving, E., et al. (1999). In search of the self: A positron emission tomography study. *Psychological Science, 10*, 26–34.

Crick, N. R. & Dodge, K. A. (1994). A review and reformulation of social information-processing mechanisms in children's social adjustment. *Psychological Bulletin, 115*, 74–101.

Crick, N. R. & Dodge, K. A. (1996). Social information-processing mechanisms in reactive and proactive aggression. *Child Development, 67*, 993–1002.

Fletcher, P. C., Happe, F., Frith, U., Baker, S. C., Dolan, R. J., Frackowiak, R. S. J., et al. (1995). Other minds in the brain: A functional imaging study of "theory of mind" in story comprehension. *Cognition, 57*, 109–128.

Gibbons, F. X. (1990). The impact of focus of attention and affect on social behavior. In R. Crozier (Ed.), *Shyness and Embarrassment* (pp. 119–143). New York: Cambridge University Press.

Goel, V., Grafman, J., Tajik, J., Gana, S., & Danto, D. (1997). A study of the performance of patients with frontal lobe lesions in a financial planning task. *Brain, 120*, 1805–1822.

Goffman, E. (1967). *Interaction Ritual* (p. 13). New York: Pantheon Books.

Grattan, L. M., Bloomer, R. H., Archambault, F. X., & Eslinger, P. J. (1994). Cognitive flexibility and empathy after frontal lobe lesion. *Neuropsychiatry, Neuropsychology, and Behavioral Neurology, 7*, 251–259.

Gray, J. A. & McNaughton, N. (2000). Neural anxiety systems: Relevant fault-lines to track and treat disorders. *European Journal of Neuroscience, 12*, 311.

Harvey, J. H. & Omarzu, J. (1997). Minding the close relationship. *Personality and Social Psychology Review, 3*, 224–240.

Heatherton, T. F. & Ambady, N. (1993). Self-esteem, self-prediction, and living up to commitments. In R. F. Baumeister (Ed.), *Self-esteem: The puzzle of low self-regard* (pp. 131–145). New York: Plenum Press.

Higgins, E. T. (1987). Self-discrepancy: A theory relating self and affect. *Psychological Review, 94*, 319–340.

Hornak, J., Rolls, E. T., & Wade, D. (1996). Face and voice expression identification in patients the emotional and behavioural changes following ventral frontal lobe damage. *Neuropsychologia, 34*, 247–261.

Kaczmarek, B. L. J. (1984). Neurolinguistic analysis of verbal utterances in patients with focal lesions of frontal lobes. *Brain and Language, 21*, 52–58.

Keenan, J. P., Wheeler, M. A., Gallup, G. G., & Pasucal-Leone, A. (2000). Self-recognition and the right prefrontal cortex. *Trends in Cognitive Sciences, 4*, 338–344.

Kelley, W. M., Macrae, C. N., Wyland, C. L., Caglar, S., Inati, S., & Heatherton, T. F. (2002). Finding the self? An event-related fMRI study. *Journal of Cognitive Neuroscience, 14*, 785–794.

Keltner, D., Capps, L., Kring, A. M., Young, R. C., & Heerey, E. A. (2001). Just teasing: A conceptual analysis and empricial review. *Psychological Bulletin, 127*, 229–248.

Lhermitte, F. (1986). Human autonomy and the frontal lobes. II. Patient behavior in complex and social situations: The "environmental dependency syndrome." *Annals of Neurology, 19*, 335–343.

Lhermitte, F., Pillon, B., & Serdaru, M. (1986). Human anatomy and the frontal lobes. I. Imitation and utilization behavior: A neuropsychological study of 75 patients. *Annals of Neurology, 19*, 326–334.

Luria, A. R. & Homskaya, E. D. (1970). Frontal lobes and the regulation of arousal process. In D. I. Mostofsky (Ed.), *Attention: Contemporary Theory and Analysis* (pp. 303–330). New York: Appleton-Century-Crofts.

MacMillan, E. (1986). A wonderful journey through skull and brains: The travels of Mr. Gage's tamping iron. *Brain and Cognition, 5*, 67–107.

Price, B. H., Daffner, K. R., Stowe, R. M., & Marsel-Mesulam, M. (1990). The comportmental learning disabilities of early frontal lobe damage. *Brain, 113*, 1383–1393.

Robins, R. W. & Beer, J. S. (2001). Positive illusions about the self: Short-term benefits and long-term costs. *Journal of Personality and Social Psychology, 80*, 340–352.

Robins, R. W. & John, O. P. (1997). The quest for self-insight: Theory and research on accuracy and bias in self-perception. In R. Hogan, J. Johnson, and S. Briggs (Eds.), *Handbook of Personality Psychology* (pp. 671–679). New York: Academic Press.

Rolls, E. T., Hornak, J., Wade, D., & McGrath, J. (1994). Emotion-related learning in patients with social and emotional changes associated with frontal lobe damage. *Journal of Neurology, Neurosurgery, and Psychiatry, 57*, 1518–1524.

Saver, J. L. & Damasio, A. R. (1991). Preserved access and processing of social knowledge in a patient with acquired sociopathy due to ventromedial frontal damage. *Neuropsychologia, 29*, 1241–1249.

Shamay-Tsoory, S. G., Tomer, R., Berger, B. D., & Aharon-Peretz, J. (2003). Characterization of empathy deficits following prefrontal brain damage: The role of the right ventromedial prefrontal cortex. *Journal of Cognitive Neuroscience, 15*, 324–337.

Stone, V. E., Baron-Cohen, S., & Knight, R. T. (1998). Frontal lobe contributions to theory of mind. *Journal of Cognitive Neuroscience, 10*, 640–656.

Stuss, D. T. & Benson, F. (1986). *The Frontal Lobes.* New York: Raven Press.

Stuss, D. T., Gallup, G. G., & Alexander, M. P. (2001). The frontal lobes are necessary for "theory of mind." *Brain, 124*, 279–286.

Symons, C. S. & Johnson, B. T. (1997). The self-reference effect in memory: A meta-analysis. *Psychological Bulletin, 121*, 371–394.

Taylor, S. E. & Brown, J. D. (1988). Illusion and well-being: A social psychological perspective on mental health. *Psychological Bulletin, 103*, 193–210.

9 A Pain by any other Name (Rejection, Exclusion, Ostracism) still Hurts the Same: The Role of Dorsal Anterior Cingulate Cortex in Social and Physical Pain

Matthew D. Lieberman and Naomi I. Eisenberger

Several chapters in this volume explore whether a particular dimension of social cognition can be reduced to more general cognitive processes by examining whether social and cognitive processes share overlapping neural bases. Some chapters argue for distinct neural processes devoted to social cognition, not shared by more general cognitive processes (Mitchell), and others identify components of social cognition that do share properties with more general cognitive systems as indicated by overlapping neural systems (Phelps, Stone). A social phenomenon that seems an unlikely candidate for this kind of cognitive reduction analysis is social pain, the distress experienced in response to rejection, exclusion, or ostracism. These experiences are profoundly social and, at first glance, there are no obvious cognitive analogues to which they can be reduced.

Nevertheless, we suggest a certain type of reduction. We propose that social pain may be profitably examined by considering its relation to physical pain. More specifically, we suggest that some of the basic neural mechanisms that support the experience of physical pain also support the experience of social pain (although it is doubtful that the overlap is as complete as the extreme position staked out by our Shakespearean title). Evidence from both animal and human research literatures suggest that the dorsal region of the anterior cingulate cortex (dACC) is similarly involved in the distressing component of both forms of pain. Moreover, once a connection is established between the experience of social pain and physical pain, we can explore the underlying computations that connect the processes. In other words, we can examine the computational function of the dACC such that it should be involved in various forms of painful experience. Thus the dACC may function as a neural alarm system, combining both detection of a problem and sounding of an alarm, typically found in any alarm system.

Finally, we reexamine the function of dACC versus rostral anterior cingulate (rACC) to expand on how our model fits with previously held views of the function of the

ACC. The current view is that dACC is involved in cognitive processes and the rACC is involved in affective processes. However, this view does not ultimately hold up to scrutiny when one takes into account pain studies, which typically activate dACC. A different dichotomy of function can account for the previous view, such that both dACC and rACC may be involved in conflict processing, but different forms. The dACC can best be understood as processing *nonsymbolic* conflict in which the conflict is not explicitly represented but is instead best characterized as the tension level in a connectionist constraint-satisfaction network. Alternatively, rACC can best be understood as processing *symbolic* conflict in which the conflict is explicitly represented with symbolic or propositional thought. The capacity to represent conflict symbolically may be a purely human attribute and has implications for how we regulate different types of affective experiences.

We should say from the outset that although we focus on the dACC as a region involved in social and physical pain, we readily admit that we do this because, for the moment, that is where the light is best. An enormous amount of work has been done examining the role of dACC in physical pain and in cognitive processing. Consequently, this region is ripe for consideration. This should not be taken to mean that that is the only region involved in both social and physical pain for it most assuredly is not; right ventral prefrontal cortex is another. Although we do not focus on other regions of the pain matrix (Peyron, Laurent, & Garcia-Larrea, 2000), such as the periacqueductal gray, insula, and somatosensory cortex, the reader should not be surprised if these regions are added in future work. Indeed, some work has already started along these lines (MacDonald & Leary, in press; Panksepp, 1998).

Linguistic Evidence Linking Social and Physical Pain

Whereas a cognitive neuroscientist might bristle at the notion of something as ethereal as social pain being similar to physical pain, nonscientists might think that the truth of this idea is obvious. A layperson might point to the fact that in our culture, we talk about social and physical pain in similar ways so that they should no doubt be related. For instance, we describe physical pain with phrases such as, "I broke my arm," and "my leg hurts." Similarly, we describe social pain with phrases such as, "she broke my heart," and "he hurt my feelings." Indeed, it is difficult to describe social pain without reference to physical pain terminology. In fact, English speakers have no other way to describe the feelings associated with social pain (MacDonald & Leary, in press).

MacDonald and Leary (in press) recently examined whether this linguistic overlap was a feature of only the English language and discovered that it was not. They asked individuals from fifteen different countries, including a number of non-European countries, to provide typical ways of describing social pain. In each country, social pain descriptions relied on physical pain words. This evidence, at the very least, suggests that social and physical pain may be universally linked in the mental lexica of humans around the world. It does not in itself, however, say much about whether common processes support the experience of the two types of pain. The relationship may simply be metaphorical, although obviously a strong metaphor for it to have spread so widely.

An Evolutionary Story for the Link

The first line of evidence to suggest that the link between social and physical pain might be more than metaphorical came from the work of Panksepp et al. (1978). They were examining the analgesic effects of opiate-based drugs in dogs when they discovered that in addition to altering the amount of pain the dogs could tolerate, opiates diminished the frequency of cries produced while in isolation. The investigators reasoned that the opiate receptor system mitigated the experience of both physical and social pain. Along similar lines, drugs that are typically prescribed to deal with social distress and depression are known to work effectively for chronic pain (Shimodozono et al., 2002). Finally, the "social attachment system" may have piggybacked onto or developed out of the physical pain system, which has older phylogenetic roots than the social pain system (Nelson & Panksepp, 1998).

Unlike other animals that are born relatively mature or have rapid developmental trajectories, infant mammals are unable to care for themselves for an extended period of time and human infants take the longest of all. Thus mammals have unprecedented dependence on their caregivers. Whereas lack of food, water, shelter, and defense against predators will lead to death for any animal, for young mammals, meeting these needs is entirely contingent on the continuing relationship with a caregiver. For mammals, then, social needs supplant all other biological needs in importance, at least in infancy, because meeting them is critical to meeting all other needs.

It is not surprising, therefore, that mammals are among the first to have a social attachment system, a system that monitors for actual or psychological distance from others and elicits distress once distance is detected so that contact can be reestablished. Attachment processes exist primarily in mammalian species and thus as the social attachment system was evolving, the physical pain system was already in place and could serve as a solid foundation for the creation of this attachment system. Whereas the physical pain system produces physical pain in response to physical injuries so that attention and other biological resources can be mobilized to prevent greater injury and promote survival, the social pain system produces social pain in response to social injuries so that attention and other biological resources can be mobilized to prevent these injuries and promote survival. It is of interest that mammals are also the first species, phylogenetically speaking, to have a cingulate gyrus (MacLean, 1993). Thus, it is plausible that this new structure may be involved in these social attachment processes.

We have chosen to focus on the role of the dACC in both social and physical pain for a number of reasons. First, the ACC has one of the highest densities of opiate receptors in the brain (Vogt, Wiley, & Jensen, 1995), and thus may have been one of the primary sites of action in Panksepp's work on the social and physical pain-alleviating properties of opiates. Second, a large literature shows the dACC to be involved in physical pain processes in humans. Finally, a number of studies with nonhuman mammals suggest that the ACC, and perhaps the dACC specifically, is involved in the experience of separation distress and in the production of distress vocalizations aimed at regaining social contact.

ACC and Physical Pain in Humans

Neural regions involved in physical pain, referred to as the pain matrix, include the dACC, rACC, somatosensory cortex, insula, periacqueductal gray, and right ventral prefrontal cortex. These regions are thought to be differentially involved in the sensory, distressing, and regulatory components of pain. Somatosensory cortex and insula are primarily, although not exclusively, linked with the sensory aspects of pain. Pain-related activity in these regions is associated with being able to identify the region of one's body that is in pain and other sensory features including intensity. Periacqueductal gray, rACC, and right ventral prefrontal cortex are more frequently associated with regulation of pain through opioid release and cognitive processing. Finally, dACC is generally associated with the subjectively distressing component of pain.

Sensory intensity and subjective distress associated with pain are often highly correlated, and thus a few illustrations of their relationship are in order to clarify the difference between the concepts. A useful metaphor is the sound of music on a radio, with sensory intensity likened to the radio's volume and subjective distress likened to the extent to which the music is experienced as unpleasant (Price, 1999). Above a certain threshold, increasing volume will usually be highly correlated with increasing unpleasantness. Nevertheless, the same volume can produce different levels of unpleasantness depending on the level of ambient noise from being in a quiet room versus outdoors at a barbeque or one's sensitivity to or tolerance of loud noises. Thus, intensity can be distinguished from unpleasantness such that under different conditions or across different people, the same degree of sensory intensity might produce different degrees of unpleasantness.

The consequences of damaging neural structures associated with the sensory and distressing aspects of pain also reveals this dissociation. In the 1960s, patients with chronic pain problems sometimes underwent cingulotomies, a procedure that involves lesioning the anterior cingulate. After the procedure, patients often experienced significant pain relief; however, the relief did not come as a result of all aspects of pain being diminished. Rather, only the subjective distress appears to have abated. Patients would report that the sensory aspects of pain continued, but it no longer seemed to bother them (Foltz & White, 1968). Alternatively, there is a reported case of a patient with damage to somatosensory cortex. When painful stimulation was applied to the body region represented by the damaged part of somatosensory cortex, the patient had difficulty reporting on the location of the stimulation but still experienced the stimulation as distressing (Nagasako, Oaklander, & Dworkin, 2003). So although intensity and distress may feel inextricably linked, they can be separated because they depend on distinct neural processes.

Furthering this conceptual separation, Rainville et al. (1997) conducted a neuroimaging study in which they used hypnotic suggestion to alter the perceived unpleasantness of painful stimulation without changing the perceived intensity. During part of the experiment, subjects were given a suggestion that they would experience the pain as more or less distressing than normal. In fact, subjects reported the pain to be more or less distressing in accord with the suggestion they received. These results were unlikely to be mere demand characteristics, because changes in reports of pain distress were highly correlated with changes in dACC activity. The changes were not correlated with activity in somatosensory cortex, suggesting that this region, unlike dACC, is not a direct contributor to the experience of pain distress.

ACC and Social Pain in Animals

Until recently little research has examined neural correlates of social pain in humans; however, some studies from the animal literature suggest a role for dACC in this regard. Care must be taken in the inferences drawn from the animal literature because similar neural regions in different animals do not always serve the same functions. Along those lines, two issues are worthy of note with respect to the anterior cingulate across animal and human brains. First, in older animal studies, ACC, which can be subdivided into rostral and dorsal sections, was sometimes referred to as rostral cingulate cortex (Smith, 1945). It is important to note that this designation refers to all of what would now be called anterior cingulate, including dACC. Second, strictly speaking, some primates such as the macaque may not have rACC at all, such that the entire macaque ACC might be functionally analogous to dACC. In humans, dACC and rACC have different morphological properties (i.e., cell layer organization). In the macaque, the entire ACC has morphological properties similar to those in dACC (Smith, 1945). This will be particularly relevant in the final section of this chapter when we suggest that rACC is associated with symbolic processes specific to humans (Lieberman et al., 2002).

One set of animal studies indicating the link between the ACC and social pain comes from work in which lesions were made to various overlapping subregions of the medial prefrontal cortex, including the ACC, in squirrel monkeys (MacLean & Newman, 1988). The only monkeys that ceased to make separation distress vocalizations after the procedure were those for which dACC had been lesioned. It should be noted that these same monkeys also had their rACC lesioned, which may or may not be functionally equivalent to rACC in humans, as rACC homology was examined in macaques, not squirrel monkeys. However, other squirrel monkeys in this study had rACC lesions without dACC lesions and did not show reduced distress vocalizations, whereas none of the monkeys had dACC lesions without reduced distress vocalizations. Moreover, the monkeys were still capable of making different kinds of vocalizations, so it is not the case that the capacity for vocalizing per se was affected by the lesions. This study suggests that dACC may be critical to the experience of social pain that would lead to distress vocalizations.

Electrical stimulation studies support this conclusion as well. Over half a century ago, Smith (1945; Jurgens & Muller-Preuss, 1977) observed that electrical stimulation of dACC produced spontaneous separation distress vocalizations in the absence of social isolation. Thus, these lesion and stimulation studies provide good converging evidence that the dACC plays an important role in social pain.

ACC and Social Pain in Humans

To examine neural correlates of social exclusion in humans, and begin to assess whether the neural basis of social pain is similar to the neural basis of physical pain, it was critical to find a manipulation that would produce a genuine episode of social pain while simultaneously seeming plausible to a subject lying inside a neuroimaging scanner. Numerous manipulations have been used successfully by social psychologists since the late 1900s to examine cognitive and behavioral responses to social exclusion (see Williams, Forgas, & von Hippel, in press).

The most common manipulation of exclusion involves subjects finding out that no one else in an experimentally created group wants to pair up with them in the upcoming task, so they will be working on their own (Leary et al., 1995; Twenge et al., 2001; see also Gaertner & Iuzzini, in press). Baumeister and colleagues also used a "future alone" manipulation by telling subjects that, on the basis of their answers on questionnaires, they are the kinds of persons who are likely to end up alone in life, even if they have close friends and loved ones now (Baumeister & De Wall, in press). Williams and Sommer (1997) used the most direct and overt manipulation of social exclusion. In their work, a subject is waiting with two same-sex confederates for an experiment to start. One of the confederates picks up a ball from the table between them and begins tossing it to the other confederate and the subject. After being included in the game for a short time, half of the subjects are then excluded by the confederates who never throw the ball to those subjects again. Each of these manipulations proved effective in producing behaviors and self-evaluations that would be expected to result from social exclusion, including lowered self-esteem, increased aggressiveness, and increased conformity to group norms (Leary, in press).

None of these manipulations is easily transferable to the context of a functional neuroimaging study. Fortunately, the ball-tossing manipulation has been converted into a virtual ball-tossing game ("cyberball"), ostensibly played on line with other players, and produces the same effects as the in-person version (Williams, Cheung, & Choi, 2000). In fact, "cyberball" works so well that even when subjects are informed that the other two "players" are really just computer players controlled by the program and that those players will exclude them part way through the game, subjects still report feeling social pain as a result of the experience (Zadro & Williams, 1998). "Cyberball" thus fit the dual constraints of producing a genuine experience of social pain while at the same time making sense in the context of a functional magnetic resonance imaging (fMRI) study.

Subjects in our study of social exclusion (Eisenberger, Lieberman, & Williams, 2003) were informed that we were working out the technical details in a new neuroimaging procedure called hyperscanning. Hyperscanning, which is a real technique being used at Johns Hopkins University, involves scanning many subjects simultaneously while they engage in some kind of coordinated activity so that the relationship between the neural patterns of the different subjects can be analyzed. We showed subjects a write-up of this work that appeared in *Nature Neuroscience* and told subjects that we wanted to do the same kind of procedure in our laboratory, but that we were still in the technical development stage. Subjects were told that two other subjects were also going to be in scanners at other locations around campus and that we would be having subjects play a simple ball-tossing game so that we would have some basic coordinated neural activity to look at, to see if we were making progress. There was no goal to the ball-tossing game, no points to be won, and no skill involved given that catching occurs automatically and subjects' only decision is whether to throw it to one person or the other (an animation of the game can be seen at www.scn.ucla.edu).

Subjects went through three functional scans. During the first scan (implicit exclusion), they were told that we were having technical difficulties making a full internet link with the other two scanners, so as a result, they would be able to watch the other two players play during the first scan, but they would not be able to receive or throw the ball. This condition is visually identical to the regular exclusion condition; however, our subjects knew that they were not being intentionally excluded. This condition conceptually replicated the study by Zadro and Williams (1998), in which subjects were informed that they were playing a computer game with animated figures and would be automatically excluded part way through the game.

For the second scan (inclusion), subjects were told we had made the full internet link so that they would be able to play. Subjects played with the other two players during this scan and were fully included such that they were thrown the ball with 50 percent probability. In the third and final scan (explicit exclusion), subjects were fully included for about thirty seconds, but then excluded by the other two players for the rest of the scan. After this scan, subjects were removed from the scanner and immediately filled out a questionnaire assessing the degree of social pain experienced during the last scan.

Our primary analysis comparing brain activations showed greater activity in dACC, right insula, and right ventral prefrontal cortex during the explicit exclusion condition, relative to inclusion. Each of these regions is commonly found in neuroimaging studies of physical pain (Petrovic et al., 2000). In a follow-up analysis, we examined the extent to which the individual differences in activity in each of these regions predicted the individual differences in self-reported social pain as measured in the postscanning questionnaire. Insula activity was not correlated with self-reported social pain, however, dACC and right ventral prefrontal activity were both strongly related to social pain. For dACC, greater activity during exclusion relative to inclusion was associated with greater reports of social pain ($r = 0.88$). For right ventral prefrontal cortex, greater activity during exclusion relative to inclusion was associated with diminished reports of social pain ($r = -0.69$).

This pattern of activity is quite similar to results in studies of physical pain. For instance, in a study of visceral pain regulation (Lieberman et al., 2004), dACC, insula, and right ventral prefrontal cortex were all more active during painful stimulation than at baseline. More important, the same pattern of relationships between neural activity and subjective pain reports was found here as in the social pain study; insula activity did not correlate with self-reported pain, dACC activity correlated positively with self-reported pain, and right ventral prefrontal cortex correlated negatively with self-reported pain. In fact, in both studies, dACC activity strongly mediated the relationship between right ventral prefrontal activity and self-reports of pain. These results, consistent with other results (Petrovic et al., 2002; Wager et al., 2004), suggest that although dACC is important in producing the subjective distress of pain, right ventral prefrontal cortex is involved in down-regulating the experience of pain (see Lieberman et al., 2005 for a discussion of the role of right prefrontal cortex in the self-regulation of negative affective states).

In a second set of analyses, we examined brain activations during implicit exclusion. As in the explicit condition, dACC activity was greater during implicit exclusion than during inclusion, and was similar to the explicit exclusion dACC activation in position, cluster size, and intensity. This may be somewhat surprising given that subjects presumably did not consciously believe that they were being excluded during this condition. However, it should be recalled that Zadro and Williams' subjects experienced social pain even when they were told ahead of time that the game was fixed. It may be the case that humans are sufficiently hardwired to respond to exclusion and that the mere appearance of exclusion is sufficient to provoke the response. If so, this would be similar to the way visual illusions function. In most cases, understanding the causes of a visual illusion in no way mitigates the effect; understanding the true shape of an Ames room does not prevent a child from looking as tall as an adult. The reliability of the visual system is either so important or so ancient that evolution has sealed its computations off from intentional modification by the owner of the visual system. Similarly, the need to detect potential exclusion may be so important or ancient that it too resists our conscious beliefs about the true meaning of an episode, in this case that it is not truly exclusionary.

Still, there ought to be some consequences of whether or not we consciously believe we are being excluded. Individuals ought to be less likely to be thinking about exclusion and its causes if they do not believe they are being excluded, and should also be less likely to try to self-regulate the negative experience. Consistent with this view, no increased right ventral prefrontal activity was evident during implicit exclusion, relative to inclusion, even at very liberal statistical thresholds. Given that right ventral prefrontal activity has been associated with explicit thought about negative affect and negatively evaluated attitude objects (Crockett, Eisenberger, & Lieberman, 2004; Cunningham et al., 2003) and is also associated with inhibition of negative affective experience, it appears that in our study, subjects were not engaged in these sorts of mental activities. In other words, because subjects had no reason to think they were being excluded, self-regulatory and attributional mechanisms were not engaged. One caveat to this account is that we cannot be sure that implicit exclusion actually produced any distress, because to have assessed social pain after the first scan would have given away our cover story.

What Is the Basic Function of the ACC?

The results thus far, from our own work as well as from the broader physical and social pain literatures, suggest that dACC activity is related to the experience of social and physical pain. The next obvious question to ask is why. What is the function of the dACC such that it should be involved in these forms of pain? It turns out that the answer to this question depends on which end of the psychology department one goes to for an answer.

If clinical and psychopathology researchers are asked, they will probably respond that the function of the dACC is to produce attention-getting affective-motivational states such as pain, anxiety, and distress. Each of these states serves important functions in motivating adaptive and appropriate behavior (Mandler, 1975). Thus, these researchers focus on phenomenological contributions of the ACC.

If cognitive researchers are asked, they will probably respond by saying that the function of the ACC is to monitor for conflict and to detect errors. Numerous studies suggested that dACC is activated when a discrepancy exists between one's goals and one's prepotent responses. For instance, during the Stroop task when a person is shown color words in different color ink (e.g., R-E-D written in blue ink), one's goal is to say the color of the ink that words are written in, but this goal conflicts with the prepotent tendency to read the words themselves. The dACC is activated in this context and "notifies" lateral prefrontal cortex that top-down control processing is necessary to promote contextually appropriate responding (Botvinick et al., 2001). One might expect the same thing to occur when one is required to give responses that appear racially biased when one does not want to appear racist (Amodio et al., 2004). Most researchers in this field agree that the dACC's role is limited to detecting conflict and alerting lateral prefrontal cortex, rather than being involved in resolving conflict directly.

Of interest in this brief tour of the psychology department is that those interested in the cognitive functions of the dACC leave the phenomenological consequences of dACC activity unexamined. At the same time, those interested in the phenomenological consequences of dACC activity have rarely expressed interest in the computations underlying that activity.

The ACC as a Neural Alarm System

We suggest that it might be profitable to consider the possibility that the dACC works as a neural alarm system (Eisenberger & Lieberman, 2004). Such a metaphor might help to bridge the explanations of ACC function that are seemingly in competition with one another. Any alarm system, whether it is a clock alarm or a smoke alarm, must integrate two functions to work effectively. First, it must be able to detect the critical environmental conditions for which it is designed. A clock must be able to detect when there is a match between the current time and the time set for the alarm to go off. A smoke detector must be able to detect when the amount of smoke in the room crosses some threshold for unacceptability. Second, the alarm must be able to notify relevant parties that the critical condition has been met. For most alarms, this means making a loud noise, either to wake people up or let them know a room is filling with smoke and possibly fire. Conceptually, these functions are separable (e.g., a clock alarm with a broken speaker), but they must be integrated and work together for the alarm to function properly.

These two functions of an alarm system sound conspicuously similarly to the two descriptions of ACC function. The cognitive account suggests that the ACC is sensitive to goal conflicts, conditions that are critical to detect. The phenomenological account suggests that the ACC can create attention-getting affective states. If these two processes operate together, they might function as a unitary alarm system, detecting conflicts and "making noise" to attract the person's attention.

There are at least two ways in which this alarm could be instantiated in dACC, and at this point the data are insufficient to differentiate them. In one account, discrepancy detection and distress may go hand in hand, such that the alarm bell sounding may be the phenomenological consequence of the detection of discrepancy. It is possible that previous cognitive studies of the ACC produced distress in their subjects but did not attempt to measure it. The second possibility is that the conflict detector and alarm bell are instantiated in nearby but distinct regions of dACC. The two functions would still be integrated with one another, but this might help explain why performing tasks such as the Stroop or oddball task do not seem to set off major alarm bells, phenomenologically speaking.

As noted, there are no compelling data yet to argue for one or the other of these accounts. However, two studies partially investigated this question by examining neural activity to pain distress and cognitive processing in the same subjects. One study (Derbyshire, Vogt, & Jones, 1998) noted that the regions in dACC activated by a pain and Stroop task were adjacent to one another, but also sometimes overlapping. The ambiguity of these results is only heightened by the small sample—six. A second study (Davis et al., 1997) also found adjacent regions activated by a pain and cognitive task, however, the task did not involve conflict detection and thus is not entirely on point.

Separate from determining whether or not these two potential subcomponents of the alarm system are in the same region of dACC, the more pressing question is whether the two subcomponents actually work together as they should if they really are part of an underlying alarm system. In other words, do phenomenological and conflict-detection processes covary with one another? If so, this would suggest that the overall alarm system is mobilizing its subcomponents en masse, rather than each performing its own process independently. Two studies examined the covariation between these subcomponents.

In a neuroimaging study (Ursu et al., 2003), individuals with obsessive-compulsive disorder (OCD), which is characterized by distress and worry, were scanned while performing a task involving response conflict. Individuals with OCD, compared with healthy controls, showed significantly more dACC activity to high-conflict trials. In addition, a trend was observed, although not significant, such that patients with more severe symptoms of OCD showed more dACC activity to conflict than those with less severe symptoms.

A second neuroimaging study (Eisenberger, Lieberman, & Satpute, in press) investigated this issue by correlating dACC activity on a conflict-detection task with neuroticism scores. Neuroticism is typically defined as a heightened tendency to experience negative affect frequently and/or intensely (Costa & McCrae, 1985). Thus, we can safely assume that neurotics tend to have alarm bells that ringer louder and/or more often than in nonneurotics. The question of interest was whether they also have more sensitive conflict-detection systems, even for conflicts that are generally nondistressing to detect. The results suggested that indeed such was the case. The magnitude of activations to conflict relative to nonconflict trials correlated strongly ($r = 0.76$) with neuroticism.

Dorsal versus Rostral ACC

Evidence presented thus far supports the notion of the dACC functioning as a neural alarm system involved both in the cognitive detection of critical conditions and the affective sounding of a phenomenological alarm. On the face of it, this characterization conflicts with an influential account of the distinct functions of dACC and rACC. Bush, Luu, and Posner (2000) reviewed various cognitive and affective task paradigms that activated the ACC. The major conclusion was that cognitive tasks tended to activate dACC and deactivate rACC, whereas affective tasks tended to activate rACC and deactivate dACC. Thus, our conclusion that one function of dACC is to sound an affective alarm conflicts with this conceptual organization of ACC function.

One limitation of that review is that it included no pain imaging studies. Once these studies are taken into account, the affective-cognitive distinction becomes muddied, because the emotional distress of pain was reliably linked to dACC rather than rACC activity in numerous studies (Rainville et al., 1997).

The fact that the affective-cognitive distinction seems insufficient to capture functional differences associated with the dorsal and rostral regions of the ACC led us to consider an alternative formulation. After a search of the literature, although admittedly not an exhaustive one, we developed the hypothesis that dACC and rACC may both be involved in conflict processing but differ with regard to the extent to which the conflict is represented symbolically or nonsymbolically.

Symbolically and nonsymbolically represented conflict should vary in a number of ways. The most critical is whether there is an explicit Intentional representation of the conflict or of the source of the conflict.[1] According to this view, when conflict is symbolic, there is not merely conflict in the system (e.g., in the ACC), rather there is awareness of the conflict as a conflict (Lieberman et al., 2002).

These two kinds of conflict processing should also vary in the kind of computational mechanisms that could support them. Nonsymbolic conflict can be thought of as tension between two or more competing representations or responses. Connectionist networks representing the combined outputs of multiple interconnected inputs naturally produce conflict maps that represent the total level of tension or conflict among various inputs. When inputs are coherent and consistent with one another, the network has a low level of tension, whereas when these inputs conflict, the network has a high level of tension (Hopfield, 1982, 1984). An elegant model of this process suggests that when the tension level in dACC is high, it automatically triggers a signal to lateral prefrontal cortex, which then exerts top-down control over the competing inputs (Botvinick et al., 2001). Nowhere in this model is conflict itself or the decision to resolve conflict modeled. Indeed, its authors value the model, in part, because it has no "ghost in the machine" regress in which an intelligent agent must be posited but left unexplained.

Although little or nothing is known about the process by which conflicts or anything else are represented symbolically (Lieberman et al., 2002), symbolic representations do have a number of features that are known. Symbolic processes involve conscious awareness of and attention to a specific symbolic representation. The resources of awareness and attention are limited (Miller, 1956) such that only a handful of symbolic representations can be attended to and processed at any one time (Schneider & Shiffrin, 1977). These representations are typically thought to be processed serially. Thus, although many conflicts may be nonsymbolically processed, only a single symbolic conflict can be processed at any one time.

We propose that symbolically represented conflict is processed primarily in rACC, whereas nonsymbolically represented conflict is processed primarily in dACC. With respect to cognitive conflict, both symbolic and nonsymbolic forms exist. For instance, connectionist networks in which tension effects naturally emerge out of competing inputs can nicely model standard oddball, go–no-go, and Stroop effects. Alternatively, error-detection tasks involve overt awareness that a particular error has been made and the nature of the error. Using a Talaraich y-coordinate of 30 to divide rostral from dorsal ACC activations, we find that most nonsymbolic forms of cognitive conflict activate dACC (Braver et al., 2001; Bush et al., 2003; Carter et al., 2000; Weissman et al., 2003). Alternatively, error-detection tasks, which involve symbolic conflict representations, tend to activate rACC (Garavan et al., 2003; Kiehl, Liddle, & Hopfinger, 2000; Rubia et al., 2003).

Emotion can also be construed within a conflict model. Appraisal models of emotion suggest that negative emotions are the result of a conflict between desired or expected outcomes and what actually occurs (Frijda, 1986; Lazarus, 1991; Mandler, 1975). Most negative emotions (anger, fear, sadness) are thought to have a specific Intentional object such that they are, symbolically speaking, about something. When someone is afraid, he or she is afraid of something in particular and knows what that something is. Anxiety, however, is a negative affect that is distinguished from fear in that it lacks an Intentional object (Kierkegaard, 1844/1981). Although anxiety may have a specific cause (e.g., someone in the room who makes us uncomfortable), it does not involve the anxious individual knowing this specific cause. Considering neuroimaging studies of anxiety-provocation versus the induction of other emotions conforms to our symbolic-nonsymbolic theory of the ACC. Sadness, anger, and fear reliably activate rACC (Damasio et al., 2000; Dougherty et al., 1999; George et al., 1995; Kimbrell et al., 1999; Liotti et al., 2000; Mayberg et al., 1999; Shin et al., 2000). Anxiety, however, tends to activate dACC (Kimbrell et al., 1999; Liotti et al., 2000). Along similar lines, perception of discrete emotional expressions activates rACC (Ueda et al., 2003) whereas perception of ambiguous emotional expressions activates dACC (Nomura et al., 2003).

Finally, pain distress can also be divided along symbolic and nonsymbolic lines. Typically, pain is a nonsymbolic bottom-up process (animals without symbolic capacities presumably experience pain). Consistent with this, the experience of pain is most often associated with dACC activation (Hsieh, Stone-Elander, & Ingvar, 1999; Ploghaus et al., 1999; Sawamoto et al., 2000; Tolle et al., 1999). However, when pain is anticipated or expected it becomes more symbolic and tends to be processed in rACC (Buchel et al., 1998; Chua et al., 1999; Ploghaus et al., 1999, 2003).

Thus, cognitive, affective, and pain processes are each distributed across dACC and rACC. They do, however, seem to be organized such that more symbolic forms of cognition, emotion, and pain are processed in rACC with less symbolic forms processed in dACC. It is worth returning to a point made earlier regarding the homology of function between human and primate ACC. Smith (1945) pointed that though macaques have ACCs that correspond in location to the human dACC and rACC, the morphology of both their rACC and dACC is consistent with the morphology of human dACC. Smith concluded that macaques do not, functionally speaking, have an equivalent to human rACC. This conclusion takes on new meaning in light of our claim that rACC is involved in symbolic representations of conflict, a capacity that humans have and may not share with any other animal (Deacon, 1997). Symbolic representations of conflict allow for unparalleled consideration of contextual factors such as time and place. If one is aware of the source of one's distress, one may choose to delay responding (one may choose to ignore the distress caused by one's boss while the boss is still in the room) or consider complex strategies of response ("I'll settle this when I have lunch with my boss's boss next week"). Nonsymbolic conflict simply produces tension and anxiety until it is resolved or a symbolic representation is generated.

Conclusion

We suggest that because of the role of social attachment in mammals, the social pain system may have piggybacked onto the physical pain system during our evolution. In addition, the dACC may have been one of the primary sites in which this overlap evolved, such that today, this region produces similar experiences of distress in response to both physical and social injuries. More generally, we propose that the dACC may be thought of as a neural alarm system that is involved both in detection of actual or potential threats as well as in sounding a phenomenological alarm that redirects our attention and motivation toward dealing with the source of the threat. Finally, we attempted to integrate our model with what previously was hypothesized regarding the function of the ACC by proposing a new conceptual distinction for the contribution of rACC versus dACC to psychological processes. Specifically, we propose that the dACC is involved in detecting nonsymbolic conflict whereas rACC is involved in detecting symbolic conflict. In future work, we hope to explore more fully the ways in which physical and social pain processes are intertwined, as well as distinct, in order to understand better in what ways "pains by any name" really do "hurt the same."

Note

1. In philosophy, when the word "Intentional" is capitalized it refers to the quality of Intentionality, which includes, but is not limited to, mental acts such as beliefs and desires. Intentionality, first described by Aristotle in *De Anima* and later by Brentano (1874/1995) and Husserl (1913) as one of the cornerstones of continental phenomenology, refers to the fact that certain reflective acts of mental life are irreducibly *directed at* or *about* something else. Physical objects are never instrinsically about anything else, serving only as representations to the extent that they are designated as such by the Intentional acts of humans. Although mental acts can possess the quality of Intentionality, many do not. For instance, most visual information that is processed by the brain at any moment is not overtly attended to. Parafoveal priming works, in part, because it is cognitively processed but not Intentionally; the individual has no thought about the primes.

References

Amodio, D. M. , Harmon-Jones, E., Devine, P. G., Curtin, J. J., Hartley, S. L., & Covert, A. E. (2004). Neural signals for the detection of unintentional race bias. *Psychological Science, 15*, 88–93.

Baumeister, R. F. & De Wall, C. N. (in press). The inner dimension of social exclusion: Intelligent thought and self-regulation among rejected persons. In K. D. Williams, J. P. Forgas, & W. von Hippel (Eds.), *The Social Outcast: Ostracism, Social Exclusion, Rejection, and Bullying*. New York: Psychology Press.

Botvinick, M. M., Braver, T. S., Barch, D. M., Carter, C. S., & Cohen. J. D. (2001). Conflict monitoring and cognitive control. *Psychological Review, 108*, 624–652.

Braver, T. S., Barch, D. M., Gray, J. R., Molfese, D. L., & Snyder, A. (2001). Anterior cingulate cortex and response conflict: Effects of frequency, inhibition and errors. *Cerebral Cortex 11*, 825–836.

Brentano, F. (1874/1995). *Psychology from an Empirical Standpoint*. London: Routledge.

Buchel, C., Morris, J., Dolan, R. J., & Friston, K. J. (1998). Brain systems mediating aversive conditioning: An event-related fMRI study. *Neuron, 20*, 947–957.

Bush, G., Luu, P., & Posner, M. I. (2000). Cognitive and emotional influences in anterior cingulate cortex. *Trends in Cognitive Sciences, 4*, 215–222.

Bush, G., Shin, L. M., Holmes, J., Rosen, B. R., & Vogt, B. A. (2003). The multi-source interference task: Validation study with fMRI in individual subjects. *Molecular Psychiatry, 8*, 60–70.

Carter, C. S., MacDonald, A. W., Botvinick, M. M., Ross, L. L., Stenger, V. A., Noll, D., et al. (2000). Parsing executive processes: Strategic vs. evaluative functions of the anterior cingulate cortex. *Proceedings of the National Academy of Sciences of The United States of America, 97,* 1944–1948.

Chua, P., Krams, M., Toni, I., Passingham, R., & Dolan, R. (1999). A functional anatomy of anticipatory anxiety. *Neuroimage, 9,* 563–571.

Costa, P. T., Jr. & McCrae, R. R. (1985). Hypochondriasis, neuroticism, and aging: When are somatic complaints unfounded? *American Psychologist, 40,* 19–28.

Crockett, M. J., Eisenberger, N. I., & Lieberman, M. D. (2004). Stereotype activation or behavior regulation? An fMRI study of automatic behavior effects. Proceedings of the 11th annual meeting of the Cognitive Neuroscience Society, San Francisco, April 16.

Cunningham, W. A., Johnson, M. K., Gatenby, J. C., Gore, J. C., & Banaji, M. R. (2003). Neural components of social evaluation. *Journal of Personality and Social Psychology, 85,* 639–649.

Damasio, A. R., Grabowski, T. J., Bechara, A., Damasio, H., Ponto, L. L., Parvizi, J., et al. (2000). Subcortical and cortical brain activity during the feeling of self-generated emotions. *Nature Neuroscience, 3,* 1049–1056.

Davis, K. D., Taylor, S. J., Crawley, A. P., Wood, M. L., & Mikulis, D. J. (1997). Functional MRI of pain- and attention-related activations in the human cingulate cortex. *Journal of Neurophysiology, 77,* 3370–3380.

Deacon, T. W. (1997). *The Symbolic Species: The Co-evolution of Language and the Brain.* New York: Norton.

Derbyshire, S. W., Vogt, B. A., & Jones, A. K. (1998). Pain and Stroop interference tasks activate separate processing modules in anterior cingulate cortex. *Experimental Brain Research, 118,* 52–60.

Dougherty, D. D., Shin, L. M., Alpert, N. M., Pitman, R. K., Orr, S. P., Lasko, M., et al. (1999). Anger in healthy men: A PET study using script-driven imagery. *Biological Psychiatry, 46,* 466–472.

Eisenberger, N. I., Lieberman, M. D., & Satpute, A. B. (in press). Personality from a controlled processing perspective: an fMRI study of neuroticism, extraversion, and self-consciousness. *Cognitive, Affective, and Behavioral Neurosciene.*

Eisenberger, N. I., Lieberman, M. D., & Williams, K. D. (2003). Does rejection hurt: An fMRI study of social exclusion. *Science, 302,* 290–292.

Foltz, E. L. & White, L. E., (1968). The role of rostral cingulotomy in "pain" relief. *International Journal of Neurology, 6,* 353–373.

Frijda, N. H. (1986). *The Emotions.* New York: Cambridge University Press.

Gaertner, L. & Iuzzini, J. (in press). Rejection and entitativity: A synergistic model of mass violence. In K. D. Williams, J. P. Forgas, & W. von Hippel (Eds.), *The Social Outcast: Ostracism, Social Exclusion, Rejection, and Bullying.* New York: Psychology Press.

Garavan, H., Ross, T. J., Kaufman, J., & Stein, E. A. (2003). A midline dissociation between error-processing and response-conflict monitoring. *Neuroimage, 20,* 1132–1139.

George, M. S., Ketter, T. A., Parekh, P. I., & Horwitz, B. (1995). Brain activity during transient sadness and happiness in healthy women. *American Journal of Psychiatry, 152,* 341–351.

Hopfield, J. J. (1982). Neural networks and physical systems with emergent collective computational abilities. *Proceedings of the National Academy of Sciences of the United States of America, 79,* 2554–2558.

Hopfield, J. J. (1984). Neurons with graded response have collective computational properties like those of two-state neurons. *Proceedings of the National Academy of Sciences of the United States of America, 81,* 3088–3092.

Hsieh, J., Stone-Elander, S., & Ingvar, M. (1999). Anticipatory coping of pain expressed in the human anterior cingulate cortex: A positron emission tomography study. *Neuroscience Letters, 262,* 61–64.

Husserl, E. (1913/1962). *Ideas: General Introduction to Pure Phenomenology.* New York: Collier Press.

Jurgens, U. & Muller-Preuss, P. (1977). Convergent projections of different limbic vocalization areas in the squirel monkey. *Experimental Brain Research, 8,* 75–83.

Kiehl, K. A., Liddle, P. F., & Hopfinger, J. B. (2000). Error processing and the rostral anterior cingulate: An event-related fMRI study. *Psychophysiology, 37,* 216–223.

Kierkegaard, S. (1844/1981). *The Concept of Anxiety.* Princeton, NJ: Princeton University Press.

Kimbrell, T. A., George, M. S., Parekh, P. I., Ketter, T. A., Podell, D. M., Danielson, A. L., et al. (1999). Regional brain activity during transient self-induced anxiety and anger in healthy adults. *Biological Psychiatry, 46,* 454–465.

Lazarus, R. S. (1991). *Emotion and Adaptation.* New York: Oxford University Press.

Leary, M. R. (in press). Varieties of interpersonal rejection. In K. D. Williams, J. P. Forgas, & W. von Hippel (Eds.), *The Social Outcast: Ostracism, Social Exclusion, Rejection, and Bullying.* New York: Psychology Press.

Leary, M. R., Tambor, E. S., Terdal, S. K., & Downs, D. L. (1995). Self-esteem as an interpersonal monitor: The sociometer hypothesis. *Journal of Personality and Social Psychology, 68,* 518–530.

Lieberman, M. D., Gaunt, R., Gilbert, D. T., & Trope, Y. (2002). Reflection and reflexion: A social cognitive neuroscience approach to attributional inference. In M. Zanna (Ed.), *Advances in Experimental Social Psychology* (pp. 199–249). New York: Academic Press.

Lieberman, M. D., Hariri, A., Jarcho, J. M., Eisenberger, N. I., & Bookheimer, S. Y. (2005). An fMRI investigation of race-related amygdala activity in African-American and Caucasian-American individuals. *Nature Neuroscience, 8,* 720–722.

Lieberman, M. D., Jarcho, J. M., Berman, S., Naliboff, B., Suyenobu, B. Y., Mandelkern, M., et al. (2004). The neural correlates of placebo effects: A disruption account. *NeuroImage, 22,* 447–455.

Liotti, M., Mayberg, H. S., Branna, S. K., McGinnis, S., Jerabek, P., & Fox, P. T. (2000). Differential limbic-cortical correlates of sadness and anxiety in healthy subjects: Implications for affective disorders. *Biological Psychiatry, 48*, 30–42.

MacDonald, G. & Leary, M. R. (in press). Why does social exclusion hurt? The relationship between social and physical pain. *Psychological Bulletin.*

MacLean, P. D. (1993). Perspectives on cingulate cortex in the limbic system. In B. A. Vogt & M. Gabriel (Eds.), *Neurobiology of Cingulate Cortex and Limbic Thalamus: A Comprehensive Handbook* (pp. 1–23). Boston: Birkhauser.

MacLean P. D. & Newman, J. D. (1988). Role of midline frontolimbic cortex in production of the isolation call of squirrel monkeys. *Brain Research, 45*, 111–123.

Mandler, G. (1975). *Mind and Emotion.* New York: Krieger.

Mayberg, H. S., Liotti, M., Brannan, S. K., McGinnis, S., Mahurin, R. K., Jerabek, P. A., et al. (1999). Reciprocal limbic-cortical function and negative mood: Converging PET findings in depression and normal sadness. *American Journal of Psychiatry, 156*, 675–682.

Miller, G. A. (1956). The magical number seven, plus or minus two: Some limits on our capactiy for processing information. *Psychological Review, 63*, 81–97.

Nagasako, E. M., Oaklander, A. L., & Dworkin, R. H. (2003). Congenital insensitivity to pain: An update. *Pain, 101*, 213–219.

Nelson, E. E. & Panksepp, J. (1998). Brain substrates of infant-mother attachment: Contributions of opioids, oxytocin, and norepinephrine. *Neuroscience and Biobehavioral Reviews, 22*, 437–452.

Nomura, M., Iidaka, T., Kakehi, K., Tsukiura, T., Hasegawa, T., Maeda, Y., et al. (2003). Frontal lobe networks for effective processing of ambiguously expressed emotions in humans. *Neuroscience Letters, 348*, 113–116.

Panksepp, J. (1998). *Affective Neuroscience.* New York: Oxford University Press.

Panksepp, J., Herman, B., Conner, R., Bishop, P., & Scott, J. P. (1978). The biology of social attachments: Opiates alleviate separation distress. *Biological Psychiatry, 13*, 607–618.

Petrovic, P., Petersson, K. M., Ghatan, P. H., Stone-Elander, S., & Ingvar, M. (2000). Pain-related cerebral activation is altered by a distracting cognitive task. *Pain, 85*, 19–30.

Peyron, R., Laurent, B., & Garcia-Larrea, L. (2000). Functional imaging of brain responses to pain. A review and meta-analysis. *Neurophysiological Clinics, 30*, 263–288.

Ploghaus, A., Becerra, L., Borras, C., & Borsook, D. (2003). Neural circuitry underlying pain modulation: Expectation, hypnosis, placebo. *Trends in Cognitive Sciences, 7*, 197–200.

Ploghaus, A., Tracey, I., Gati, J. S., Clare, S., Menon, R. S., Matthews, P. M., et al. (1999). Dissociating pain from its anticipation in the human brain. *Science, 284*, 1979–1981.

Price, D. D. (1999). *Psychological Mechanisms of Pain and Analgesia.* Seattle: IASP Press.

Rainville, P., Duncan, G. H., Price, D. D., Carrier, B., & Bushnell, M. D. (1997). Pain affect encoded in human anterior cingulate but not somatosensory cortex. *Science, 277*, 968–971.

Rubia, K., Smith, A. B., Brammer, M. J., & Taylor, E. (2003). Right inferior prefrontal cortex mediates response inhibition while mesial prefrontal cortex is responsible for error detection. *Neuroimage, 20*, 351–358.

Sawamoto, N., Honda, M., Okada, T., Hanakawa, T., Kanda, M., Fukuyama, H., et al. (2000). Expectation of pain enhances responses to nonpainful somatosensory stimulation in the anterior cingulate cortex and parietal operculum/posterior insula: An event-related functional magnetic resonance imaging study. *Journal of Neuroscience, 20*, 7438–7445.

Schneider, W. & Shiffrin, R. M. (1977). Controlled and automatic human information processing. I. Detection, search, and attention. *Psychological Review, 84*, 1–66.

Shimodozono, M., Kawahira, K., Kamishita, T. Ogata, A., Tohgo, S., & Tanaka, N. (2002). Reduction of central poststroke pain with the selective reuptake inhibitor fluvoxamine. *International Journal of Neuroscience, 112*, 1173–1181.

Shin, L. M., Dougherty, D. D., Orr, S. P., Pitman, R. K., Lasko, M., Macklin, M. L., et al. (2000). Activation of anterior paralimbic structures during guilt-related script-driven imagery. *Biological Psychiatry, 48*, 43–50.

Smith, W. (1945). The functional significance of the rostral cingular cortex as revealed by its responses to electrical excitation. *Journal of Neurophysiology, 8*, 241–255.

Tolle, T. R., Kaufmann, T., Siessmeier, T., Lautenbacher, S., Berthele, A., Munz, F., et al. (1999). Region-specific encoding of sensory and affective components of pain in the human brain: A positron emission tomography correlation analysis. *Annals of Neurology, 45*, 40–47.

Twenge, J. M., Baumeister, R. F., Tice, D. M., & Stucke, T. S. (2001). If you can't join them, beat them: Effects of social exclusion on aggressive behavior. *Journal of Personality and Social Psychology, 81*, 1058–1069.

Ueda, K., Okamoto, Y., Okada, G., Yamashita, H., Hori, T., & Yamawaki, S. (2003). Brain activity during expectancy of emotional stimuli: An fMRI study. *Neuroreport, 14*, 41–44.

Ursu, S., Stenger, V. A., Shear, M. K., Jones, M. R., & Carter, C. S. (2003). Overactive action monitoring in obsessive-compulsive disorder: Evidence from functional magnetic resonance imaging. *Psychological Science, 14*, 347–353.

Vogt, B. A., Wiley, R. G., & Jensen, E. L. (1995). Localization of mu and delta opioid receptors to anterior cingulate afferents and projection neurons and input/output model of mu regulation. *Experimental Neurology, 135*, 83–92.

Wager, T. D., Riling, J. K., Smith, E. E., Sokolik, A., Casey, K. L., Davidson, R. J., et al. (2004). Placebo-induced changes in fMRI in the anticipation and experience of pain. *Science, 303*, 1162–1167.

Weissman, D. H., Giesbrecht B., Song A. W., Mangun G. R., & Woldorff M. G. (2003). Conflict monitoring in the human anterior cingulate cortex during selective attention to global and local object features. *Neuroimage, 19,* 1361–1368.

Williams, K. D. & Sommer, K. L. (1997). Social ostracism by coworkers: Does rejection lead to loafing or compensation? *Personality and Social Psychology Bulletin, 23,* 693–706.

Williams, K. D., Cheung, C. K. T., & Choi, W. (2000). Cyberostracism: Effects of being ignored over the Internet. *Journal of Personality and Social Psychology, 79,* 748–762.

Williams, K. D., Forgas, J. P., & von Hippel, W. (in press). *The Social Outcast: Ostracism, Social Exclusion, Rejection, and Bullying.* New York: Psychology Press.

Zadro, L. & Williams, K. D. (1998). Impact of ostracism on social judgments and decisions: Explicit and implicit responses. In J. P. Forgas, K. D. Williams, & W. von Hippel (Eds.), *Social Judgments: Implicit and Explicit Processes* (pp. 325–342). Cambridge: Cambridge University Press.

10 The Social Neuroscience of Stereotyping and Prejudice: Using Event-Related Brain Potentials to Study Social Perception

Tiffany A. Ito, Geoffrey R. Urland, Eve Willadsen-Jensen, and Joshua Correll

In his book on prejudice, Gordon Allport (1954/1979) stated, "The human mind must think with the aid of categories. . . . We cannot possibly avoid this process. Orderly living depends on it," (p. 20). Allport followed this assertion with some relatively benign implications of categorization, such as carrying an umbrella when perception of a darkening sky and falling barometric pressure produces a categorization of "rain likely." As Allport recognized, though, categorization becomes potentially more problematic when applied to the perception of people. This is because we often have strong beliefs and evaluative reactions associated with social groups, and these can be easily activated after categorization. Allport noted, "The category saturates all that it contains with the same ideational and emotional flavor" (p. 21).

Social psychologists have not surprisingly devoted a great deal of attention to these issues. A variety of experimental methods have been used to do so (self-report, nonverbal behavior, response latencies), but with the continued development of neuroscience methods and theories, we now also have the opportunity to approach these questions from a more explicitly neuroscientific perspective. In this chapter, we review the line of research we have pursued using event-related brain potentials (ERPs) to study aspects of social perception.

Using ERPs to Study Social Perception

ERPs are changes in electrocortical activity that occur in response to discrete events such presentation of a visual stimulus or commission of a behavior. They can be recorded noninvasively from the surface of the scalp, and are thought to reflect summated postsynaptic potentials from large sets of synchronously firing neurons in the cerebral cortex (Fabiani, Gratton, & Coles, 2000). The recorded waveform is composed of a series of positive and negative deflections whose polarity is determined by the polarity of the electrical potential at that particular time relative to a reference electrode or collection of electrodes. These time-locked deflections are referred to as *components* and are assumed to reflect one or more information-processing operations (Gehring et al., 1992). They are typically named with reference to both their polarity and time course, so N100 would refer to a component with a negative polarity that peaks at around 100 msec after stimulus onset. Other naming schemes refer to the presumed psychological meaning of the component (a selection negativity refers to a component with a negative polarity associated with the selection of attention). The amplitude of a component is thought to reflect the extent to which the associated psychological operation has been engaged, and latency of the component is taken as the time by which the operation has been completed.

ERPs are attractive for the study of social perception for several reasons. First, they have been examined in response to a wide range of psychological operations, providing a large corpus of research to draw on in linking observed electrical activity to its underlying psychological meaning. They also possess high temporal resolution (on the order of milliseconds). They consequently provide access to processes that occur very quickly after stimulus onset, and allow for inferences about the time course and ordering of mental operations. Because of this temporal resolution, ERPs also provide access to complex mental phenomena about which participants might be unaware. As the research reviewed in this chapter will illustrate, perceiving other people is likely to be composed of many information-processing operations, and there is no reason to expect that perceivers are explicitly aware of all aspects of this process. In addition, because they do not rely on conscious awareness or willingness to report internal states accurately, ERPs are useful in measuring socially sensitive topics such as processes related to stereotyping and prejudice. Finally, although ERPs have a lower spatial resolution than some imaging techniques, the scalp distribution of observed activity can be used to obtain estimates of neuroanatomical location.

Based on these considerations, we used ERPs to study many aspects of person perception. Two main associated issues are how social category information relevant to classifying individuals into meaningful social groups is perceived, and how this information, in conjunction with stereotypes, influences behavior.

Perception of Social Category Information

A very simply model of social perception is as follows: if we categorize an individual as belonging to a particular social group, stereotypical beliefs and prejudicial evaluative reactions associated with the entire group can become activated, and once activated, this information can influence how we respond to the individual. Although much is known about some aspects of this process (application of stereotypes), the earliest aspect, involving encoding cues indicative of social category membership, is less well understood. It has generally been assumed that an individual's membership in different social categories is perceived relatively automatically (Brewer, 1988; Bruner, 1957; Fiske & Neuberg, 1990; Macrae & Bodenhausen, 2000; Stangor et al., 1992), but it was not known how quickly this information could be perceived and whether the perceptions could be easily modified.

In two initial studies, we investigated the degree to which race and gender are encoded when perceiving faces, and whether attention to these factors requires perceivers to attend to these dimensions explicitly (Ito & Urland, 2003). To do so, we had (primarily white) participants view faces while making either race or gender judgments. Regardless of task, all participants viewed the same pictures of black and white males and females. This design allowed us to look at both explicit and implicit social-categorization processes. For instance, for participants attending to race, we could examine ERP responses as a function of target race to assess explicit race categorization effects and responses as a function of target gender to examine implicit gender categorization effects. Resulting waveforms revealed four distinct deflections: a negative-going component with a mean latency of 122 msec after face onset, a positive-going component with a mean latency of 176 msec, a negative-going component with a mean latency of 256 msec, and a positive-going component with a mean latency of 485 msec. We refer to these components based on their polarity and latency as N100, P200, N200, and P300[1], respectively.

Consistent with models of automatic social categorization, target race modulated ERP responses early in processing. Specifically, racial category information affected N100 responses. As can be seen in the top panel of figure 10.1, which shows waveforms as a function of target race, N100s were larger to blacks than whites. This occurred both for participants explicitly attending to race as well as those explicitly attending to gender.

Gender category information had effects slightly later, beginning in the P200. As can be seen in the bottom panel of figure 10.1, which shows waveforms as a function of target gender, P200s were larger to males than to females. As with race effects, this occurred for participants explicitly attending to both race and gender, and also for both male and female participants. In addition to gender effects, the larger responses to blacks than whites continued in P200 as well.

Of interest, the direction of the target race and gender effects reversed in the N200, with larger responses to whites than blacks and to females than males (see also Ito,

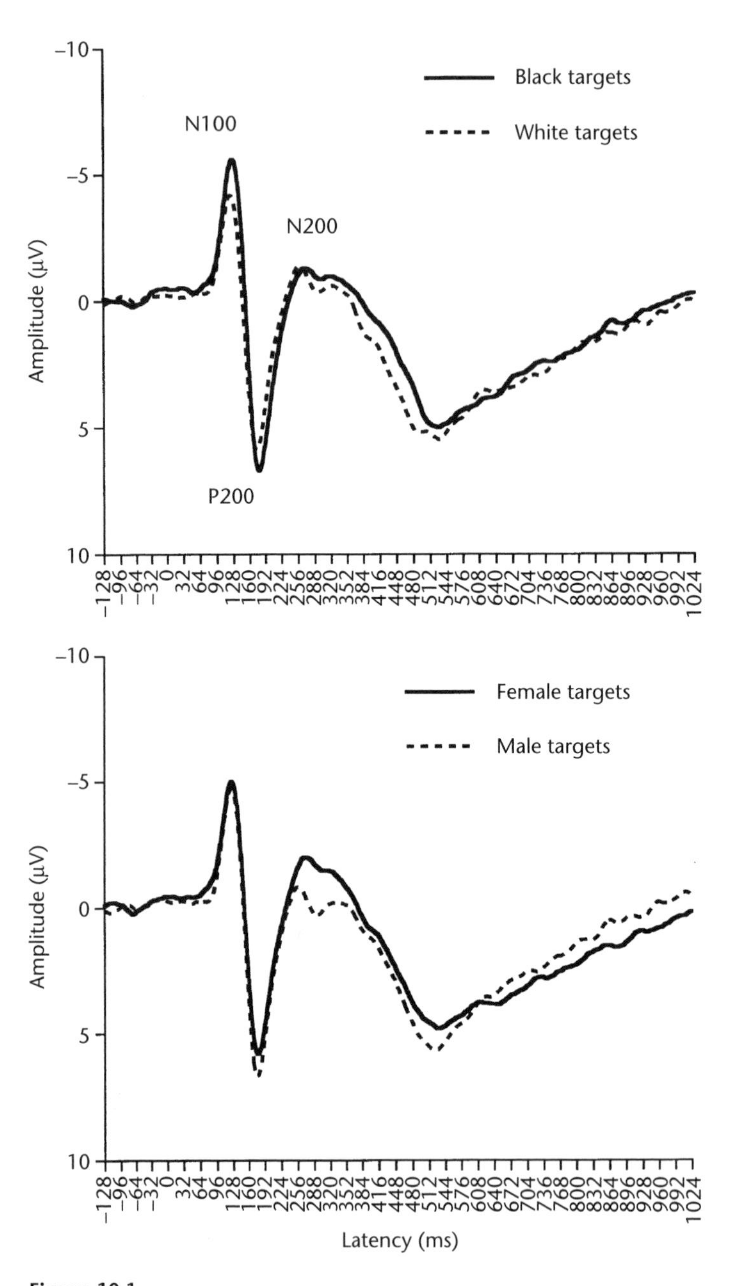

Figure 10.1

N100, P200, and N200 effects as a function of target race (top) and gender (bottom). Waveforms are from the central area (Cz) in response to viewing of color photographs of faces. (From "Race and gender on the brain: Electrocortical measures of attention to race and gender of multiply categorizable individuals," by T. A. Ito and G. R. Urland, 2003, *Journal of Personality and Social Psychology*, *85*, p. 619. Copyright 2003 by American Psychological Association. Adapted with permission.)

Thompson, & Cacioppo, 2004; James, Johnstone, & Hayward, 2001). Again, this occurred in both task conditions and for both male and female participants.

The N100, P200, and N200 have all been strongly associated with nonspatial visual attention. The amplitude of these early components is taken as an index of covert orienting to relevant and/or salient features (Czigler & Geczy, 1996; Eimer, 1997; Kenemans, Kok, & Smulders, 1993; Naatanen & Gaillard, 1983; Wijers et al., 1989). The results of Ito and Urland (2003) therefore suggest that attention is directed to race and gender cues at early processing stages regardless of the dimension to which participants are explicitly attending.

We can also ask what the direction of target race and gender effects reflects. This is an issue we are still investigating, but we do have preliminary hypotheses. In American culture, stereotypes of blacks include aspects of aggressiveness and violence and stereotypes of men include more reference to power and strength than those about women. Larger N100s and P200s to blacks and larger P200s to males may therefore represent initial covert orienting to targets heuristically associated with the greater potential for threat. In addition, racial contact disparities often result in more experience with racial ingroup members, which could produce orienting to more novel targets. Thus, early race and gender effects may reflect threat or novelty detection. The time course of these effects is consistent with other types of affective modulation, which can occur as early as 100 msec (Pizzagalli, Regard, & Lehmann, 1990; Smith et al., 2003).

Although attention is apparently initially oriented to blacks and males, effects shift by the time the N200 peaks, with larger N200s to whites and females. Larger N200s to whites may represent greater, more individuated processing of racial in-group members and/or members of the more culturally dominant racial group. This is consistent with the large body of research showing that whites and racial in-group members are spontaneously processed more deeply than other racial groups (Anthony, Cooper, & Mullen, 1992; Levin, 2000). The meaning of larger N200s to females is less clear. Processing differences in favor of female over male targets have been obtained (Lewin & Herlitz, 2002; McKelvie, 1981; McKelvie et al., 1993; O'Toole et al., 1998), but these effects are more variable than differences in processing as a function of race. We are therefore continuing our examination of what these particular patterns of attentional focus represent, and whether race and gender effects are governed by common attentional mechanisms (early orientation to threat) or different mechanisms that simply manifest with the same time course (different mechanisms that later promote greater attention to whites and females).

In addition to components occurring earlier in processing that are associated with attention, we have examined a component linked with working memory processes in these studies. The amplitude of this component, the P300, increases when a stimulus differs from preceding stimuli along task-relevant dimensions. Thus, P300 amplitude is determined by the relation of a given stimulus to the stimuli that precede it. This sensitivity led to the belief that P300 amplitude reflects updates to working memory that maintain an accurate mental model of the external environment (Donchin, 1981). To examine the effect that social category information has on working memory, we systematically varied stimuli in these same studies so that we could analyze responses in terms of race and gender not only of the target picture, but also of the face that preceded it (Ito & Urland, 2003).[2]

Consistent with past P300 research on nonsocial stimuli (Donchin, 1981), P300s were larger when a target individual's social category membership differed from that of preceding individuals on the task-relevant dimension compared with when it matched the preceding context. For instance, for a participant attending to race and viewing a majority of faces of blacks, P300s were larger to pictures of whites than blacks. This shows, not surprisingly, that working memory processes are sensitive to the social category dimension along which categorization was explicitly occurring. However, we also obtained evidence of implicit categorization effects on working memory. P300 amplitude also increased when a target picture differed from individuals pictured in preceding pictures along the task-irrelevant dimension. For instance, for a participant attending to race and viewing a majority of male faces, P300s were larger to females than males. Both explicit and implicit working memory effects can be seen in figure 10.2. The top panel shows waveforms from participants attending to race, and the bottom panel from participants attending to gender. In both panels, data are represented in terms of similarities and differences between race and gender of target and preceding pictures.

Our two initial studies in this area differed in physical features of stimuli, with one study showing faces in full color and the other in gray scale (Ito & Urland, 2003). This was done to examine the degree to which low-level perceptual features such as luminance and contrast were responsible for the results. Of importance, N100, P200, and N200 results were the same in both studies (in the P300, there was some greater sensitivity to race that was seen only with color stimuli). This suggests that participants were responding to social cues per se, rather than to general perceptual features (luminance) that happen to covary with the social cues.

Together, these studies provide relatively direct evidence that when perceivers encounter individuals who can be categorized into several social groups, race and

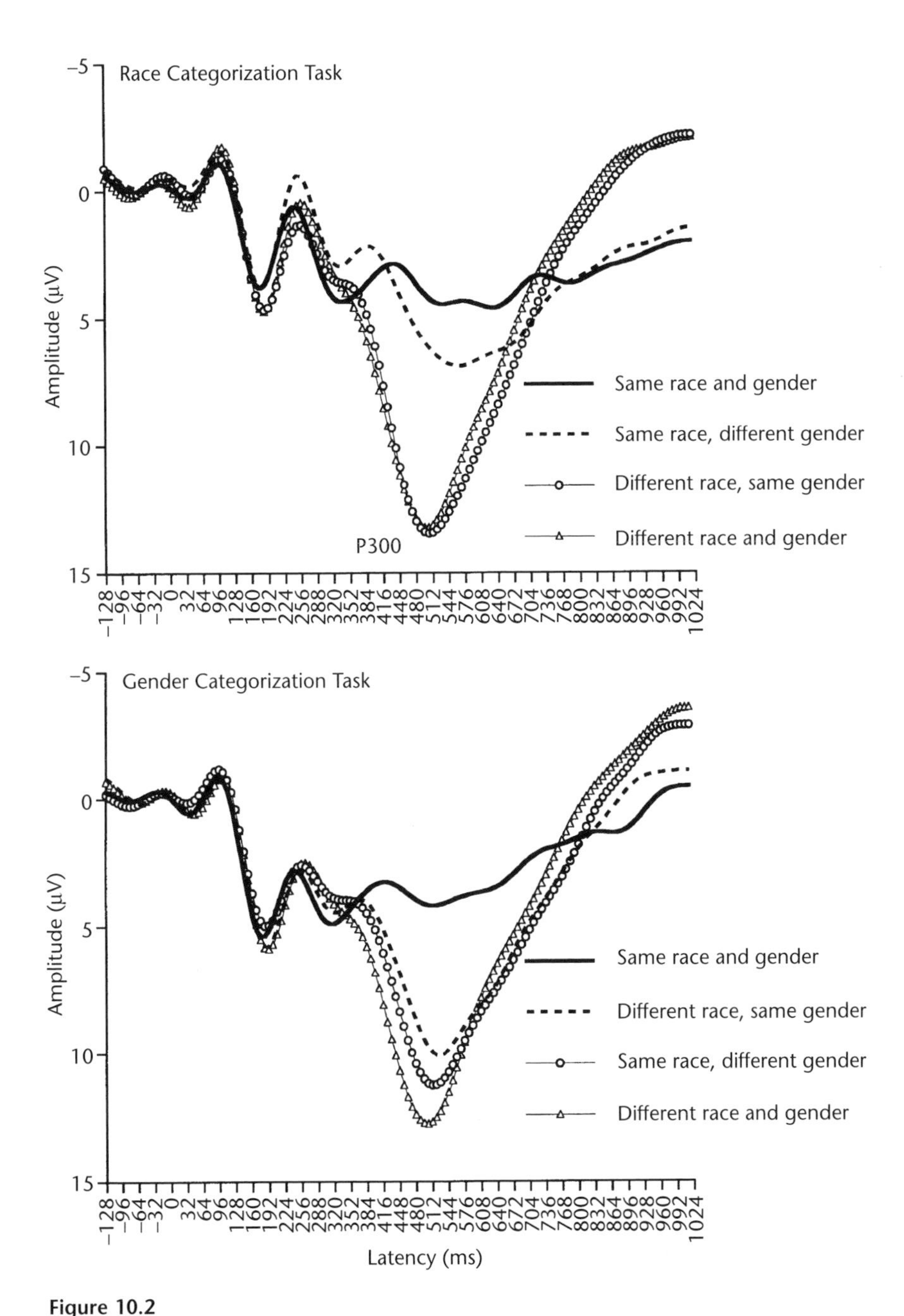

Figure 10.2
P300 effects as a function of a target individual's race and gender relative to the race and gender of preceding individuals. The top panel shows responses from participants categorizing faces by race and the bottom panel shows responses from participants categorizing by gender. Waveforms are from the parietal area (Pz) in response to viewing of color photographs of faces. (From "Race and gender on the brain: Electrocortical measures of attention to race and gender of multiply categorizable individuals," by T. A. Ito and G. R. Urland, 2003, *Journal of Personality and Social Psychology, 85*, p. 619. Copyright 2003 by American Psychological Association. Adapted with permission.)

gender information are both activated at very early stages in processing. That race and gender modulated ERP responses even when those dimensions were not relevant to the task is consistent with theories that attention to race and gender information is automatic (Brewer, 1988; Fiske & Neuberg, 1990; Macrae & Bodenhausen, 2000). However, it is the case that all participants were performing some type of social-categorization task. It might be that this sensitized participants to attend to all social category information, even dimensions that were not directly task relevant.

Boundary Conditions on Attention to Race and Gender

Tasks that focus attention on a target's individual characteristics and foster person-based (as opposed to category-based) processing have proved successful in reducing stereotyping and prejudice (Fiske & Neuberg, 1990). Similarly, focusing attention at lower levels of analysis, for example, on a nonsocial physical cue such as a circle on a picture, decreases stereotype activation (Macrae et al., 1997). These results could occur for three different reasons: (1) encoding of category membership has been attenuated, thereby blocking activation and application of stereotypes and prejudice; (2) category membership is still encoded, but activation of stereotypes-prejudice is attenuated; or (3) category membership is encoded and stereotypes-prejudice are activated, but their application is attenuated. To the degree that categorization is afforded unconditional automatic status (Brewer, 1988; Fiske & Neuberg, 1990; Macrae & Bodenhausen, 2000), option 1 would seem unlikely. To assess this, and to examine the effects obtained in Ito and Urland (2003) using tasks that did not require simple social categorization, two additional studies were done in which participants viewed pictures of black and white males and females while performing two different individuating tasks: having participants make introversion-extroversion judgments about each person and having them indicate whether each individual would like various kinds of vegetables (Ito & Urland, 2005). The introversion task was chosen because we assumed it would be engaging and easy for participants to perform, thereby easily directing attention away from social category cues. The vegetable task was chosen because it decreases stereotype activation and the typically greater amygdala activation to racial out-group than in-group faces that is thought to reflect greater negativity toward the out-group (Wheeler & Fiske, 2005). We also had some participants perform the visual feature-detection task from Macrae et al. (1997), in which they determined for each picture whether a white dot was or was not present (Ito & Urland, 2005). This task has also been associated with decreased stereotype activation (Macrae et al., 1997).

We observed the same four components as in Ito and Urland (2003), with very similar latencies and nearly identical effects as when participants were explicitly attending to social category. Specifically, sensitivity of the P200, N200, and P300 was replicated even under these very different processing goals. This indicates that directing participants to attend to individuals at a deeper (more individuated) or more shallow (downplaying their social nature altogether) level does not inhibit most aspects of racial and gender perception observed under conditions of explicit attention to race and gender, further supporting the obligatory nature of social category encoding. It also suggests that prior instances in which processing manipulations attenuated stereotyping and prejudice effects are due to changes in subsequent stages of processing, after at least elementary aspects of social category information have been processed.

The one main departure compared with when participants were explicitly attending to race and gender was absence of N100 target race effects when participants were attending to dots or making the vegetable judgments. For both of these tasks, stimulus presentation was more complex. The vegetable task was accomplished by presenting the name of the vegetable about which the preference judgment was to be made before each face. The dot task required placing dots on some of the faces. The simple increase in visual complexity may have been sufficient to slow visual processing of racial cues observed in more visually simple contexts (Ito & Urland, 2003). It is notable that for participants performing the introversion-extroversion task, stimulus presentation was identical to that in our previous studies, and N100s here were significantly larger to blacks than to whites. This leads us to conclude that increasing the visual complexity of the task can slow race-sensitive aspects of face processing, delaying them from the N100 to the P200, a conclusion consistent with general effects of attentional load in reducing visual attention (for a review, see Rees & Lavie, 2001). Nevertheless, it is still the case that race effects emerge relatively early in processing in all cases. Even when effects are delayed to the P200, they occurr within 190 msec after stimulus onset.

Looking across these studies examining the initial perception of racial and gender cues, the ease with which social category information can be encoded is consistently demonstrated. Access to very quick aspects of processing afforded by ERPs reveals that early perceptual aspects of social categorization are relatively obligatory, driven more by properties of the individual being perceived than by goals and intentions of the perceiver. Such encoding occurs even when attention is explicitly directed to another social category dimension (Ito & Urland, 2003), when targets are processed in a more individuated manner (Ito & Urland, 2005), and when they are processed in a presemantic manner (Ito & Urland, 2005). The last two manipulations have been shown to decrease stereotype and prejudice activation, so the present results suggest that attention to category membership need not invariantly result in stereotyping and prejudice (cf. Wolsko et al., 2000).

Stereotyping and Behavior

Understanding how social category information is processed is an important piece of the puzzle that is social perception, but it is also important to understand how the stored information perceivers have about social groups (stereotypes, prejudice) comes to influence behavior. We found that ERPs are also well suited to addressing this issue. We particularly focused on understanding how race influences the detection of weapons in a first-person shooter video game paradigm developed by Correll et al. (2002; see also Amodio et al., 2004; Greenwald, Oakes, & Hoffman, 2003; Payne, 2001). The original work using this paradigm was spurred by several instances in which unarmed black men were erroneously shot to death by police (Amadou Diallo, Timothy Thomas, Anthony Dwain Lee). In some cases, officers reported fearing that the black individual was reaching for a weapon. This raised the issue of whether officers' behavior was influenced by personal and cultural stereotypes linking blacks with violence and aggression, and whether the outcome would have differed if the individuals in question had been white.

To assess this experimentally, Correll et al. (2002) developed a simulation in which participants viewed pictures of black and white men who were holding either guns or innocuous objects such as wallets and cell phones. Participants were asked to make speeded decisions to "shoot" armed targets by pressing one button and "not shoot" unarmed targets by pressing another. Behavioral results showed a consistent bias against blacks relative to whites. Participants were faster and more accurate in "shooting" armed blacks compared with armed whites. By contrast, they were faster and more accurate in "not shooting" unarmed whites than unarmed blacks. Behavioral bias was also associated with stereotypical beliefs, with bias increasing as associations between blacks and violence-aggression grew stronger.

At the neuroscience level, the issue of how race affects decisions to shoot can be recast as an issue of behavioral regulation and executive control, which is thought to be governed by a two-part system (Botvinick et al., 2001; Carter et al., 1998). The first involves monitoring conflict in continuing information processing. Such conflict might take the form of a prepotent but incorrect response activated at the same time as a less potent but correct response. The second component responds to identification of conflict with implementation of cognitive control to reduce discrepancies in information processing. Given the ease with which race is processed, and strong cultural and sometimes personal stereotypes associating violence with blacks more strongly than with whites, stereotypes relevant to violence and aggression are likely to be activated easily and quickly in this task. They may, however, conflict with other representations activated by the task, leading to cognitive conflict.

Erroneous behavior is particularly likely to involve activation of incongruent representations and may therefore particularly benefit from a behavior-regulation analysis. In the shooter task, erroneously shooting someone may be associated with some activation promoting shooting (this person looks dangerous) as well as activation inhibiting shooting (eventual determination that the individual is unarmed). But whereas it seems reasonable to expect that conflict monitoring is occurring in this situation, it might not be occurring equally for black and white targets. It seems likely that the greater association between blacks and violence will diminish conflict between activated representations of "he's black" and "I've shot him" compared with the representations of "he's white" and "I've shot him." Similarly, less conflict seems likely between the representations of "he's white" and "I did not shoot him" than between "he's black" and "I did not shoot him."

This analysis suggests that assessments of conflict monitoring during performance of the shooter task should differ as a function of target race. Such an assessment can be accomplished by measuring an ERP component associated with the degree of detected conflict, that is, the error-related negativity (ERN). As conflict increases, as when a prepotent response is inconsistent with an accurate response, ERN amplitude generated shortly after the response associated with conflict increases (Falkenstein, Hohnsbein, & Hoormann, 1995; Falkenstein, Hoormann, & Hohnsbein, 1999; Gerhing et al., 1995; Gerhing et al., 1993; Scheffers & Coles, 2000). Consistent with localization of functions ascribed to conflict monitoring to the anterior cingulate cortex (ACC; Botvinick et al., 2001; Carter et al.,1998; MacDonald et al., 2000), dipole modeling has located the source of the ERN to medial prefrontal areas (Dehaene, Posner, & Tucker, 1994).

Conflict Monitoring and Erroneous Decisions to Shoot

Integrating across these findings, we have evidence that (1) processing of race occurs quickly and easily (Ito & Urland, 2003, 2005); (2) behavior in the shooter paradigm is racially biased (Correll et al., 2002); (3) models of behavior regulation may be relevant to understanding racial bias; and (4) the ERN is sensitive to conflict monitoring after error commission. We therefore examined ERNs in response to errors as participants played the shooter videogame developed by Correll et al. (2002) to determine if conflict monitoring differs for black and white targets. Participants playing the videogame were required to respond within a relatively short response window (580–600 msec after stimulus onset) to ensure a relatively large number of errors. Replicating Correll et al.'s initial behavioral studies, the pattern of errors participants committed was racially biased, with more errors to unarmed blacks than whites (Ito & Correll, unpublished data). That is, unarmed black men were more likely to be erroneously "shot" than were unarmed white men.

ERNs were observed after erroneous responses, but their amplitude differed depending on gun status. They were larger after mistakenly "shooting" unarmed targets compared with mistakenly "not shooting" armed targets. This effect is interesting, given the explicit point contingencies of the game. To motivate performance, a cumulative score based on performance was displayed after every trial. Correct detection of an unarmed target earned five points, correct detection of an armed target earned ten, incorrect detection of an unarmed target cost twenty, and incorrect detection of an armed target cost forty points. This last penalty was made the most severe to model real-life outcomes in which failing to fire on an armed target can have fatal consequences. Within this point contingency, failing to "shoot" an armed target is the most undesirable response, but the largest ERNs were generated by an outcome whose explicit point implications were only half as severe.

Rather than reflecting our externally imposed payoff matrix, larger ERNs to erroneous firing on unarmed individuals appear to reflect participants' subjective interpretation of their behavior (Holroyd, Larsen, & Cohen, 2004), with greater conflict over erroneous firing. In this regard, results are consistent with the well-documented tendency to be more bothered by errors of commission than of omission immediately after the behavior (Gilovich & Medvec, 1995). They also confirm the psychological realism of the game; participants were disturbed when they "shot" an innocent person.

More important to understanding racial bias, the ERN difference for erroneously "shooting" versus "not shooting" also interacted with target race. The largest ERNs were associated with erroneously "shooting" unarmed whites, which resulted in significantly larger ERNs than failing to "shoot" armed whites. By contrast, when the target was black, ERNs after erroneous "shoot" and "not shoot" responses did not differ. Thus, consistent with stereotypes, "shooting" someone who is not associated with violence generated the greatest conflict, but "shooting" someone stereotypically associated with violence was not more problematic than "not shooting" him. Because conflict monitoring is thought to trigger the implementation of cognitive control, these results suggest that behavior toward unarmed whites is most strongly regulated, and that behavioral regulation for white targets is most responsive to their weapon status. Although we cannot know what transpired in recent situations in which unarmed black men were shot by police, our results suggest that the police officers' behavior was not as strongly regulated by higher-order processes at it might have been if the victims were white.

Conflict Monitoring during Successful Behavior Regulation

These initial results implicate cognitive control processes, in particular, failure to implement control equally toward white and black targets in racial bias (Ito & Correll, unpublished data). This was revealed by analyzing conflict monitoring during the commission of errors, but conflict monitoring and cognitive control are also at work when behavior is successfully regulated. This suggests that even when participants correctly detect who to "shoot" and "not shoot," differences in the implementation of cognitive control should be observed. To examine this, we conducted a second study in which ERP responses were recorded during performance of the shooter video task. Participants were given a slightly longer window within which to respond (850 msec), which allows for variability in response latency and results in primarily correct responses (Correll et al., 2002). This increase had the intended effect, as participants made few errors, but showed a biased pattern of response latencies. They were faster to "shoot" armed blacks than whites, but faster to "not shoot" unarmed whites than blacks (Correll, Urland, & Ito, in press).

Several anteriorally distributed negative-going ERP components are associated with conflict monitoring and cognitive control during successful behavior regulation. The N200 in particular is associated with the magnitude of response conflict (Nieuwenhuis et al., 2003), whereas temporally later negativities (N400, negative slow wave) are associated with cognitive control process that successfully resolve the detected conflict (Curtin & Fairchild, 2003; West & Alain, 1999). Source modeling of these ERP effects is consistent with involvement of the ACC and other areas of prefrontal cortex thought to be involved in cognitive control (Liotti et al., 2000; Nieuwenhuis et al., 2003). Based on these results, we expected evidence of conflict monitoring and cognitive control when participants correctly detected who to "shoot" and "not shoot." From the pattern of ERP results obtained when participants made errors in this task (Ito & Correll, unpublished data), we expected that cognitive control would differ as a function of target race, with greater sensitivity of cognitive control mechanisms for whites than for blacks.

The ERP waveforms recorded during shoot–not-shoot decisions showed five clear deflections (N100, P200, N200, P300, N400), and can be characterized in terms of two different effects. First, weapon status was the first variable to modulate ERP responses. This occurred first in the N100, which peaked at a mean latency of 159 msec, and continued for all subsequent components. This indicates that although searching for a weapon may be a relatively novel task, participants can orient to cues indicative of gun status very early in processing. Although correct decisions to "shoot" and "not shoot" occur much later (at around 550–650 msec), these effects reflect the early point at which perception begins to differentiate between armed and unarmed individuals. The second effect observed in the ERP waveforms was a stereotype-consistent differentiation on the basis of both weapon status and race. Starting in the P200 and continuing through to the N400, unarmed whites were associated with a negative shift in the waveform that produced larger N200s and N400s to unarmed whites than to armed whites and blacks (given the negative shift for unarmed whites, this effect manifested as a smaller P200 to unarmed whites). Of interest, unarmed blacks were not differentiated from armed targets, but were significantly dissociated from unarmed whites. Said differently, starting as early as 200 msec, the "safeness" of unarmed whites was being perceived, but unarmed blacks were being processed in a manner similar to that for individuals holding guns.

As noted, the N200 is associated with conflict monitoring during successful behavior regulation in situations of conflict between prepotent but incorrect and less potent but correct responses (Nieuwenhuis et al., 2003). The N200 results we obtained therefore map onto ERN responses after error commission obtained in Ito and Correll (unpublished data). Conflict monitoring is greatest in response to unarmed whites, a pattern consistent with the presumed weaker association of whites with violence than of blacks with violence. That this pattern was also seen in the preceding P200 may indicate that this processing difference began slightly earlier than in some other paradigms. Negativities later in the waveform are associated with the implementation of cognitive control (Curtin & Fairchild, 2003; West & Alain, 1999), suggesting that the later effect we obtained in the N400, in which responses were largest to unarmed whites, reflects implementation of control in response to conflict detection.

The purpose of conflict detection and executive control is to regulate behavior. To the degree that the ERP effects we observed reflect the operation of behavior-regulation mechanisms, we would expect ERP effects to predict subsequent behavior. This was in fact the case. As participants' N100, P200, and N200 responses more strongly differentiated unarmed whites from all other targets, they showed more bias in their response latencies (they were faster to "shoot" blacks compared with whites and faster to "not shoot" whites compared with blacks). Note that N100s did not differentiate between unarmed whites and blacks at the average level (there was only a gun main effect in the N100). Nevertheless, the significant correlation with behavior indicates individual variability in the sensitivity to combined race and gun cues at this point in processing that predicts behavior. Contrary to expectations, N400 effects did not predict behavior, which therefore leaves us to question whether they did reflect cognitive control processes. This lack of correlation with the N400 notwithstanding, the results on balance were consistent with those from our first study and support the executive control analysis of decisions to shoot. Participants on average showed a biased pattern of differentiation in their ERPs as early as the P200 that was consistent with implementation of greater executive control in response to unarmed white targets than to unarmed black targets. They also on average showed a biased pattern of response latencies, and bias in earlier parts of the ERP waveform predicted behavioral bias.

Conclusion

The field of social psychology has seen many changes since 1954, when Gordon Allport wrote his seminal book on prejudice from which this chapter's opening quotation was taken. Integration of neuroscience into the study of social phenomena exemplified by this volume is one particularly salient change. But in many ways, social psychology is really not so different from the field Allport knew. We still grapple with many of the same issues, and we still seek interdisciplinary perspectives to address them. The two core issues addressed in this chapter—how social category information is processed and how racial stereotypes influence behavior—would be very much at home in the social psychology of Allport, but their investigation with a neuroscience perspective has revealed what we think are new conclusions and avenues for inquiry.

Consistent with Allport's concern over the ease and fluency of social categorization, and subsequent suggestions that social category information is processed automatically (Brewer, 1988; Fiske & Neuberg, 1990; Macrae & Bodenhausen, 2000), our research with ERPs reveals perceivers' sensitivity to racial and gender cues. Perceivers orient to this information early in perception and process it in a relatively obligatory manner across a range of processing goals known to attenuate later activation of stereotypes and prejudice (Ito & Urland, 2003, 2005). Moreover, early social perception processes appear to be multifaceted. Initially larger N100s and P200s shown to blacks and males, and subsequently larger N200s to whites and females suggest that different types of processing occur across time as individuals are processed.

Allport (1954/1979) noted that the social category "saturates all that it contains with the same ideational and emotional flavor," (p. 21), but how exactly is this achieved?

Integration of neuroscience into the investigation of social perception suggests that cognitive control models may be helpful in answering this question. The parallel, distributed nature of cognitive processing makes it likely that many representations are often activated when processing social targets. Stereotypes and prejudices associated with automatically encoded social category cues, for instance, may be simultaneously activated together with information based on more individuated processing. When these numerous sources of representations are consistent with each other, behavioral implications are in concert and behavior can follow from any of these activations with little need for higher-order cognitive control. But when many activated representations are in conflict, cognitive control processes have to be recruited to regulate behavior successfully. This analysis suggests that understanding behavior thought to be influenced by a target's social category membership requires consideration of activated representations, how these influence conflict-monitoring processes, and the degree to which cognitive control is being implemented. To some degree, cognitive control can be inferred if a perceiver's behavior and some likely activated representations are known, but ERPs provide a means for observing these processes as they occur on-line (see also Amodio et al., 2004).

When this analysis was applied to the understanding of racially biased behavior related to perceptions of threat and danger, differences in conflict monitoring and cognitive control between white and black targets was important (Correll et al., in press; Ito & Correll, unpublished data). To some extent, emphasis on behavior regulation in understanding bias is not new. Other researchers noted that motivations to control prejudice can result in behavior regulation (Dunton & Fazio, 1997; Monteith, 1993; Plant & Devine, 1998), but neuroscientific models of cognitive control offer a more detailed model of the processes involved in regulation. Moreover, whereas regulation of behavior toward stigmatized groups has often been emphasized, our results show the importance of also understanding how regulation occurs in response to members of nonstigmatized groups in the sense that *differences* in conflict monitoring toward white and black targets in the shooter videogame task seem important to understanding the behavioral effects.

There are many challenging aspects in studying social perception, such as perceivers' potential lack of awareness of all component processes and the influence of social desirability concerns. ERPs are remarkably well suited for addressing many of these complicated aspects, and we hope continued development of social neuroscience will increase our understanding of the dynamic interplay between the mind and the neural processes that underlie it.

Acknowledgments

Supported by NIMH grants R21 MH66739 and R03 MH61327 to Tiffany A. Ito, a National Science Foundation Graduate Research Fellowship to Geoffrey R. Urland, and NIMH grant F31 MH069017 to Joshua Correll.

Notes

1. Although the P300 had a latency longer than 300 msec, we use the P300 name because its scalp distribution and response to psychological processes mirrors the classic P300 component. The N100, P200, and N200 were typically largest at frontal and central scalp sites, whereas the P300 had a parietal-maximal distribution.

2. The N100, P200, and N200 amplitudes were insensitive to the preceding stimuli and differed only as a function of the face currently being processed, as previously described.

References

Allport, G. W. (1954/1979). *The Nature of Prejudice*. Reading, MA: Perseus Books.

Amodio, D. M., Harmon-Jones, E., Devine, P. G., Curtin, J. J., Hartley, S. L., & Covert, A. E. (2004). Neural signals for the detection of unintentional race bias. *Psychological Science, 15*, 88–93.

Anthony, T., Copper, C., & Mullen, B. (1992). Cross-racial facial identification: A social cognitive integration. *Personality and Social Psychology Bulletin, 18*, 296–301.

Botvinick, M. M., Carter, C. S., Braver, T. S., Barch, D. M., & Cohen, J. D. (2001). Conflict monitoring and cognitive control. *Psychological Review, 108*, 624–652.

Brewer, M. C. (1988). A dual process model of impression formation. In R. Wyer & T. Scrull (Eds.), *Advances in Social Cognition* (vol. 1, pp. 1–36). Hillsdale, NJ: Erlbaum.

Bruner, J. S. (1957). On perceptual readiness. *Psychological Review, 64*, 123–151.

Carter, C. S., Braver, T. S., Barch, D. M., Botvinick, M. M., Noll, D., & Cohen, J. D. (1998). Anterior cingulate cortex, error detection, and the online monitoring of performance. *Science, 280*, 747–749.

Correll, J., Park, B., Judd, C. M., & Wittenbrink, B. (2002). The police officer's dilemma: Using ethnicity to disambiguate potentially threatening individuals. *Journal of Personality and Social Psychology, 83*, 1314–1329.

Correll, J., Urland, G. R., & Ito, T. A. (in press). Shooting straight from the brain: Early attention to race promotes bias in the decision to shoot. *Journal of Experimental Social Psychology*.

Curtin, J. J. & Fairchild, B. A. (2003). Alcohol and cognitive control: Implications for regulation of behavior during response conflict. *Journal of Abnormal Psychology, 112*, 424–436.

Czigler, I. & Geczy, I. (1996). Event-related potential correlates of color selection and lexical decision: Hierarchical processing or late selection? *International Journal of Psychophysiology*, *22*, 67–84.

Dehaene, S., Posner, M. I., & Tucker, D. M. (1994). Localization of a neural system for error detection and compensation. *Psychological Science*, *5*, 303–305.

Donchin, E. (1981). Surprise! . . . Surprise? *Psychophysiology*, *18*, 493–513.

Dunton, B. C. & Fazio, R. H. (1997). An individual difference measure of motivation to control prejudiced reactions. *Personality and Social Psychology Bulletin*, *23*, 316–326.

Eimer, M. (1997). An event-related potential (ERP) study of transient and sustained visual attention to color and form. *Biological Psychology*, *44*, 143–160.

Fabiani, M., Gratton, G., & Coles, M. G. H. (2000). Event-related brain potentials. In J. T. Cacioppo, L. G. Tassinary, & G. G. Berntson (Eds.), *Handbook of Psychophyiology*, 2nd ed. (pp 53–84). Cambridge: Cambridge University Press.

Falkenstein, M., Hohnsbein, J., & Hoormann, J. (1995). Event-related potential correlates of errors in reaction tasks. *Perspectives of Event-Related Potentials Research* (EEG Suppl. 44), 287–296.

Falkenstein, M., Hoormann, J., & Hohnsbein, J. (1999). ERP components in go/nogo tasks and their relation to inhibition. *Acta Psychologica*, *101*, 267–291.

Fiske, S. T. & Neuberg, S. L. (1990). A continuum of impression formation, from category-based to individuating processes: Influences of information and motivation on attention and interpretation. *Advances in Experimental Social Psychology*, *23*, 1–73.

Gehring, W. J., Coles, M. G. H., Meyer, D. E., & Donchin, E. (1995). A brain potential manifestation of error-related processing. *Perspectives of Event-Related Potentials Research* (EEG Suppl. 44), 261–272.

Gehring, W. J., Goss, B., Coles, M. G. H., Meyer, D. E., & Donchin, E. (1993). A neural system for error detection and compensation. *Psychological Science*, *4*, 385–390.

Gehring, W. J., Gratton, G., Coles, M. G. H., & Donchin, E. (1992). Probability effects on stimulus evaluation and response processes. *Journal of Experimental Psychology: Human Perception and Performance*, *18*, 198–216.

Gilovich, T. & Medvec, V. H. (1995). The experience of regret: What, when, and why. *Psychological Review*, *102*, 379–395.

Greenwald, A. G., Oakes, M. A., & Hoffman, H. G. (2003). Targets of discrimination: Effects of race on responses to weapons holders. *Journal of Experimental Psychology*, *39*, 399–405.

Holroyd, C. B., Larsen, J. T., & Cohen, J. D. (2004). Context dependence of the event-related brain potential associated reward and punishment. *Psychophysiology*, *41*, 245–253.

Ito, T. A., Thompson, E., & Cacioppo, J. T. (2004). Neural mechanisms of social perception: The effects of racial cues on ERPs. *Personality and Social Psychology Bulletin*, *30*, 1267–1280.

Ito, T. A. & Urland, G. R. (2003). Race and gender on the brain: Electrocortical measures of attention to race and gender of multiply categorizable individuals. *Journal of Personality and Social Psychology, 85,* 616–626.

Ito, T. A. & Urland, G. R. (2005). The influence of processing objectives on the perception of faces: An ERP study of race and gender perception. *Cognitive, Affective, and Behavioral Neuroscience,* 5, 21–36.

James, M. S., Johnstone, S. J., & Hayward, W. G. (2001). Event-related potentials, configural encoding, and feature-based encoding in face recognition. *Journal of Psychophysiology, 15,* 275–285.

Kenemans, J. L., Kok, A., & Smulders, F. T. Y. (1993). Event-related potentials to conjunctions of spatial frequency and orientation as a function of stimulus parameters and response requirements. *Electroencephalography and Clinical Neurophysiology, 88,* 51–63.

Levin, D. T. (2000). Race as a visual feature: Using visual search and perceptual social perception 25 discrimination tasks to understand face categories and the cross-race recognition deficit. *Journal of Experimental Psychology 129,* 559–574.

Lewin, C. & Herlitz, A. (2002). Sex differences in face recognition—Women's faces make the difference. *Brain and Cognition, 50,* 121–128.

Liotti, M., Woldorff, M. G., Perez, R., & Mayberg, H. S. (2000). An ERP study of the temporal course of Stroop color-word interference effect. *Neuropsychologia, 38,* 701–711.

MacDonald III, A. W., Cohen, J. D., Stenger, V. A., & Carter, C. S. (2000). Dissociating the role of the dorsolateral prefrontal and anterior cingulate cortex in cognitive control. *Science, 288,* 1835–1838.

Macrae, C. N. & Bodenhausen, G. V. (2000). Social cognition: Thinking categorically about others. *Annual Review of Psychology, 51,* 93–120.

Macrae, C. N., Bodenhausen, G. V., Milne, A. B., Thorn, T. M. J., & Castelli, L. (1997). On the activation of social stereotypes: The moderating role of processing objectives. *Journal of Experimental Social Psychology, 33,* 471–489.

McKelvie, S. J. (1981). Sex differences in memory for faces. *Journal of Psychology, 107,* 109–125.

McKelvie, S. J., Standing, L., St. Jean, D., & Law, J. (1993). Gender differences in recognition memory for faces and cars: Evidence for the interest hypothesis. *Bulletin of the Psychonomic Society, 31,* 447–448.

Monteith, M. J. (1993). Self-regulation of prejudiced responses: Implications for progress in prejudice-reduction efforts. *Journal of Personality and Social Psychology, 65,* 469–485.

Naatanen, R. & Gaillard, A. W. K. (1983). The orientating reflex and the N2 deflection of the event-related potential (ERP). In A. W. K. Gaillard & W. Ritter (Eds.), *Tutorials in ERP Research: Endogenous Components* (pp. 119–141). New York: North-Holland.

Nieuwenhuis, S., Yeung, N., Van Den Wildenberg, W., & Ridderinkhof, K. R. (2003). Electrophysiological correlates of anterior cingulate function in a go/no-go task: Effects of response conflict and trial type frequency. *Cognitive, Affective and Behavioral Neuroscience, 3,* 17–26.

O'Toole, A. J., Deffenbacher, K. A., Valentin, D., McKee, K., & Abdi, H. (1998). The perception of face gender: The role of stimulus structure in recognition and classification. *Memory and Cognition, 26,* 146–160.

Payne, B. K. (2001). Prejudice and perception: The role of automatic and controlled processes in misperceiving a weapon. *Journal of Personality and Social Psychology, 81,* 181–192.

Pizzagalli, D., Regard, M., & Lehmann, D. (1999). Rapid emotional face processing in the human right and left brain hemispheres: An ERP study. *Neuroreport, 10,* 2691–2698.

Plant, E. A. & Devine, P. G. (1998). Internal and external motivation to respond without prejudice. *Journal of Personality and Social Psychology, 75,* 811–832.

Rees, G. & Lavie, N. (2001). What can functional imaging reveal about the role of attention in visual awareness? *Neuropsychologia, 39,* 1343–1353.

Scheffers, M. K. & Coles, M. G. H. (2000). Performance monitoring in a confusing world: Error-related brain activity, judgments of response accuracy, and types of errors. *Journal of Experimental Psychology: Human Perception and Performance, 26,* 141–151.

Smith, N. K., Cacioppo, J. T., Larsen, J. T., & Chartrand, T. L. (2003). May I have your attention, please: Electrocortical responses to positive and negative stimuli. *Neuropsychologia, 41,* 171–183.

Stangor, C., Lynch, L. Duan, C., & Glass, B. (1992). Categorization of individuals on the basis of multiple social features. *Journal of Personality and Social Psychology, 62,* 207–218.

West, R. & Alain, C. (1999). Event-related neural activity associated with the Stroop task. *Cognitive Brain Research, 8,* 157–164.

Wheeler, M. E. & Fiske, S. T. (2005). Controlling racial prejudice: Social cognitive goals affect amygdala and stereotype activation. *Psychological Science, 16,* 56–63.

Wijers, A., Mulder, G., Okita, T., Mulder, L. J. M., & Scheffers, M. (1989). Attention to color: An analysis of selection, controlled search, and motor activation, using event-related potentials. *Psychophysiology, 26,* 89–109.

Wolsko, C. V., Park, B., Judd, C. M., & Wittenbrink, B. (2000). Framing interethnic ideology: Effects of multicultural and color-blind perspectives on judgments of groups and individuals. *Journal of Personality and Social Psychology, 78,* 635–654.

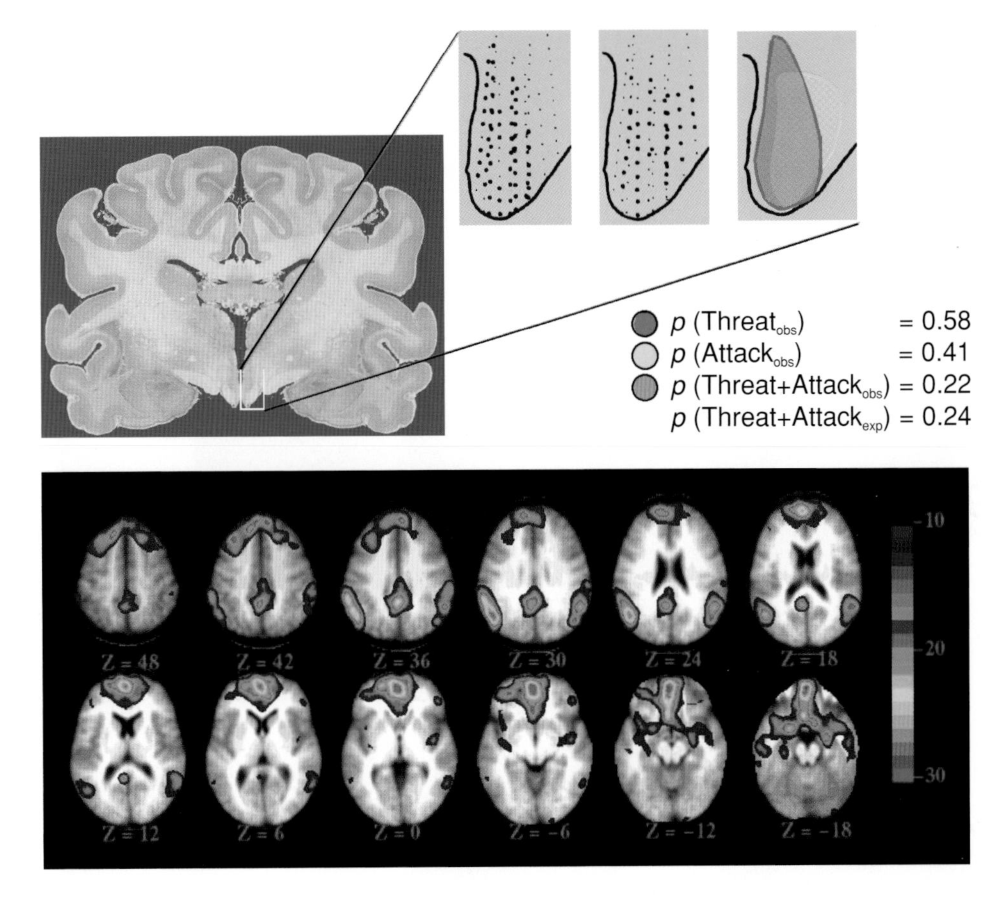

Plate 1

Stimulation-induced behaviors. Distribution of effective (large dots) and ineffective (small dots) electrode points yielding defensive threat and predatory attacklike responses in a hypothalamic-stimulation mapping study in animals. Coronal section of the brain on the left illustrates the relevant area, which is expanded to the right. Observed probabilities of threat [p (Threat$_{obs}$)] and attack [p (Attack$_{obs}$)] are illustrated, together with the probability of both threat and attack responses elicited by the same electrode [p (Threat+Attack$_{obs}$)]. The latter compares closely with the expected probability [p (Threat+Attack$_{exp}$)] based on the joint occurrence of independent events. Results suggest an overlapping distribution of specific differentiated behavioral systems. Results further show that stimulation of a single locus can induce more than one response and that stimulation of divergent points within a distribution field can induce the same behavior. See chapter 1.

Plate 2

Transverse images of PET data averaged across nine cognitive tasks (n = 134) demonstrating the pattern of the common (impersonal task-induced) decreases. See chapter 3.

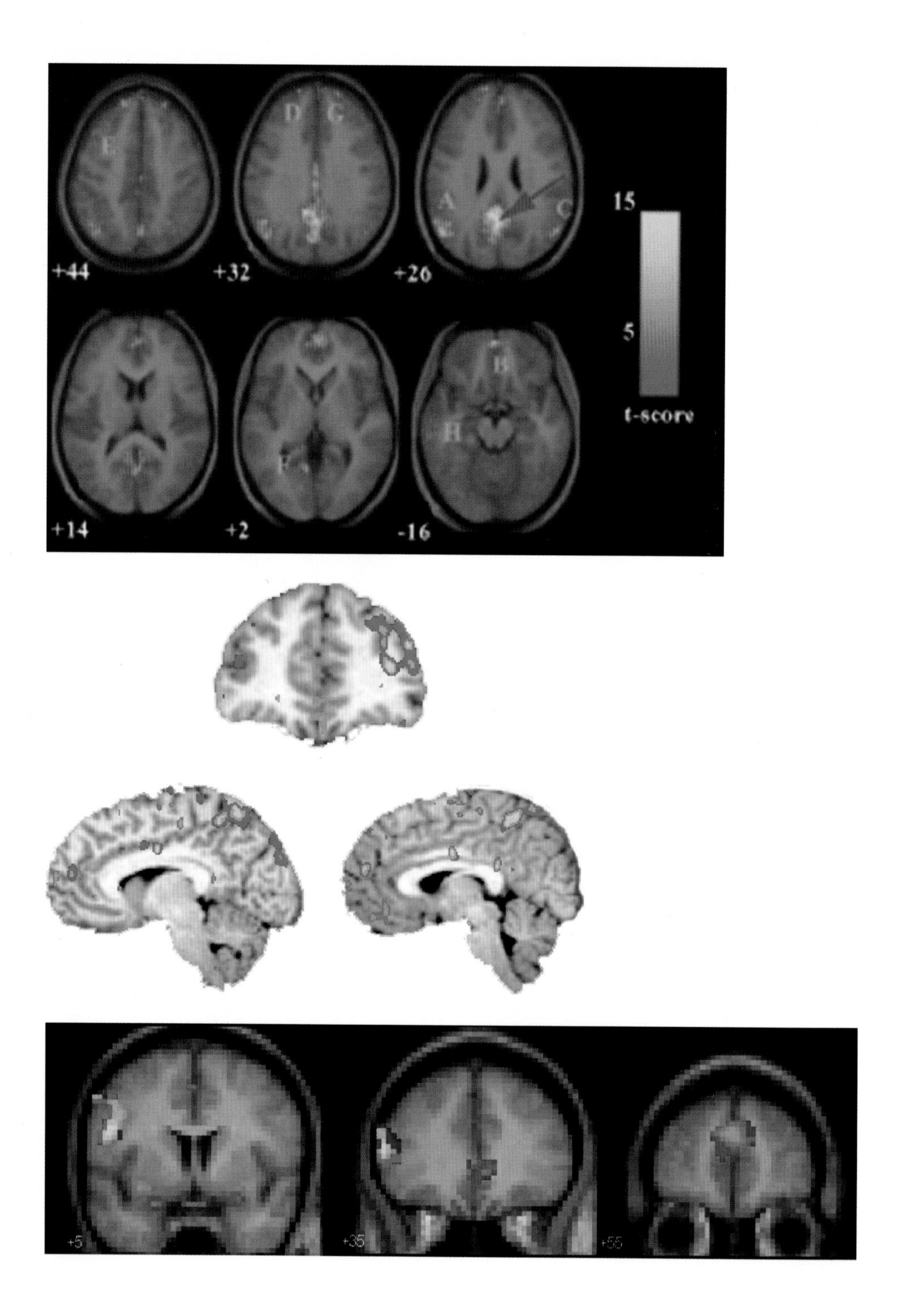
D
G
E
A
C
15
5
t-score
B
H
F
+44
+32
+26
+14
+2
-16
+5
+35
+55

Plate 3

Map of the resting-state neural connectivity for the posterior cingulate cortex (PCC) (blue arrow indicates the PCC peak used for the analysis). Labels A–H designate the significant clusters in order of descending *t* score (Greicius et al., 2003). Note correspondence of this map with the pattern of the common decreases (plate 2). See chapter 3.

Plate 4

In the single subject whose image data is displayed here (contrasting the condition of task with camera on to that with camera off), one observes increased activity in portions of medial prefrontal and medial parietal (precuneus) cortices, consistent with enhanced attention to self-evaluative processes in this behavioral context. Activations are also seen in the dorsal anterior cingulate, which has been implicated in the affective processing of physical pain (Rainville et al., 1997) and social pain (Eisenberger, Lieberman, & Williams, 2003), as well as in the left dorso-lateral prefrontal cortex (typically associated with aspects of working memory) and left ventral prefrontal cortex, both of which are implicated in control strategies, including those involved in emotional self-regulation (Ochsner et al., 2002), which are presumably engaged in this context of becoming more self-conscious, as the subject tries to persist in successful task performance. See chapter 3.

Plate 5

Regions of the prefrontal cortex that were differentially engaged by semantic judgments of objects and of people (Mitchell, Heatherton, & Macrae, 2002). Replicating earlier studies of neural systems that subserve semantic knowledge, object judgments engaged an extensive region of left ventrolateral PFC (red-orange scale). In contrast, person judgments were associated with activity in qualitatively distinct regions of the prefrontal cortex, specifically, medial PFC (blue-green scale). T-maps from comparisons of object and person judgments are overlaid on coronal slices (y values = 5, 35, and 55, respectively) of participants' mean normalized brain. See chapter 4.

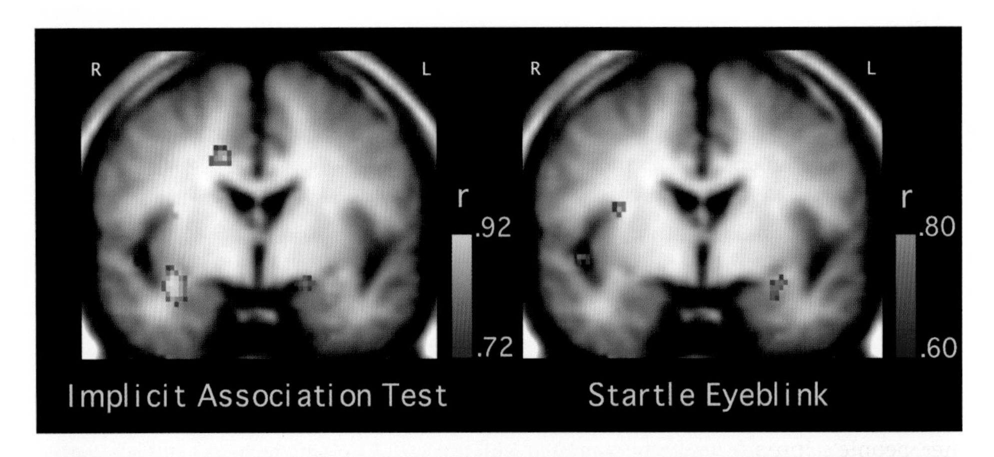

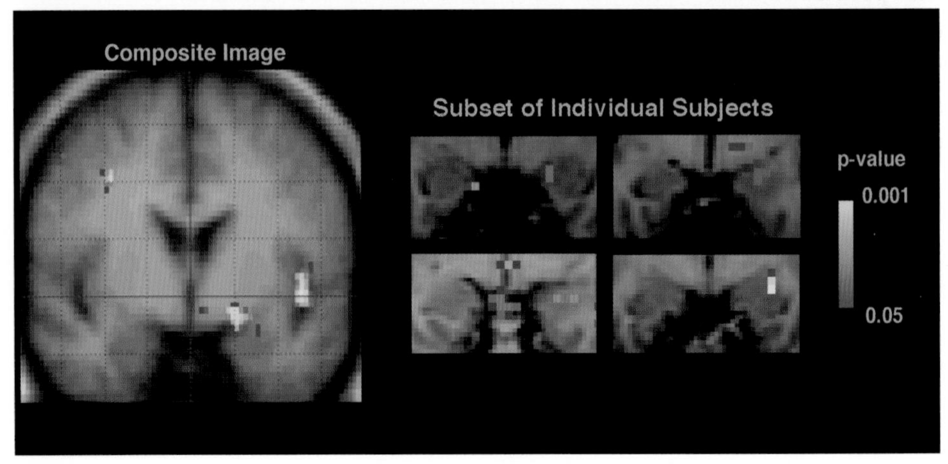

Plate 6

Correlation maps indicating regions where the differential BOLD response to Black versus White was significantly correlated with the strength of implicit race bias as assessed by the IAT (left) and startle eyeblink (right). (Reprinted from Phelps et al., 2000.) See chapter 12.

Plate 7

Activation to instructed fear. A group composite map (left) and selected individual subjects (right) indicate regions with a significantly greater BOLD response to a "threat" stimulus (paired with possibility of shock) versus a "safe" stimulus (no possibility of shock). (Reprinted from Phelps et al., 2001.) See chapter 12.

11 Race and Emotion: Insights from a Social Neuroscience Perspective

Nalini Ambady, Joan Y. Chiao, Pearl Chiu, and Patricia Deldin

Successful human social interaction relies on each individual's ability to understand other people's intentions, beliefs, and desires (Baron-Cohen, 1988). An important building block for understanding others' mental states is being able to recognize how they feel. We can infer this in many ways, for example, from a person's tone of voice or social context. Another important way is through deciphering facial expressions. Endowed with a flexible, sophisticated set of facial muscles, human and nonhuman primates use these expressions as a primary means of communicating emotions (Darwin, 1872).

Although the ability to express and recognize emotions is to a large extent universally shared across human and nonhuman primates, social experience, and in particular racial or cultural experience, moderates how well an individual recognizes these emotions in others. For example, emotional expressions expressed by members of a different species or racial group are often inaccurately recognized (Elfenbein & Ambady, 2002a). Cultural differences in display rules indicating when it is appropriate to communicate certain kinds of emotions are another example of how social experience can shape the ability to recognize emotions. Differences in the capacity to recognize emotions accurately may explain social misunderstandings and communication difficulties that can arise during interracial interactions.

The Origins of Emotion Recognition

Recognizing and expressing emotions through the face is a robust and innate human ability. From the beginning of life, infants show proclivity to express facial emotions, imitate the facial movements of others (Oster & Ekman, 1978; Meltzoff & Moore, 1983), and discriminate between positive and negative expressions (Nelson, Morse, & Leavitt, 1979). By age five years children learn to make finer-grained distinctions among facial expressions and recognize them beyond the positive-negative dimension

(Widen & Russell, 2003). Even children who are blind and deaf and have no visual or auditory notion of what it is like to communicate an emotion create spontaneous facial expressions that greatly resemble those displayed by normal children (Eibl-Eibesfeldt, 1970). Their ability to generate emotional expressions recognizable by others is compelling evidence of an innate mechanism for communicating emotions through the face. Darwin (1872) first noted over 100 years ago that humans share distinct facial morphology and expressions with nonhuman primates. For example, rhesus macaques display facial threat by widening their eyes and mouth in a fashion very similar to a human expression of fear (Knapp & Hall, 1997). Finally, people from various cultures are able to recognize emotional expressions such as happiness, fear, anger, sadness, disgust, and surprise at above-chance levels of accuracy (Ekman, 1992, 1994; Izard, 1994).

The potent ability to recognize emotions expressed by the face is supported by a complex but relatively discrete network of bodily systems including brain regions within and connected to the limbic system. Numerous brain imaging and patient studies show that the amygdala plays a role in the perception, detection, and subsequent recognition of fearful facial expressions (Phan et al., 2002; Adolphs, 2002). Studies using neuroimaging and tensor magnetic stimuli provide converging evidence that medial frontal gyri are critically involved in recognizing angry expressions (Phan et al., 2002; Harmer et al., 2001). Other imaging studies suggest that the basal ganglia is engaged during recognition of happy expressions and other pleasant visual stimuli (Phan et al., 2002).

How do these brain regions interact during emotion recognition? Evaluation of emotional content in faces is a rapid process that occurs very early in the perceptual processing stream. Affective evaluation of faces along positive and negative dimensions occurs as early as 160 msec after stimulus presentation (Pizzagalli et al., 2002). Perceiving fearful facial expressions, in particular, modulates neural responses in frontocentral regions even earlier, at approximately 120 msec (Eimer & Holmes, 2002). Thus, it appears that the neural processes associated with emotion recognition occur early.

In summary, substantial evidence indicates that the human capacity to express and recognize emotions through the face has evolutionary roots, is shared across human cultures, and has dedicated neural machinery.

Emotion Recognition and the Role of Social Experience

In addition to emotion recognition through facial expressions, social experience influences this process. The face conveys a multitude of social information, not only about

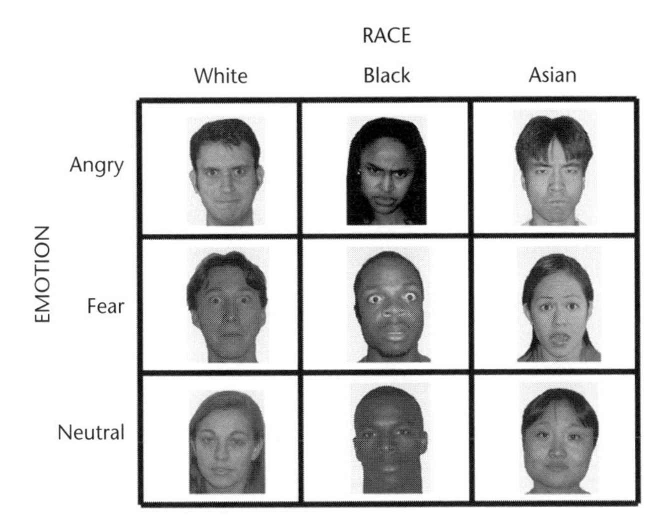

Figure 11.1
Sample stimuli of different racial emotional expressions in functional magnetic resonance imaging and event-related potential experiments on emotion recognition.

internal attributes such as how someone feels, but also external features (age, gender, race) that shape a person's identity by influencing how one sees oneself and how others construe one (Fiske & Neuberg, 1990). Indeed, behavioral findings suggest that such social attributes as race and cultural group membership can influence emotion recognition.

For instance, people can recognize emotions in faces from all cultures at striking levels of accuracy (Ekman, 1992). However, they recognize emotions most accurately in members of the same cultural group relative to other cultural groups (Elfenbein & Ambady, 2002b; figure 11.1). Several factors are thought to moderate this recognition advantage, including differences in attitude toward one's own cultural group relative to other groups, and quantity and quality of familiarity with other cultures[1] (Elfenbein & Ambady, 2002a). Moreover, cultural standards for when it is appropriate to convey a particular emotion can influence when and how much an emotion is publicly expressed, which may in turn affect the internal experience of that emotion (Ekman & Friesen, 1969; Hochschild, 1979).

Race and the Brain

The investigation of neural and behavioral responses to in-group and out-group stimuli has to date focused exclusively on reactions to neutral faces of different races, to imagined partners of a different race within a neutral context, or to race-specific words (Hart et al., 2000; Golby et al., 2001; Phelps et al., 2000; Richeson et al., 2003). However, given the impact of emotion on social perception and interaction (Hess, Barry, & Kleck, 2000; Keltner, Ellsworth, & Edwards, 1993; Vrana & Rollock, 1998, 2002), and the particularly rapid, possibly unconscious, processing of negative expressions (Fox et al., 2000; Dimberg & Oehman, 1996; White, 1996), both the racial and emotional salience of a target face are likely to affect neural and behavioral responses.

Despite substantial evidence for the impact of emotional states on prejudiced attitudes and stereotypical judgments (Jackson et al., 2001; Asuncion & Mackie, 1996; Bodenhausen, Kramer, & Susser, 1994; Bodenhausen, Sheppard, & Kramer, 1994; Lambert et al., 1997), and evidence that physiological reactivity of perceivers during social interaction varies according to both racial and emotional contexts (Vrana & Rollock, 1998, 2002), scant attention has been given to the emotion expressed by targets of such judgment (Vaes et al., 2003). Indeed, little previous work to our knowledge has examined perceivers' responses to in-group and out-group members expressing different emotions (Hugenberg & Bodenhausen, 2003). This is a surprising oversight given considerable evidence indicating rapid and efficient processing of facial expressions, particularly negative expressions (Dimberg & Oehman, 1996; Fox et al., 2000; White, 1996).

We conducted functional magnetic resonance imaging (fMRI) and event-related brain potential[2] (ERP) studies in our laboratory that emphasize not only the impact of race on emotional processing, but also the influence of emotional expression on evaluation of in-group and out-group members. Our data underscore both the importance of emotional expression on how a target is appraised and also the utility of using converging measures to clarify social phenomena and processes that contribute to social behavior.

Does Race Affect Brain Processes during Emotion Recognition?

To examine how racial group membership affects brain processes during emotion recognition, Chiao and colleagues (2004) conducted an event-related fMRI in eight Caucasian participants (4 men) while they explicitly identified fear, anger, and neutral expressions in faces of Caucasian, Asian-American, and African-American men and women. Participants were shown each facial expression for 750 msec and responded within 2500 msec, pressing an appropriate button to indicate which emotion the face was expressing. We predicted that participants would recognize all expressions but that they would be most accurate at recognizing those in faces of their own race.

Consistent with previous behavioral work, all emotional expressions were recognized at better-than-chance accuracy levels; however, Caucasian participants recognized neutral faces better than fearful and angry faces. They were also best at recognizing emotions in Caucasian and Asian-American faces relative to African-American faces, specifically, fear and anger.

Neuroimaging results[3] revealed greater amygdala activity in response to Caucasian and Asian-American faces showing fear relative to African-American faces. Caucasian expressions of anger elicited increased signal change in medial frontal cortex relative to Asian-American and African-American anger. These findings suggest that neural regions specifically involved in recognizing fear and anger show differences in signal change depending on the race of the person expressing the emotion.

Chiao, Lowenthal, and Ambady (unpublished data) examined when neural processes involved in emotion recognition are influenced by racial group membership. The vertex positive potential (VPP) of ERPs was used to study fourteen Caucasian participants while they viewed expressions and judged whether or not each face was angry, fearful, or neutral. We hypothesized that race of the facial target would influence basic structural face processing approximately 170 msec after stimulus onset and this would be observable in the amplitude of the VPP, an ERP that is critical to face processing (Bentin et al., 1996; Jeffreys, 1989, 1996). Moreover, we predicted that the emotional expression being processed would affect the extent to which race influenced neural processing. To investigate this hypothesis, participants were asked to self-report their overall exposure to Caucasians, Asian-Americans, and African-Americans on a Likert scale of 1 to 7; they reported most exposure to Caucasians and least to African-Americans. Participants detected angry expressions most accurately in African-American and Caucasian faces relative to Asian-American faces, but they recognized fear most accurately in Caucasian faces relative to African-American and Asian-American faces. Finally, neutral expressions were recognized equally well across the three racial groups.

The VPP amplitude was sensitive to both race and emotion of the face. It was greatest for all African-American faces, regardless of the emotion being expressed. Furthermore, angry expressions yielded the most positive amplitude for African-American and Asian-American faces relative to Caucasian faces. For fear expressions, African-American faces yielded the greatest positive amplitude in the VPP relative to Caucasian and Asian-American faces. For anger expressions, African-American and Asian-American faces yielded greater amplitude relative to Caucasian faces. For neutral faces, VPP amplitude was similar across all different race faces.

Taken together, these neuroimaging and ERP data suggest that race affects brain processes involved in recognizing of fear and anger. First, regions important in the successful recognition of fear and anger show modulation of signal change based on the race of the expressor. Second, behavioral and neural evidence suggests that not all outgroup faces are processed alike. The amygdala and VPP respond differently to African-American fear faces relative to Caucasian and Asian-American fear faces. However, Asian-American and African-American anger faces are processed more similarly relative Caucasian anger faces.

Differences in behavioral and neural responses to out-group faces may be a result of several factors. Caucasians and Asian-Americans may have greater exposure to each other, thus leading to better recognition of emotions expressed by those groups relative to African-Americans (Elfenbein & Ambady, 2002a). Furthermore, social groups often vary in social status relative to each other. This observation was first made in reference to the power differentials witnessed between the sexes; however, social psychologists and sociologists have also applied this subordination hypothesis to racial groups as they vary not only in socioeconomic status but also their historical progressions, which influence the degree to which these minority out-groups are stigmatized by the majority. Of importance, neuroimaging and ERP evidence shows that differences seen at the group level are also detectable at the neural level within individuals. It has yet to be determined definitively whether these neural differences in processing of emotion expressed by individuals of different races arise from variations in racial exposure or social status, or additional factors.

Together, our data are among the first to indicate that the emotional expression of a racially salient target influences processing of different out-group members at not only the behavioral but also the physiological level. As we describe below, using neuroscience techniques to investigate the influence of race and emotion on the processing of one another may also illuminate such costly social phenomena as prejudice. Specifically, theories of prejudice emphasize that a combination of factors that make up a stereotype may in turn influence behavior toward out-group members (Fiske et al., 2002; Fiske, 1998). That is, the degree to which prejudiced behavior is manifested may depend on not just group membership of the target (out-group African-American) but also the extent to which the target possesses other qualities (warmth or competence; Fiske et al., 2002). Given that facial expressions are processed differently depending on the valence of the expression, that prejudiced behavior is influenced by the particular composition of qualities portrayed by an out-group target, and that emotion clearly influences social behavior, an investigation of the interactions among race, emotion, and prejudice is warranted.

ERP Differentiation to In-Group and Out-Group Facial Expressions: Insights for Understanding Prejudice

Intriguing work on the neural responses to outgroup faces evaluated correlations with implicit measures of prejudice on passive viewing tasks of neutral faces (Phelps et al., 2000; Richeson, et al., 2003). A series of studies from our laboratory examined cortical and behavioral responses of high- and low-prejudiced individuals to in-group and out-group emotional stimuli. In contrast with previous investigations that used passive viewing tasks to examine neural responses to out-group stimuli, we employed an active evaluation task in which participants were asked to make a socially relevant judgment (do I want to work with this person?) regarding in-group and out-group members (Chiu, Ambady, & Deldin, 2004).

Both behavioral responses and the contingent negative variation (CNV) component of the ERP were measured in high- and low-prejudiced individuals (selected on the basis of their responses on the modern racism scale; McConahay, Hardee, & Batts, 1981) who were asked to make evaluative judgments of emotionally and racially salient facial stimuli (figure 11.2). The CNV is typically elicited by a warning

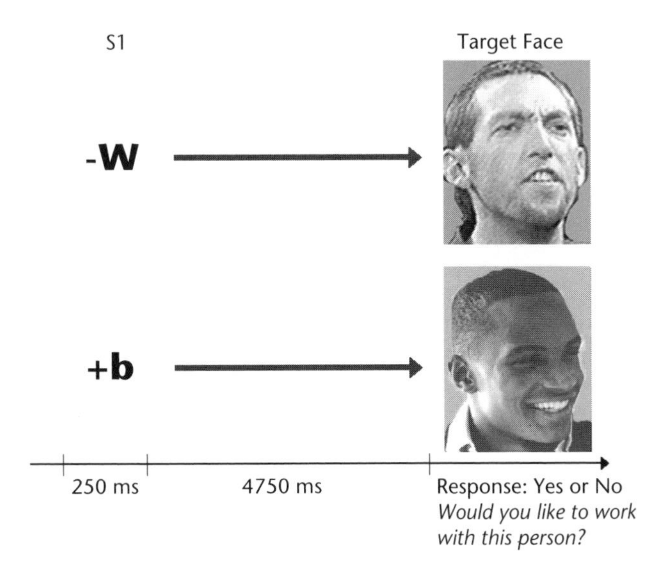

Figure 11.2
Schematic representation of CNV task. Each trial consisted of a warning stimulus (–w, +w, –b, +b, r) presented for 250 msec followed 4750 msec later by a corresponding target face (happy white, angry white, etc.). Participants were asked to respond yes or no according to their preference for working with the individual.

stimulus that requires anticipation of a target stimulus (Walter et al., 1964; Picton & Hillyard, 1988). The early component of the CNV is thought to index initial attention to information carried by the warning stimulus, the expected degree of expenditure of cognitive effort to respond to the target stimulus, and the degree of motivation to respond to the target stimulus (Low & McSherry, 1968; Forth & Hare, 1989; Hamon & Seri, 1987). Moreover, the presence of the early CNV is generally thought to be a cortical reflection of controlled, rather than automatic, psychological processes in response to an S1 that requires anticipation of a subsequent S2 (Picton & Hillyard, 1988; Shiffrin & Schneider, 1977). Several groups demonstrated the sensitivity of the CNV to the anticipation of affective stimuli and successfully used the CNV to identify individual and group differences in distinct components of information processing that reflect the subjective significance of anticipated stimuli (Rockstroh et al., 1979; Klorman & Ryan, 1980; Yee & Miller, 1988; Regan & Howard, 1995).

Our behavioral and ERP data indicate not only that high- and low-prejudiced individuals are differently influenced by the affective relevance of in-group and out-group members, but also that the affective nature of target stimuli may be especially salient for low-prejudiced individuals (figure 11.3). Specifically, low-prejudiced individuals showed an increased CNV not only to angry out-group stimuli, but also in anticipation of angry faces more generally, compared with happy faces. Longer behavioral response latencies of the low-prejudiced group in evaluating angry black targets further indicate this enhanced processing. Together, these data support, and extend to include emotion, theories of prejudice proposing that these individuals monitor automatic

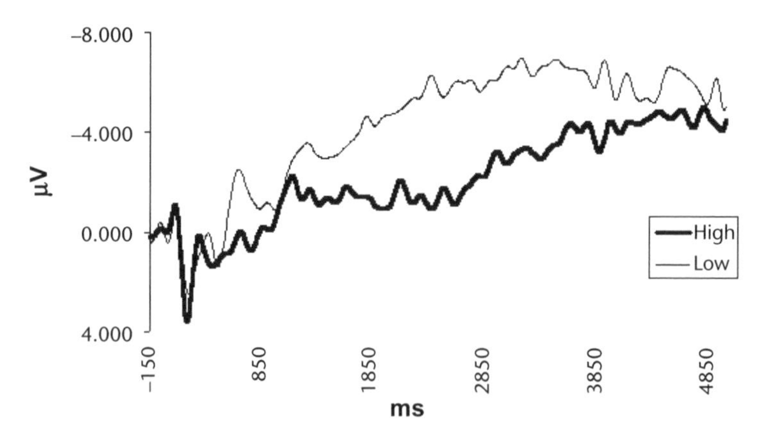

Figure 11.3
Contingent negative variation amplitude at site Cz of high- and low-prejudiced individuals to S1, indicating the subsequent presentation of an angry black American face (S1 = "–b").

reactions to negative stereotypes elicited by out-group stimuli (Bodenhausen & Macrae, 1998; Devine, 1989; Monteith, Devine, & Zuwerink, 1993; Plant & Devine, 1998).

In comparison, the high-prejudiced group showed decreased CNV in anticipation of angry black targets compared with all other targets, supporting theories that individuals high in explicit prejudice may be characterized by a decreased tendency, or motivation, to monitor automatic prejudiced responses to negative stereotypes (Bodenhausen & Macrae, 1998; Monteith et al., 1993; Plant & Devine, 1998). The shorter behavioral response latencies of the high-prejudiced group to angry black targets further reflects absence of effortful suppression of prejudiced behavior. In further comparison, the high-prejudiced group showed an enhanced CNV in anticipation of happy white targets; this suggests a greater recruitment of cognitive resources to respond to happy white stimuli, and is consistent with theories proposing that prejudiced individuals may expend extra effort to make individuating responses when required to evaluate in-group stimuli, and that less effort in individuating out-group members may contribute to the expression of prejudice (Brewer, 1999; Miller & Brewer, 1986). Indeed, high-prejudiced individuals showed enhanced CNV in anticipation of a specific nonthreatening in-group stimulus (happy white), but not a more general response to happy faces.

Our data augment the small but growing literature investigating neural concomitants of race perception and race bias. In the one fMRI study (Phelps et al., 2000) that employed measures of both implicit and explicit race bias, positive correlations between amygdala activity and race bias were found only on implicit (implicit association test; startle potentiation) and not explicit measures of race evaluation (modern racism scale, MRS). In contrast, in our work (Chiu et al., 2004), CNV amplitudes distinguished participants with high and low scores on an explicit measure of racial bias (MRS) such that low-, compared with high-prejudiced individuals showed greater cortical activity to angry black targets. At first glance, these data may seem at odds with those of Phelps et al. However, most of their participants scored below 2 on the MRS, suggesting that their participants may be comparable with our low-prejudice group. Indeed, within this group, the data show greatest cortical resources to angry black stimuli, compared with all other targets.

A study investigating possible mechanisms by which race bias may be suppressed provides further evidence for the role of controlled processing in prejudice (Richeson et al., 2003). Briefly, participants with high scores on the implicit association test of racial bias had greater activity in brain regions associated with cognitive control when viewing out-group faces. The authors noted that this pattern of results was counterintuitive, and suggested that it may not be race bias per se that correlates with activity in brain regions associated with cognitive control, but rather greater concern with exhibiting overt signs of prejudice that is subsequently reflected in the recruitment of cognitive control in order to suppress prejudicial behavior (Richeson et al., 2003; Richeson & Shelton 2003; see also Gehring, Karpinkski, & Hilton, 2003). Such an interpretation indicates that these individuals may score low on explicit measures of prejudice and, as proposed by current perspectives on prejudice, may be more motivated to control prejudiced reactions to negative stereotypes, devote more cognitive resources to monitoring their behavior toward out-group members (Blascovich et al., 1997; Devine, 1989; Devine et al., 1991; Fazio et al., 1995), and thus show greatest activity in brain regions associated with cognitive control, as reported by Richeson et al. Clearly, it would be of theoretical interest to explicate further the nature of prejudice by examining whether individuals who score high on explicit measures of racial prejudice show differentiable subcortical activation to in-group and out-group emotional faces, and also to investigate the relationship between implicit and explicit measures of race evaluation and cortical processes.

It is further notable that in our sample the high- and low-prejudiced groups exhibited no differences in behavioral or physiological responses to the evaluation of happy out-group stimuli (Hugenberg & Bodenhausen, 2003, did not specifically investigate participants' identification of happy expressions). Indeed, the differences between these groups were enhanced or attenuated by simply varying the anticipated and actual valence of facial expression from angry to happy. The pattern of null group differences to happy out-group stimuli, in conjunction with the robust between-group differences to angry out-group stimuli, is striking and suggests at least that prejudice and stereotyping are not unitary phenomena and may be malleable. It should be emphasized, however, that no analyses within race or emotion alone yielded significant between-group effects; thus, both race and emotion, and likely an interaction between them, appear critical in influencing behavioral, attitudinal, and physiological responses toward in-group and out-group members.

Preliminary data from two other ERP studies from our laboratory supplement these findings. First, high- and low-prejudiced individuals appear to engage different neural resources and cognitive processes when confronted with racial stereotypes and violation of these race-based expectancies. Briefly, we measured N400 and late positive components (LPCs) of the ERP as high- and low-prejudiced individuals viewed sentences that confirmed or violated racial stereotype-based expectancies ("Jamaal's favorite sport is lacrosse," versus "Jamaal's favorite sport is basketball"). The N400 is typically enhanced in response to the presentation of context incongruent words relative to context congruent words, and is thus thought to index both lexical and conceptual-level information processing (Halgren, 1990; Kutas & Van Petten, 1994; Osterhout & Holcomb, 1995; Rugg, 1990). In sentence-comprehension paradigms, the component temporally after the N400, variously termed the LPC, P600, or syntactic positive shift, is not only sensitive to syntactic violations, but also reflects subjective probability and the ease with which new information is integrated into one's representation of the environment (as part of the P300 family; Donchin, 1981; Donchin & Coles, 1988; Coulson, King, & Kutas, 1998; Gunter, Stowe, & Mulder, 1997). Our data indicate that out-group stereotypes may be equally represented in the cognitive representations of high- and low-prejudiced individuals, and, for all participants, more salient than in-group stereotypes (greater N400 to violation of out-group versus in-group expectancies; lack of between-prejudice differences). Notably, although the representation of in-group and out-group stereotypes, respectively, may be equivalent between high- and low-prejudiced individuals, the cognitive processes involved in resolving these stereotypes appear to be quite different. Specifically, low-prejudiced individuals exhibited enhanced LPC to violations of out-group race-based expectancies, whereas high-prejudiced individuals exhibited enhanced LPC to violations of in-group expectancies (Chiu et al., unpublished observations).

In another investigation of the influence of race and emotion on the processing of each, we examined the cognitive processes and temporal resolution of cortical physiology as high- and low-prejudiced individuals viewed images of in-group and out-group emotional faces. Preliminary analyses indicate that racial and emotional differentiation occur as early as 120 msec after stimulus presentation (as evidenced by more positive voltages for out-group relative to in-group faces, and delayed latencies to angry faces compared with happy faces), and effects of target gender are seen even earlier (~80 msec; female targets elicited more negative voltage than male targets). Of importance, interaction effects including race, emotion, and level of prejudice are evident at all stages of processing, beginning as early as 80 msec after stimulus presentation. Although data analyses from this study are still under way, preliminary findings unambiguously highlight the mutual and respective influence of race and emotion on person perception and social behavior (Chiu et al., unpublished observations).

Conclusion

Studies in our laboratory emphasize not only the impact of emotion on the perception of members of in-groups and out-groups but also, we believe, the utility of converging measures to clarify the impact of race and emotion on social behavior. Indeed, self-report measures are vulnerable to selective distortion by self-presentational and experimenter biases, especially with regard to socially questionable phenomena such as prejudice. A neuroscience approach to understanding social phenomena allowed progress relatively unhindered by such biases. Nevertheless, we concur with others (Cacioppo et al., 2003) who caution against the use of brain imaging tools simply to show that there are biological concomitants of social phenomena; indeed, that changes in social behavior correspond with physiological differences is expected. Instead, we draw on the long tradition of using ERPs as quantifiable measures of the temporal manifestation of cortical resources and specific cognitive processes that may not be readily accessible by isolated use of traditional behavioral and self-report measures. We believe that these may provide both converging evidence for existing theories of social behavior and also supply fuel for new conceptualizations. For example, our data suggesting that high- and low-prejudiced individuals do not differ at the N400 component but do differ at the LPC in response to violations of racial expectancies, provide evidence not only that race-based information may trigger a societally constrained conceptual representation of race, but also that the manifestation of prejudice depends how individuals process this information. Moreover, the specificity of our between-group findings to the early, but not the late, component of the CNV provides intriguing insight about when and how in the processing of social information high- and low-prejudiced individuals may begin to diverge. That is, substantial evidence indicates that the late CNV is the sum of motor (response preparation) and nonmotor (cognitive anticipation of a task-relevant stimulus) components (Damen & Brunia, 1994; van Boxtel & Brunia, 1994). Thus, given that the early CNV is thought to reflect sustained processing that extends to the late CNV and also the greater early CNV in the low-prejudiced group, a relatively smaller contribution of motoric preparation to the late CNV in anticipation of angry black targets is evidenced in this group compared with the high-prejudiced group.

The brain's functional anatomy places important constraints on psychological theories of racial and emotional processing, and their relationship in social behavior. Thus, a social neuroscience approach integrating the spatial resolution of fMRI with the temporal resolution of ERPs along with traditional behavioral and self-report measures not only facilitates a comprehensive description of the functional anatomy of racial and emotional processing, but also encourages development of more comprehensive theoretical models of prejudice and other social phenomena, uses current knowledge of brain function to test hypotheses about the processes underlying social phenomena that may not otherwise not be assessed, and allows an assessment of independent and interactive contributions of emotional, race, and cognitive factors to social discrimination. Our series of studies emphasizes not only the impact of emotion on the perception of members of in-groups and out-groups but also, we believe, the utility of converging measures to clarify the impact of race and emotion on social behavior.

The main purpose of this work is to show that facial expressions of emotion affect both neural and behavioral responses to in-group and out-group faces, and that a social neuroscience approach is of utility for understanding factors that contribute to social behavior. Indeed, findings from our laboratory are among the first to indicate that the emotional expression of a racially salient target influences processing of in-group and out-group members at not only the behavioral, but also the physiological level, and the first to show neural differentiation between individuals who score high and low on explicit measures of racial prejudice. Together, these data underscore both the importance of emotional expression on how social targets are appraised and also the utility of using converging measures to clarify processes that may contribute to social behavior. These investigations are, however, a first step toward understanding the respective contributions of racially and emotionally salient features of individuals to how we perceive and interact with each other.

Acknowledgments

Supported by NSF PECASE grant to N.A. and NSF predoctoral fellowship to J.Y.C. We thank Hillary Anger-Elfenbein, Y. Susan Choi, Heather Gray, Abigail Marsh, Ashli Owen-Smith, and Jennifer Steele for their insightful feedback on this work.

Notes

1. As indexed by geographical distance and self-report.

2. An extensive and rapidly growing literature in cognitive psychophysiology suggests that event-related brain potentials (ERPs), voltage changes time-locked to stimulus presentation, may be a particularly useful tool for exploring the cognitive and emotional processes that may be associated with social behavior. The amplitude and latency of these voltages changes are thought to reflect the cognitive processing associated with the presentation, or pending presentation, of discrete events. Relative immunity to demand characteristics renders ERPs of particular utility for exploring phenomena, such as prejudice, that in purely behavioral paradigms may be especially sensitive to experimenter effects and self-presentation biases. Moreover, since ERPs are considered the gold standard among noninvasive imaging methods for measuring the temporal resolution of the physiological manifestation of psychological processes (Fabiani, Gratton, & Coles, 2000), the temporal pattern of social impression formation and reaction can be examined.

3. Whole-brain analyses were conducted using a general linear model in SPM'99. All contrast analyses were conducted at the $p < 0.001$ level corrected. Due to the limited number of incorrect trials per given condition, all events were included in the contrast analyses regardless of whether or not the event response was correct.

References

Adolphs, R. (2002). Neural systems for recognizing emotion. *Current Opinion in Neurobiology, 12,* 169–177.

Adolphs, R., Tranel, D., & Damasio, A. R. (2003). Dissociable neural systems for recognizing emotions. *Brain Cognition, 52*(1), 61–69.

Asuncion, A. & Mackie, D. (1996). Undermining social stereotypes: Impact of affect-relevant and behavior-relevant information. *Basic and Applied Social Psychology, 18,* 367–386.

Baron-Cohen, S. (1988). Without a theory of mind one cannot participate in a conversation. *Cognition, 29*(1), 83–84.

Bentin, S., Allison, T., Puce, A., Perez, A., & McCarthy, G. (1996). Electrophysiological studies of face perception in humans. *Journal of Cognitive Neuroscience, 8,* 551–565.

Blascovich, J., Wyer, N., Swart, L., & Kibler, J. (1997). Racism and racial categorization. *Journal of Personality and Social Psychology, 72,* 1364–1372.

Bodenhausen, G., Kramer, G., & Susser, K. (1994). Happiness and stereotypic thinking in social judgment. *Journal of Personality and Social Psychology, 66,* 621–632.

Bodenhausen, G. & Macrae, N. (1998). Stereotype activation and inhibition. In R. Wyer (Ed.), *Advances in Social Cognition* (pp. 1–52). Hillsdale, NJ: Erlbaum.

Bodenhausen, G., Sheppard, L., & Kramer, G. (1994). Negative affect and social judgment. *European Journal of Social Psychology, 24,* 45–62.

Brewer, M. (1999). The psychology of prejudice: Ingroup love or outgroup hate? *Journal of Social Issues, 55,* 429–444.

Brunia, C. & Damen, E. (1988). Distribution of slow-potentials related to motor preparation and stimulus anticipation in a time estimation task. *Electroencephalography and Clinical Neurophysiology, 69,* 234–243.

Cacioppo, J., Berntson, G., Lorig, T., Norris, C., Rickett, E., & Nusbaum, H. (2003). Just because you're imaging the brain doesn't mean you can stop using your head: A primer and set of first principles. *Journal of Personality and Social Psychology, 85,* 650–661.

Chiu, P., Ambady, N., & Deldin, P. (2004). Contingent negative variation to emotional in- and out-group stimuli differentiates high- and low-prejudiced individuals. *Journal of Cognitive Neuroscience, 16,* 1830–1839.

Coulson, S., King, J. W., & Kutas, M. (1998). ERPs and domain specificity: Beating a straw horse. *Language and Cognitive Processes, 13*(6), 653–672.

Damen, E. & Brunia, C. (1994). Is a stimulus conveying task-relevant information a sufficient condition to elicit a stimulus-preceding negativity. *Psychophysiology, 31*, 129–139.

Darwin, C. (1872). *The Expression of Emotion in Man and Animals*. New York: Appleton.

Devine, P. (1989). Stereotypes and prejudice: Their automatic and controlled components. *Journal of Personality and Social Psychology, 56*, 5–18.

Devine, P., Monteith, M., Zuwerink, J., & Elliot, A. (1991). Prejudice with and without compunction. *Journal of Personality and Social Psychology, 60*, 817–830.

Dimberg, U. & Oehman, A. (1996). Behold the wrath: Psychophysiological responses to facial stimuli. *Motivation and Emotion, 20*, 149–182.

Donchin, E. (1981). Surprisep . . . Surprise? *Psychophysiology, 18*(5), 493–513.

Donchin, E. & Coles, M. G. (1998). Is the P300 component a manifestation of context updating? *Behavioral and Brain Sciences, 11*(3), 357–427.

Eibl-Eibesfeldt, I. (1970). *Ethology: The Biology of Behavior*. Oxford, England: Holt, Rinehart.

Eimer, M. & Holmes, A. (2002). An ERP study on the time course of emotional face processing. *Neuroreport, 13*(4), 427–431.

Ekman, P. & Oster, H. (1979). Facial Expressions of Emotion. *Annual Review of Psychology, 30*, 527–554.

Ekman, P. (1992). Are there basic emotions? *Psychololgical Review, 99*(3):550–553.

Ekman, P. (1994). Strong evidence for universals in facial expressions: A reply to Russell's mistaken critique. *Psychological Bulletin, 115*(2), 268–287.

Elfenbein, H. A. & Ambady, N. (2002a). On the universality and cultural specificity of emotion recognition: A meta-analysis. *Psychological Bulletin, 128*, 203–235.

Elfenbein, H. A. & Ambady, N. (2002b). Is there an ingroup advantage in emotion recognition? *Psychological Bulletin, 128*, 243–249.

Fabiani, M., Gratton, G., & Coles, M. G. H. (2000). Event-Related Brain Potentials. In J. T. Cacioppo & L. G. Tassinary et al. (Eds.), *Handbook of psychophysiology (2nd ed.)*. (pp. 52–84). New York, NY: Cambridge University Press.

Fazio, R. H., Jackson, J. R., Dunton, B. C., & Williams, C. J. (1995). Variability in automatic activation as an unobstrusive measure of racial attitudes: A bona fide pipeline? *Journal of Personality and Social Psychology, 69*, 1013–1027.

Fiske, S. (1998). Stereotyping, prejudice, and discrimination. In D. Gilbert, S. Fiske, & G. Lindzey (Eds.), *Handbook of Social Psychology* (vol. 2, pp. 357–411), Boston: McGraw-Hill.

Fiske, S., Cuddy, A., Glick, P., & Xu, J. (2002). A model of (often mixed) stereotype content: Competence and warmth respectively follow from perceived status and competition. *Journal of Personality and Social Psychology, 82,* 878–902.

Fiske, S. & Neuberg, S. L. (1990). A continuum of impression formation, from category-based to individuating processes: Influences of information and motivation of attention and interpretation. *Advances in Experimental Social Psychology, 23,* 1–73.

Forth, A. & Hare, R. (1989). The contingent negative variation in psychopaths. *Psychophysiology, 26,* 676–682.

Fox, E., Lester, V., Russo, R., Bowles, R. J., Pichler, A., & Dutton, K. (2000). Facial expressions of emotion: Are angry faces detected more efficiently? *Cognition and Emotion, 14,* 61–92.

Gehring, W., Karpinksi, A., & Hilton, J. (2003). Thinking about interracial interactions. *Nature Neuroscience, 6,* 1241–1243.

Golby, A. J., Gabrieli, J. D. E., Chiao. J. Y., & Eberhardt, J. L. (2001). Differential responses in the fusiform region to same-race and other-race faces. *Nature Neuroscience, 4,* 845–850.

Gunter, T. C., Stowe, L. A., & Mulder, G. (1997). When syntax meets semantics. *Psychophysiology, 34*(6), 660–676.

Halgren, E. (1990). Insights from evoked potentials into the neuropsychological mechanisms of reading. In A. B. Scheibel & A. F. Wechsler (Eds.), *Neurobiology of Higher Cognitive Function. UCLA Forum in Medical Sciences,* No. 29 (pp. 103–150). New York, NY: Guilford Press.

Hamon, J. & Seri, B. (1987). Relation between warning stimuli and contingent negative variation in man. *Activitas Nervosa Superior, 29,* 249–256.

Harmer, C. J., Thilo, K. V., Rothwell, J. C., & Goodwin, G. M. (2001). Transcranial magnetic stimulation of medial-frontal cortex impairs the processing of angry facial expressions. *Nature Neuroscience, 4*(1), 17–18.

Hart, A. J., Whalen, P. J., Shin, L. M., McInerney, S. C., Fischer, H., & Rauch, S. L. (2000). Differential response in the human amygdala to racial outgroup vs. ingroup face stimuli. *Neuroreport, 11,* 2351–2355.

Hess, U., Barry, S., & Kleck, R. (2000). The influence of facial emotion displays, gender, and ethnicity on judgments of dominance and affiliation. *Journal of Nonverbal Behavior, 24,* 265–281.

Hochschild, A. R. (1983). *The managed heart: Commercialization of human feeling.* Berkeley: University of California Press.

Hugenberg, K. & Bodenhausen, G. (2003). Facing prejudice: Implicit prejudice and the perception of facial threat. *Psychological Science, 14,* 640–643.

Izard, C. E. (1994). Innate and universal facial expressions: Evidence from developmental and cross-cultural research. *Psychological Bulletin, 115*(2), 288–299.

Jackson, L., Lewandowski, D., Fleury, R., & Chin, P. (2001). Effects of affect, stereotype consistency and valence of behavior on causal attributions. *Journal of Social Psychology, 141*, 31–48.

Jeffreys, D. A. (1989). A face-responsive potential recorded from the human scalp. *Experimental Brain Research, 78*, 193–202.

Jeffreys, D. A. (1996). Evoked studies of face and object processing. *Visual Cognition, 3*, 1–38.

Keltner, D., Ellsworth, P., & Edwards, K. (1993). Beyond simple pessimism: Effects of sadness and anger on social perception. *Journal of Personality and Social Psychology, 64*, 740–752.

Klorman, R. & Ryan, R. M. (1980). Heart rate, contingent negative variation, and evoked potentials during anticipation of affective stimulation. *Psychophysiology, 17*(6), 513–523.

Knapp, M. L. & Hall, J. A. (1997). *Nonverbal Communication and Human Interaction*, 5th ed. Belmont, CA: Wadsworth.

Kutas, M. & Van Petten, C. K. (1994). Psycholinguistics electrified: Event-related brain potential investigations. In M. A. Gernsbaoher (Ed.), *Handbook of Psycholinguistics* (pp. 83–143). San Diego, CA: Academic Press, Inc.

Lambert, A., Khan, S., Lickel, B., & Fricke, K. (1997). Mood and correction of positive versus negative stereotypes. *Journal of Personality and Social Psychology, 72*, 1002–1016.

Low, M. & McSherry, J. (1968). Further observations of psychological factors involved in CNV genesis. *Electroencephalography and Clinical Neurophysiology, 25*, 203–207.

McConahay, J., Hardee, B., & Batts, V. (1981). Has racism declined in America? It depends on who is asking and what is asked. *Journal of Conflict Resolution, 25*, 563–579.

Meltzoff, A. N. & Moore, M. K. (1983). The origins of imitation in infancy: Paradigm, phenomena, and theories. In L. P. Lipsitt (Ed.), *Advances in Infancy Research* (vol. 2, pp. 265–301). Norwood, NJ: Ablex.

Miller, N. & Brewer, M. (1986). Categorization effects on ingroup and outgroup perception. In J. Dovidio & S. Gaertner (Eds.), *Prejudice, Discrimination, and Racism* (pp. 209–230). San Diego: Academic Press.

Monteith, M., Devine, P., & Zuwerink, J. (1993). Self-directed versus other-directed affect as a consequence of prejudice-related discrepancies. *Journal of Personality and Social Psychology, 64*, 198–210.

Nelson, C. A., Morse, P. A., & Leavitt, L. A. (1979). Recognition of facial expressions by seven-month-old infants. *Child Development, 50*(4), 1239–1242.

Oster, H. & Eleman, P. (1978). Facial behavior in child development. In W. A. Collins (Ed.), *Minnesota Symosium or Child Development* (pp. 231–276). Hillsdale, NJ: Erlbaum.

Osterhout, L. & Holcomb, P. J. (1995). Event related potentials and language comprehension. In M. D. Rugg & M. G. H. Coles (Eds.), *Electrophysiology of Mind: Event-related Brain Potentials and Cognition. Oxford psychology series, No. 25.* (pp. 171–215). London: Oxford University Press.

Phan, K. L., Wager, T., Taylor, S. F., & Liberzon, I. (2002). Functional neuroanatomy of emotion: A meta-analysis of emotion activation studies in PET and fMRI. *Neuroimage, 16,* 331–348.

Phelps, E., O'Connor, K., Cunningham, W., Funayama, E., Gatenby, J., Gore, J., et al. (2000). Performance on indirect measures of race evaluation predicts amygdala activation. *Journal of Cognitive Neuroscience, 12,* 729–738.

Picton, T. & Hillyard, S. (1988). Endogenous components of the event-related brain potential. In T. Picton (Ed.), *Human Event-Related Potentials: EEG Handbook* (pp. 361–426). Amsterdam: Elsevier.

Pizzagalli, D. A., Lehmann, D., Hendrick, A. M., Regard, M., Pascual-Marqui, R. D., & Davidson, R. J. (2002). Affective judgments of faces modulate early activity (approximately 160 ms) within the fusiform gyri. *Neuroimage, 16*(3 Pt 1), 663–677.

Plant, A. & Devine, P. (1998). Internal and external motivation to respond without prejudice. *Journal of Personality and Social Psychology, 75,* 811–832.

Regan, M. & Howard, R. (1995). Fear conditioning, preparedness, and the contingent negative variation. *Psychophysiology, 32*(3), 208–214.

Richeson, J., Baird, A., Gordon, H., Heatherton, T., Wyland, C., Trawalter, S., et al. (2003). An fMRI investigation of the impact of interracial contact on executive function. *Nature Neuroscience, 6,* 1323–1328.

Richeson, J. & Shelton, N. (2003). When prejudice does not pay: Effects of interracial contact on executive function. *Psychological Science, 14,* 287–290.

Rockstroh, B., Elbert, T., Lutzenberger, W., & Birbaumer, N. (1979). Slow cortical potentials under conditions of uncontrollability. *Psychophysiology, 16*(4), 374–380.

Rugg, M. D. (1990). Event-related brain potentials dissociate repetition effects of high- and low-frequency words. *Memory and Cognition, 18*(4), 367–379.

Shriffin, R. & Schneider, W. (1977). Controlled and automatic human information processing. II. Perceptual learning, automatic attending, and a general theory. *Psychological Review, 84,* 127–190.

Vaes, J., Paladino, M., Castelli, L., Leyens, J., & Giovanazzi, A. (2003). On the behavioral consequences of infrahumanization: The implicit role of uniquely human emotions in intergroup relations. *Journal of Personality and Social Psychology, 85,* 1016–1034.

van Boxtel, G. & Brunia, C. (1994). Motor and non-motor aspects of slow brain potentials. *Biological Psychology, 38,* 37–51.

Vrana, S. R. & Rollock, D. (1998). Physiological response to a minimal social encounter: Effects of gender, ethnicity, and social context. *Psychophysiology, 35*(4), 462–469.

Vrana, S. R. & Rollock, D. (2002). The role of ethnicity, gender, emotional content, and contextual differences in physiological, expressive, and self-reported emotional responses to imagery. *Cognition and Emotion, 16*(1), 165–192.

Walter, W., Cooper, R., Aldridge, V., McCallum, W., & Winter, A. (1964). Contingent negative variation: An electric sign of sensori-motor association and expectancy in the human brain. *Nature, 203*, 380–384.

White, M. (1996). Anger recognition is independent of spatial attention. *New Zealand Journal of Psychology, 25*, 30–35.

Widen, S. C. & Russell, J. A. (2003). A closer look at preschoolers' freely produced labels for facial expressions. *Developmental Psychology, 39*, 114–128.

Yee, C. M. & Miller, G. A. (1988). Emotional information processing: Modulation of fear in normal dysthymic subjects. *Journal of Abnormal Psychology, 97*(1), 54–63.

12 Animal Models of Human Attitudes: Integrations Across Behavioral, Cognitive, and Social Neuroscience

Elizabeth A. Phelps and Mahzarin R. Banaji

The development of brain imaging techniques has led to rapid advancement in our understanding of the human brain. Initially, this growth relied on detailed models of neural systems in nonhuman animals to help interpret the function and significance of brain signals observed in humans (Cohen, Noll, & Schneider, 1993). However, as we move from investigating the neural mechanisms of simple perceptual, motor, and cognitive behaviors to complex social interactions, the usefulness of referring to animal models may seem less apparent. This is primarily because human social behavior differs fundamentally from the types of behavior typically observed in laboratory animals. Evidence suggests, however, that even in the study of complex social behavior and culture-specific learning, animal models may provide a useful starting point as we attempt to identify the neural systems underlying these behaviors.

The basic processes of emotional learning and expression are similar across species, although the content and complexity of stimuli may vary widely. By using animal models of emotional learning as a basis for studying complex social responses, we can take advantage of previously identified neural and behavioral mechanisms to inform our understanding of human social interaction. In our experiments, we have benefited from using the mechanisms of classical fear conditioning, which have been investigated across species, to explore affective states in humans that reveal race bias.

Classical conditioning was first described by Pavlov over a century ago, but more recently aversive classical conditioning paradigms, or fear conditioning, have been used to delineate the neural systems of emotional learning (see LeDoux, 2000, for a review). In fear conditioning, a neutral stimulus, such as a tone, is paired with an aversive event, such as a mild shock. After a few pairings, the subject learns that the tone, or conditioned stimulus (CS), predicts the shock, or unconditioned stimulus (UCS), and the tone itself begins to elicit a fear response, called the conditioned response (CR). Studies examining fear conditioning in rats have traced the pathways of emotional learning from stimulus input to response output (Davis, 1997; Kapp et al., 1979;

LeDoux, 1996). These models identified the amygdala as a critical structure in the acquisition, storage, and expression of conditioned fear (LeDoux, 2002; see also Cahill et al., 1999).

Although it is not possible to study neural mechanisms in humans with the same level of specificity as in other animals, both human brain imaging and lesion studies are consistent with animal models. Using functional magnetic resonance imaging (fMRI), amygdala activation was observed during fear conditioning (Buchel et al., 1998; LaBar et al., 1998). Moreover, the strength of that activation is correlated with the CR, as assessed with skin conductance (SCR), an indication of physiological arousal (LaBar et al., 1998). Furthermore, consistent with animal models suggesting the critical involvement of the amygdala in the acquisition and expression of conditioned fear, humans with damage to this region fail to demonstrate a CR as assessed with SCR (Bechara et al., 1995; LaBar et al., 1995).

Human beings, however, can also be asked to report on their experience of emotion. It appears that even though amygdala damage in humans impairs the physiological expression of conditioned fear, these same patients are able to report explicitly on the events that constitute fear conditioning. For example, patient SP, who suffers from bilateral amygdala damage, was given several presentations of a blue square paired with a mild shock to the wrist. Normal control subjects show an increased SCR to the blue square after a few pairings with the shock, an indication of conditioned fear. SP, however, never demonstrated a SCR to the blue square, even though her SCR to the mild shock was normal. When shown her data indicating a lack of conditioned fear, SP commented:

> I knew that there was an anticipation that the blue square, at some particular point in time, would bring in one of the volt shocks. But even though I knew that, and I knew that from the very beginning, except for the very first one when I was surprised. That was my reaction—I knew it was going to happen. So I learned from the very beginning that is was going to happen: blue-shock. And it happened. I turned out to be right! (Phelps, 2002, p. 559).

Thus, in spite of the lack of a CR as assessed with a physiological measure, SP had explicit awareness and understanding of the events of fear conditioning. Patients with damage to the hippocampus, whose amygdalas are intact, show the opposite pattern (Bechara et al., 1995). That is, they have a normal CR as measured implicitly through physiological arousal, but they are unable to report explicitly the events of fear conditioning. This dissociation between an explicit understanding of the emotional properties of the CS and an implicit assessment of the emotional significance of the CS suggests that the mechanisms of explicit and implicit emotional evaluation rely on distinct neural circuits. This discovery has obvious implications for dissociations that are observed in the behavioral responses of humans to complex situations in which some behaviors, such as thoughts and feelings, can be disjointed from action.

From Animal Models of Fear Conditioning to the Neural Systems of Race Bias

In particular, animal data on fear conditioning indicating a dissociation between implicit and explicit responses are reminiscent of findings from humans that illustrate several dissociations. Group data show small to no preferences on explicit measures of attitudes, whereas implicit measures of attitudes reveal large preferences. Or, explicit attitudes demonstrate large preferences that are not mimicked on implicit measures of the same. For example, white Americans express small to no preference (depending on the type of sample) for their own group compared to African-Americans. In the implicit association test (IAT), subjects see items that belong to one of four categories: photos of Black and White individuals and words for good and bad concepts (joy, love, friend versus devil, vomit, agony). For half the trials, the task involves classifying White + good using one computer keyboard key while also classifying Black + bad using another key. For the remainder of the trials, the opposite pairing is requested, with classification on White + bad using a single key and Black + bad using another. The difference in response latencies in these two conditions constitutes the IAT effect and is regarded to be a measure of evaluative strength of association regarding these race groups. Responses on the IAT (Greenwald & Banaji, 1995) using large numbers of White American subjects showed stronger association of White + good–Black + bad than to Black + good–White + bad.

On the other hand, African-Americans verbally report strong preference for their own group compared to white Americans, but their IAT data show a much weaker preference for their own group compared to that of white Americans. About half of African-Americans, as opposed to 72 percent of whites, favor their own group.

Animal data on conditioning had led to hypotheses concerning humans with damage to either the hippocampus or amygdala. Similarly, these models offered a starting point to infer the mechanisms that underlie the dissociations in attitudes observed in normal humans. Moreover, human data on implicit-explicit dissociations in race and other group attitudes had raised doubts about the validity of the measure. What is the meaning of the associative strength observed in response latencies that reveal faster or slower responses to some pairings over others? Many objections were raised (the measure reflects an effect of familiarity, not attitude; the measure taps something that has no predictive validity), but the one for which animal models proved particularly useful concerned a specific objection regarding the construct validity of the IAT.

The claim was that the IAT response was not a measure of evaluation or affect at all; in other words, it was not a measure of a preference or attitude. It was, instead, a measure of colder associations that are not typically viewed as revealing an attitude—a warm if not hot construct that was devised to capture the evaluative or affective portion of one's responses to any object.

The animal data and its convergence with patient and fMRI data provided the setting to conduct the first study of this sort to investigate the construct validity of the IAT while also shedding light on the dissociation itself. To test this hypothesis, we investigated whether the amygdala, which mediates implicit physiological indications of conditioned fear, might also mediate implicit expressions of race bias as measured by both the IAT and a physiological assessment (Phelps et al., 2000). To the extent that this is observed, behavioral data resulting from the IAT can be assumed to be tapping into an affective process.

Phelps et al. (2000) used fMRI to examine amygdala activation in White Americans while they observed pictures of Black and White unfamiliar male faces and correlated observed brain activity with two behavioral measures, the IAT and startle eyeblink. During imaging, subjects were asked to indicate by button press whether a face was repeated. Afterward, they were given the two implicit measures of race bias; the IAT, a reaction time measure of cognitive conflict, and startle eyeblink, a physiological assessment of evaluation, as well as an explicit measure of race bias, the Modern Racism Scale (MRS; McConahay, 1986). The IAT effect was computed as a difference score between the speed to respond to White + good–Black + bad pairs and the opposite, White + bad–Black + good pairs. Social groups were represented using faces that could be clearly classified as belonging to one or the other group. Good and bad stimuli were words that again, could be clearly classified as good or bad (love, joy, friend versus agony, vomit, devil).

The physiological assessment was a measure of the startle reflex, a natural response to a startling stimulus, such as a loud noise. One of the first components of this reflex response is an eyeblink, the strength of which can be assessed in the laboratory by measuring the reaction of muscles around the eyes. Startle responses are enhanced or potentiated in the presence of negative stimuli relative to neutral or positive stimuli (Lang, Bradley, & Cuthbert, 1990). In our assessment of implicit race bias, subjects were startled with a loud white noise while viewing pictures of unfamiliar Black and White male faces. The difference in strength of the startle eyeblink response while viewing Black versus White faces was the measure of race bias.

Variability among the White American subjects was seen in the amygdala response to Black versus White faces. Although most subjects showed greater amygdala activation while viewing black compared to white faces, the overall effect was not significant. We had expected overall greater amygdala activation to Black than to White faces, but the more important idea was the relationship between the magnitude of amygdala activation and the IAT, startle reflex, and MRS. When amygdala activation to Black versus White faces was correlated with the three measures of race bias, only the implicit measures predicted amygdala activation. On region of interest analysis, amygdala activation correlated significantly with implicit race bias as measured by the IAT ($r = 0.576$) and startle eyeblink ($r = 0.556$), with more negative assessments of Black relative to White on the measure of behavior predicting greater amygdala activation. No correlation was seen between amygdala activation and the MRS. An additional exploratory analysis generated correlation maps indicating brain regions where behavioral measures of race bias were significantly correlated with the strength of activation to black versus white faces. These correlation maps indicated that a region of the right amygdala correlated with both the IAT and startle eyeblink measures of bias (figure 12.1, plate 5).

This study revealed a variety of results that can be the basis of future research. First, learning what a social group means, along the evaluative dimension, involves mechanisms that are similar across species. For example, in Black Americans, greater amygdala activation was evident to White versus Black unfamiliar male faces (Hart et al.,

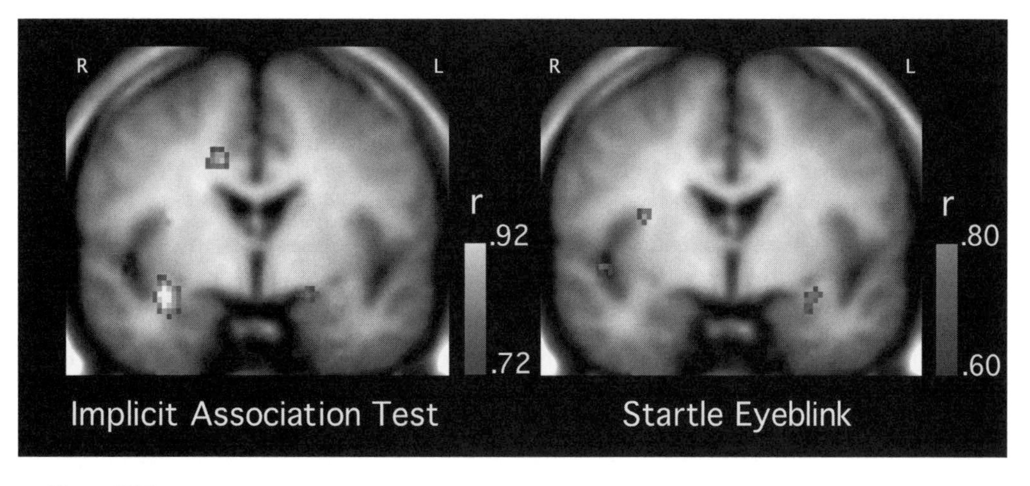

Figure 12.1
Correlations maps indicating regions where the differential BOLD response to Black versus White was significantly correlated with the strength of implicit race bias as assessed by the IAT (left) and startle eyeblink (right). (Reprinted from Phelps et al., 2000.) See plate 5 for color version.

2000), suggesting that this effect extends to outgroup faces in general. Another study observed amygdala activation in White Americans to Black versus White faces that are presented subliminally, so quickly that subjects are unaware of their presentation and report seeing only a flicker (Cunningham et al., 2004).

The amygdala is implicated in learning the emotional significance of an event, such as fear. It also is implicated in fear extinction, in which a CS is no longer paired with a UCS and, eventually, the CS fails to elicit a CR. During initial extinction learning, the amygdala appears to play an active role in learning that the CS no longer predicts an aversive event that it previously did (Falls, Miserendino, & Davis, 1992; Phelps et al., 2004). That is, after a period of time of no pairing, the amygdala response to an extinguished CR is diminished, and other brain regions, such as the prefrontal cortex (PFC), appear to play a role in the retention of extinction learning and inhibiting an amygdala response (Morgan, Romanski, & LeDoux, 1993; Phelps et al., 2004; Quirk et al., 2000).

In terms of race bias, we might expect two types of extinction learning. The first is extinction to specific individual members of a race group who are familiar. A further question of interest concerns the extent to which such effects with specific individuals generalize beyond those instances to include new members of Black or White race categories who were not previously involved in the "learning" of extinction.

To assess the effect of familiarity and exposure on the neural systems of race bias, we repeated the fMRI study described above, with one exception. The Black and White male faces were not unfamiliar, but rather were well known figures from the popular media with relatively positive public images (Martin Luther King, John F. Kennedy, Denzel Washington, Harrison Ford). With these stimuli, we failed to find consistent amygdala activation. Much like the lack of amygdala activation to CSs that had been previously extinguished (Phelps et al., 2004), the amygdala response in White subjects did not differentiate same and other race group faces. There was also no accompanying correlation between amygdala activation and implicit or explicit measures of race bias (Phelps et al., 2000), suggesting that individuals who are well known may be subtyped so as not to evoke favorable-unfavorable feelings themselves, whereas the social category as whole continues to elicit them.

In the Phelps et al. (2000) study, we did not explore responses to the entire brain, so we do not know if the PFC was involved in retention of this extinction learning, or inhibition of the amygdala response. However, Cunningham et al. (2004) found that the magnitude of activation in regions of the PFC is accompanied by lower activity in the amygdala in the Black versus White comparison, similar to extinction learning (Phelps et al., 2004). However, this was observed only when subjects were aware of the presentation of Black and White faces, suggesting a consciously mediated mechanism of cognitive control that may be specific to humans. In addition, Richeson et al. (2003) reported that activation of the PFC may be related to the inhibition of race bias. The precise regions of the PFC observed in these studies differ from those observed in extinction learning (Phelps et al., 2004), but they are consistent with studies of emotion regulation (Ochsner et al., 2002), suggesting that cognitive control of emotion, like extinction learning, results in the interaction of the PFC and amygdala. Although not able to speak directly to the issue of communication between subcortical and cortical regions, these studies suggests the potential that conscious values and beliefs may have an impact on the subcortical mechanisms of race bias. They speak to the question of the role of individual differences in conscious attitudes and, in this case, the ability to modulate the negative response to Black versus White race groups through the work of mechanisms that are more directly under conscious control.

Although we have not explored how the amygdala response to all members of another race group (familiar and unfamiliar) might change with extensive positive exposure to a number of individuals from that group, it was suggested that intergroup contact is a factor that predicts positive outgroup attitudes and interactions (Pettigrew & Tropp, 2000). These results are consistent with the hypothesis that exposure, without aversive consequences, may help extinguish culturally acquired race bias.

From Animal Models of Fear Conditioning to Social Learning

Thus far, we have argued that the neural systems of classical fear conditioning might be similar to the neural systems of culturally acquired race bias. However, a fundamental difference between conditioned fear and race bias is the means of acquisition. In fear conditioning, the subject learns the emotional significance of a CS through direct, personal aversive experience (UCS) paired with the CS. Culturally dependent race bias is learned through social means, that is, interaction and communication between individuals, however subtle the exchanges are. Although on rare occasions race bias may be acquired through aversive experience, in most cases, it is learned without personal, aversive consequences. Given this, it is important to show that social means of fear learning rely on the same neural mechanisms as learning through direct aversive experience.

A primary means of social learning in humans is language—a symbolic form of communication between individuals. Studies on the acquisition of phobias, suggest that linguistic communication about the potential aversive consequences of stimuli can generate potent fears (King, Eleonora, & Ollendick, 1998). For example, individuals suffering from a phobia of germs have extreme reactions to stimuli they have been told about—microscopic organisms they have neither seen nor will see. Some individuals go to great lengths to avoid these situations. An experimental paradigm to investigate the verbal communication of fears is instructed fear. This is similar to fear conditioning, except that subjects are simply told that they might receive a shock paired with a stimulus (the Instructed CS). No shocks are ever actually delivered. Instructed fear results in similar physiological expressions of fear as actual fear conditioning involving an aversive stimulus (Hugdahl & Ohman, 1977).

There is little reason to expect that the amygdala would play a significant role in the acquisition of verbally acquired fears. Simply being told that a neutral stimulus may, in the future, predict a possible aversive consequence is unlikely to elicit a strong emotional response at the time this information is learned. However, it is possible that the expression of these verbally acquired fears may rely on similar mechanisms as fears acquired with direct aversive experience with fear conditioning. In an effort to determine if the amygdala is involved in verbally acquired fears, we investigated the neural systems of instructed fear using fMRI (Phelps et al., 2001). Subjects were informed that they would see several presentations of a blue square, and that paired with some of these presentations they would also receive a mild shock to the wrist. None of the subjects actually received a shock at any time. Afterward, all subjects indicated awareness that a shock would be presented with a blue square. Consistent with this, the SCR response to the blue square was increased, a physiological indication of fear. Similar to fear conditioning, there was significant activation of the left amygdala when subjects believed they might receive a shock (figure 12.2, plate 6). As with the earlier study with fear conditioning and implicit measures of race bias, the strength of amygdala activation was correlated with the strength of the SCR response to the blue square.

These results suggest that the amygdala responds to verbally instructed fears, but brain imaging data cannot indicate if the amygdala plays a critical role in the expression of these fears. This was shown in a second study in patients with amygdala damage (Funayama et al., 2001). Using a similar paradigm, there patients failed to show a physiological indication of instructed fear, although they were able to report verbally that the blue square indicated the possibility of shock. Although studies examining the neural systems of instructed fear differ from fear conditioning in that only the left amygdala is involved, perhaps because of the verbal nature of the learning and representation, in both paradigms the amygdala was critical for the

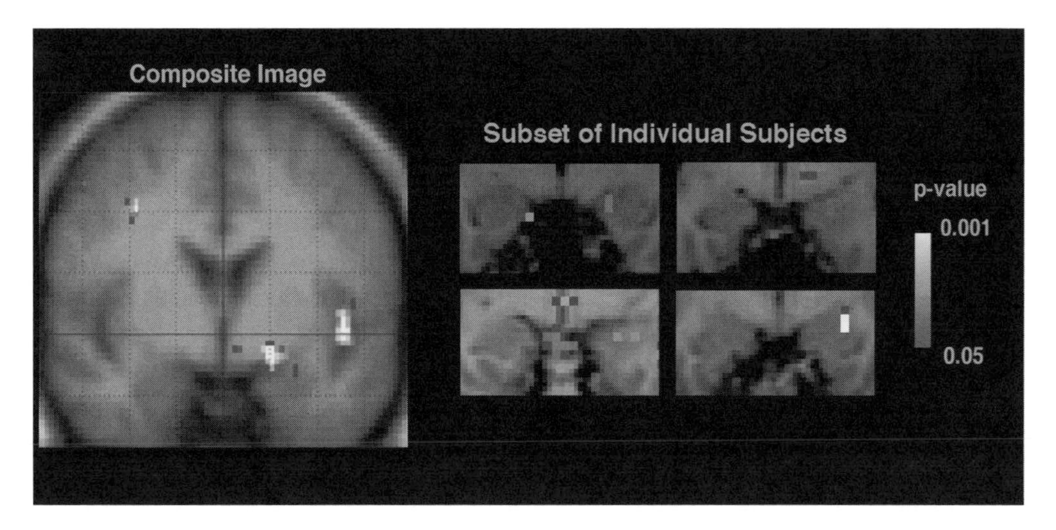

Figure 12.2
Activation to instructed fear. A group composite map (left) and selected individual subjects (right) indicate regions with a significantly greater BOLD response to a "threat" stimulus (paired with possibility of shock) versus a "safe" stimulus (no possibility of shock). (Reprinted from Phelps et al., 2001.) See plate 6 for color version.

physiological expression of fear learning. These studies indicate that fears acquired through social symbolic means, that are imagined and anticipated but never actually experienced, depend on similar mechanisms as fears acquired through direct, aversive experience.

Another powerful means of social learning is observation (Bandura, Ross, & Ross, 1961; Hygge & Ohman, 1978). By watching other individuals, we can learn about situations that should be approached or avoided ourselves. Emotional learning through observation has been demonstrated in infants viewing the emotional reactions of their mothers (Gerull & Rapee, 2002), as well as in monkeys that observed reactions of other monkeys to specific stimuli (Mineka & Cook, 1993). A recent study explored the physiological expression of fear acquired through social (instruction and observation) and nonsocial (classical conditioning) means (Olsson & Phelps, 2004). In the observation condition, subjects watched a video of another individual responding to a stimulus (observational CS) that was paired with a mild shock to the wrist (observational UCS). They were told they would experience a similar procedure immediately after viewing the video. The subjects were then presented the observational CS, but never actually received a shock. All three methods of fear learning resulted in similar levels of physiological fear expression when the stimulus (conditioned, instructed, or observational CS) that was paired with the aversive event was presented supraliminally, so that subjects were aware of its presentation (Olsson & Phelps, 2004). Of interest, both fear conditioning and observational fear resulted in the physiological expression of fear when the CS was presented subliminally—so quickly subjects were unaware it was presented. When an instructed CS was presented subliminally, there was no indication of a fear response (Olsson & Phelps, 2004). Thus, observing another individual's response to a CS paired with an aversive UCS results in a representation that is similar to direct experience with the aversive event.

The similarity in expression of fear to a subliminal CS in fear conditioning and observational fear suggests overlapping neural circuits. Although the acquisition of verbally instructed fears is unlikely to involve the amygdala, it is possible that the vicarious experience of observing another individual's aversive reaction to a UCS paired with a CS evokes an amygdala response. To investigate this, an fMRI study (Olsson et al., 2004) examined amygdala activation when observing a confederate receiving a mild shock paired with a CS (acquisition stage) and then participating in the same study (test stage), although shocks were never actually delivered. Significant amygdala activation occurred to both observing a confederate receive a shock and expecting to receive shock oneself. Thus the amygdala may be involved both in the acquisition and expression of observational fear learning, similar to fear conditioning.

Studies on the neural systems of instructed and observational fear learning indicate that socially acquired negative evaluations may not be fundamentally different than those acquired through classical conditioning. The amygdala plays a role in the acquisition (observation) and expression (observation and verbal instruction) of fears learned through social means. Given this, one might expect culturally acquired preferences, which are learned through social communication and observation, to rely on similar neural mechanisms as classical conditioning, as our previous studies suggested.

From Animal Models of Fear Conditioning to the Dynamics of Race Bias and Learning

If classical fear conditioning is an appropriate model for socially acquired race bias, we should be able to use this model to generate novel predictions concerning responses to race group stimuli. Previous studies of classical conditioning indicated that not all CSs are created equal. Some stimuli, when paired when an aversive event, will result in stronger expression and slower extinction of a CR. These stimuli have been called "prepared," to capture the idea that they are evolutionarily prepared to support fear learning. For instance, in both humans and monkeys, when a stimulus representing a snake (such as a picture or a toy snake) is paired with a UCS it will quickly acquire a CR that is stronger and less prone to extinction than a CS representing something more neutral, such as a flower or mushroom (see Ohman & Mineka, 2001, for a review). Although it may seem that more potent CRs to stimuli representing snakes, which already have a negative connotation for most people, may simply reflect a previously acquired negative evaluation, two lines of evidence suggest that this is not the case. Monkeys raised in the laboratory that never had experience with snakes will show a prepared CR to a toy snake through fear conditioning or observational fear learning (Mineka et al., 1984).

In humans, discrimination fear conditioning paradigms, in which one stimulus (CS+) is paired with the UCS and another stimulus (CS–) is not and serves as a baseline, suggest that instances of some categories can serve as prepared stimuli. A specific representation of a snake when used as a CS+ will show a prepared CR, even when the baseline is responses to another representation of a snake that is used as a CS–; that is, one that was not paired with an additional aversive stimulus. Even if a class of stimuli has an overall negative evaluation, a specific instance of that class, when used as a CS and paired with aversive consequences, will acquire a CR that is stronger and more resistant to extinction than a specific CS from a class of stimuli that is not prepared (Ebert et al., 2005). We asked to what extent can social groups that vary in race or ethnicity be such prepared stimuli?

In an effort to examine if the race of a face (as a CS) will alter the characteristics of a CR, we examined fear conditioning to pictures of White and Black unfamiliar males faces in White and Black American subjects. We hypothesized that an outgroup face, when used as a CS, would result in a CR that was more resistant to extinction than an ingroup face CS. A discrimination fear conditioning paradigm was used in which there were four CSs, two pictures of Black faces and two of White faces. One picture from each race group served as a CS+ and was paired with a mild shock to the wrist (UCS) during acquisition, while the other served as a CS– and was never paired with shock. For each race group, we compared the SCR to the CS+ to the same race CS– as an indication of fear conditioning. Regardless of subject's race, both Black and White face CS+ stimuli showed evidence of acquisition of a CR. However, the CR to an outgroup did not show the same rate of extinction as the CR to an ingroup face. In other words, an outgroup race face when paired with an aversive event is resistant to extinction learning, indicating that fear learning to a specific member of an outgroup race will be less likely to diminish than fear learning to a specific member of an ingroup race (Ebert et al., 2005). These results parallel those obtained when comparing extinction to CRs to specific instances of snakes versus specific instances of flowers. In fear conditioning, outgroup race faces look like prepared stimuli.

The difference in rate of extinction to another race face CS suggests that a learned negative evaluation to a specific member of an outgroup race may be especially difficult to change, even with additional exposure and familiarity with no aversive consequences. Although extinction learning may help diminish negative bias to outgroup race members that have not been specifically linked to aversive consequences, if an outgroup race member is linked to a negative event, this learned response appears to be more durable than an acquired negative response to an ingroup race member. Using classical fear conditioning as a model, we have been able to generate novel predictions about the impact of race group membership on differential levels of emotional learning.

Conclusion

As the research outlined in this chapter indicates, animal models can be quite useful in understanding a wide range of human behaviors, including complex social interactions. In our studies we relied on models of conditioned fear as a starting point in our investigations of the neural systems of race bias and even used these models to derive direct hypotheses of social learning and race bias. In this sense, animal models are useful in that they provide a way to detect those aspects of social learning and memory that may be extensions of the processes that are common to all animals. On the other hand, what makes humans *human* is the ability for conscious awareness, intention, and control to shape behavior. A human being, unlike other animals, can regulate behavior from deciding to delay eating gratification, even when hungry, to dine with a friend, to deciding to recode stimuli that produce negativity into those that are neutral or positive. As studies on the conscious control of race bias (Cunningham et al., 2004; Richesonn et al., 2003) and the cognitive control of emotion (Oschner et al., 2002) indicate, different neural mechanisms within the PFC may be specifically linked to these human characteristics. How these human mechanisms interact with more general neural systems of emotion inhibition that are common across species (Phelps et al., 2004) is unclear. It is clear, however, that animal models can be useful tools in our investigations of the neural systems underlying human social behavior. They serve as the starting point from where we can learn about what links us to other animals and distinguishes us from them.

Acknowledgments

Supported by grants from the NIMH and the McDonnell Foundation to E. A. Phelps, and NIMH and Rockefeller Foundation's Bellagio Study Center to M. R. Banaji. We thank Molain Saintilus for help preparing the manuscript.

References

Bandura, A., Ross, D., & Ross, S. A. (1961). Transmission of aggression through imitation of aggressive models. *Journal of Abnormal Social Psychology, 63*, 575–582.

Bechara, A., Tranel, D., Damasio, H., Adolphs, R., Rockland, C., & Damasio, A. R. (1995). Double dissociation of conditioning and declarative knowledge relative to the amygdala and hippocampus in human. *Science, 269*(5227), 1115–1118.

Buchel, C., Morris, J., Dolan, R. J., & Friston, K. J. (1998). Brain systems mediating aversive conditioning: An event-related fMRI study. *Neuron, 20*(5), 947–957.

Cahill, L., Weinberger, N. M., Roozendaal, B., & McGaugh, J. L. (1999). Is the amygdala a locus of "conditioned fear"? Some questions and caveats. *Neuron, 23*(2), 227–228.

Cohen, J. D., Noll, D. C., & Schneider, W. (1993). Functional magnetic resonance imaging: Overview and methods for psychological research. *Behavioral Research Methods, Instruments and Computers, 25*(2), 101–113.

Cunningham, W. A., Johnson, M. K., Raye, C. L., Gatenby, J. C., Gore, J. C., & Banaji, M. R. (2004). Separable neural components in the processing of black and white faces. *Psychological Science, 15*, 806–813.

Davis, M. (1997). Neurobiology of fear responses: The role of the amygdala. *Journal of Neuropsychiatry and Clinical Neuroscience, 9*(3), 382–402.

Ebert, J., Olsson, A., Phelps, E. A., & Banaji, M. R. (2005). Classical conditioning effects during extinction as a measure of race bias. Poster presented at the meeting of Society for Personality and Social Psychology.

Falls, W. A., Miserendino, M. J., & Davis, M. (1992). Extinction of fear-potentiated startle: Blockade by infusion of an NMDA antagonist into the amygdala. *Journal of Neuroscience, 12*(3), 854–863.

Funayama, E. S., Grillon, C., Davis, M., & Phelps, E. A. (2001). A double dissociation in the affective modulation of startle in humans: Effects of unilateral temporal lobectomy. *Journal of Cognition Neuroscience, 13*(6), 721–729.

Gerull, F. C. & Rapee, R. M. (2002). Mother knows best: Effects of maternal modelling on the acquisition of fear and avoidance behaviour in toddlers. *Behaviour Research and Therapy, 40*(3), 279–287.

Greenwald, A. G. & Banaji, M. R. (1995). Implicit social cognition: Attitudes, self-esteem, and stereotypes. *Psychological Review, 102*(1), 4–27.

Hart, A. J., Whalen, P. J., Shin, L. M., McInerney, S. C., Fischer, H., & Rauch, S. L. (2000). Differential response in the human amygdala to racial outgroup vs ingroup face stimuli. *Neuroreport, 11*(11), 2351–2355.

Hugdahl, K. & Ohman, A. (1977). Effects of instruction on acquisition and extinction of electrodermal responses to fear-relevant stimuli. *Journal of Experimental Psychology, 3*(5), 608–618.

Hygge, S. & Ohman, A. (1978). Modeling processes in the acquisition of fears: Vicarious electrodermal conditioning to fear-relevant stimuli. *Journal Personal Social Psychology, 36*(3), 271–279.

Kapp, B. S., Frysinger, R. C., Gallagher, M., & Haselton, J. R. (1979). Amygdala central nucleus lesions: Effect on heart rate conditioning in the rabbit. *Physiology and Behavior, 23*(6), 1109–1117.

King, N. J., Eleonora, G., & Ollendick, T. H. (1998). Etiology of childhood phobias: Current status of Rachman's three pathways theory. *Behavior Research and Therapy, 36*(3), 297–309.

LaBar, K. S., Gatenby, J. C., Gore, J. C., LeDoux, J. E., & Phelps, E. A. (1998). Human amygdala activation during conditioned fear acquisition and extinction: A mixed-trial fMRI study. *Neuron, 20*(5), 937–945.

LaBar, K. S., LeDoux, J. E., Spencer, D. D., & Phelps, E. A. (1995). Impaired fear conditioning following unilateral temporal lobectomy in humans. *Journal of Neuroscience, 15*(10), 6846–6855.

Lang, P. J., Bradley, M. M., & Cuthbert, B. N. (1990). Emotion, attention, and the startle reflex. *Psychological Review, 97*(3), 377–395.

LeDoux, J. E. (1996). *The Emotional Brain.* New York: Simon & Schuster.

LeDoux, J. E. (2002). *Synaptic Self: How Our Brains Become Who We Are.* New York: Viking.

McConahay, J. P. (1986). Modern racism, ambivalence, and the modern racism scale. In J. F. Dovidio & S. L. Gaertner (Eds.), *Prejudice, Discrimination and Racism* (pp. 91–125). Orlando: Academic Press.

Mineka, S. & Cook, M. (1993). Mechanisms involved in the observational conditioning of fear. *Journal of Experimental Psychology, 122*(1), 23–38.

Mineka, S., Davidson, M., Cook, M., & Keir, R. (1984). Observational conditioning of snake fear in rhesus monkeys. *Journal of Abnormal Psychology, 93*(4), 355–372.

Morgan, M. A., Romanski, L. M., & LeDoux, J. E. (1993). Extinction of emotional learning: Contribution of medial prefrontal cortex. *Neuroscience Letter, 163*, 109–113.

Ochsner, K. N., Bunge, S. A., Gross, J. J., & Gabrieli, J. D. (2002). Rethinking feelings: An fMRI study of the cognitive regulation of emotion. *Journal of Cognitive Neuroscience, 14*(8), 1215–1229.

Ohman, A. & Mineka, S. (2001). Fears, phobias, and preparedness: Toward an evolved module of fear and fear learning. *Psychological Review, 108*(3), 483–522.

Olsson, A., Nearing, K., Zheng, J., & Phelps, E. A. (2004). Learning by observing: Neural correlates of fear learning through social observation. Presented at the 34th annual meeting of the Society for Neuroscience, San Diego.

Olsson, A. & Phelps, E. A. (2004). Learned fear of "unseen" faces. *Psychological Science, 15*, 822–828.

Pettigrew, T. F. & Tropp, L. R. (2000). Does intergroup contact reduce prejudice: Recent meta-analytic findings. In S. Oskamp (Ed.), *Reducing Prejudice and Discrimination* (pp. 93–114). Mahwah, NJ: Erlbaum.

Phelps, E. A. (2002). The cognitive neuroscience of emotion. In M. S. Gazzaniga, R. B. Ivry, & G. R. Magnum (Eds.), *Cognitive Neuroscience: The Biology of Mind*, 2nd ed. (pp. 537–576). New York: Norton.

Phelps, E. A., Delgado, M. R., Nearing, K. I., & LeDoux, J. E. (2004). Extinction learning in humans: Role of the amygdala and vmPFC. *Neuron, 43*, 897–905.

Phelps, E. A., O'Connor, K. J., Cunningham, W. A., Funayama, E. S., Gatenby, J. C., Gore, J. C., et al. (2000). Performance on indirect measures of race evaluation predicts amygdala activation. *Journal of Cognitive Neuroscience, 12*(5), 729–738.

Phelps, E. A., O'Connor, K. J., Gatenby, J. C., Gore, J. C., Grillon, C., & Davis, M. (2001). Activation of the left amygdala to a cognitive representation of fear. *Nature Neuroscience, 4*(4), 437–441.

Quirk, G. J., Russo, G. K., Barron, J. L., & Lebron, K. (2000). The role of ventromedial prefrontal cortex in the recovery of extinguished fear. *Journal of Neuroscience, 20*(16), 6225–6231.

Richeson, J. A., Baird, A. A., Gordon, H. L., Heatherton, T. F., Wyland, C. L., Trawalter, S., & Shelton, J. N. (2003). An fMRI investigation of the impact of interracial contact on executive function. *Journal of Neuroscience, 6*, 1323–1328.

13 Characterizing the Functional Architecture of Affect Regulation: Emerging Answers and Outstanding Questions

Kevin N. Ochsner

Marcus Aurelius was a thinking person's Roman emperor, and his *Meditations* has been counted among the most introspective of ancient philosophical works. Among the many observations about society, leadership, life, and death that fill this diarylike text is one whose importance might be overlooked by the casual reader. Aurelius observed, "If you are distressed by anything external, the pain is not due to the thing itself, but to your estimate of it; and this you have the power to revoke at any moment." A careful reader might catch the twofold impact of these words. First, they convey the insight that the world is what we make of it, and that our mental machinery constructs our experience of the world. Second, they betray Aurelius's role as Rome's commander in chief, who knew full well the specific power of pain, and of negative affect more generally, to cripple even the heartiest of soldiers. Aurelius sought to understand how one might be protected from the consequences of injury, both psychic and physical.

A full appreciation of Aurelius's insights may have never been more timely. Our modern landscape may lack Roman emperors and Roman legions, but it is full of everyday battles both personal and societal. We pursue professional success. We wrestle with romantic relationships. We confront conflicted feelings. We intervene in intergroup conflicts. We wage war on terrorism. Over two millennia ago a philosophical general echoed sentiments shared by philosophers before him, and presaged the message of countless scholars and psychologists ever since: the mind is a double-edged sword. Swung one way, it serves your enemy, causing distress and pain. Swung another, it serves your champion, cutting a swath through fear, through grief, through false belief. Indeed, your mind is the best weapon you can carry into any battle, for with it you can conquer that most pernicious and inescapable enemy: your own self, and your painful estimate of the world.

How is it that the way we think about and interpret the world determines the way in which we respond to it? How does thinking control negative feeling? Modern multidisciplinary research can provide answers to these questions that move beyond philosophical musings (Cacioppo et al., 2002; Ochsner & Lieberman, 2001). Using the tools of modern neuroscience in combination with the methodology of social psychology, researchers have begun constructing theories that link the experience of negative affect to a set of psychological processes implemented by neural systems that together comprise a functional architecture for affect regulation.

Brain Bases of Affective Evaluation

Emotion and Affect

Although the terms emotion and affect often are used interchangeably by psychologists to refer to valenced feeling states that may be accompanied by changes in physiology and behavior, a useful distinction between them can be made. Emotion refers to a coordinated set of behavioral, physiological, and experiential responses that ready an organism to adapt to specific environmental challenges (Cacioppo & Berntson, 1999; Feldman Barrett, Ochsner, & Gross, in press; Lazarus, 1991). Emotions are episodic in the sense that they have an eliciting stimulus or set of eliciting conditions that trigger a response whose trajectory can be described and demarcated by an end point. Although debates have raged over the existence of a more or less finite set of basic emotion types, there is consensus concerning a core set of appraisal patterns that determine which emotions are generated as a function of the relationship between external events and internal goals, wants, and needs (Scherer, Schorr, & Johnstone, 2001). By contrast, affect often is considered to be an umbrella term that encompasses not just emotional responses, but enduring moods that lack specific eliciting conditions and consciously perceived referents, as well as moment-to-moment valenced (good-bad, or affective) evaluations of external stimuli or internal mental contents.

A full consideration of the merits of differentially inclusive definitional boundaries is beyond the scope of this chapter (for discussion, see (Feldman Barrett et al., in press; Scherer et al., 2001). For present purposes, the term affect will be used for two reasons. The first is that it can usefully refer to various types of phenomena that involve thinking to regulate a valenced response. As considered below, thinking or cognition can be used to regulate one's positive or negative perception of another person or attitude object more generally, as well as to regulate one's own personal experience of emotion. The second is that neuroscience research is increasingly suggesting that a common set of brain systems supports the evaluative processing that colors both person perception and personal experience (Anderson et al., 2003; Anderson &

Phelps, 2001; Cunningham et al., 2003; Ochsner & Feldman Barrett, 2001; Phelps et al., 1998).

Thus the term affect regulation refers either to cognitive control of judgments and interpretations that lead an individual to perceive another person in a favorable, prejudicial, suspicious, or friendly light, or to cognitive control of situation-based appraisals that lead an individual to feel happy, sad, angry, or glad. In keeping with Aurelius's recognition of the pernicious consequences of psychological distress, in recognition of the maladaptive mental and physical consequences of failures to regulate negative emotion (Davidson, Putnam, & Larson, 2000; Gross, 1998), and in light of the fact that much more is known about the neural bases of negative than of positive affect, the focus here is on regulating negative affective evaluations.

Affective Evaluation and the Amygdala

Since the late 1990s, evidence from both human and animal research has implicated the amygdala, an almond-shaped subcortical set of nuclei near the tip of the temporal lobe's medial wall, in a number of functions intimately related to affective evaluation. Studies of fear conditioning in rats, for example, initially revealed that the amygdala was critical for associating neutral unconditioned stimuli with aversive conditioned stimuli such as electric shocks (LeDoux, 2000). These findings were confirmed by studies of humans with amygdala damage (LaBar et al., 1995) who failed to show autonomic evidence of fear conditioning, as well as imaging studies of healthy participants who showed amygdala activation during acquisition and expression of conditioned fear (Buchel & Dolan, 2000).

The amygdala seems to play a special role in the automatic, rapid, early, and even preattentive detection of affectively relevant stimuli. This hypothesis was initially based on the discovery that a subcortical pathway from sensory organs to amygdala, which bypassed the cortex, was sufficient to support conditioned fear in rats (LeDoux, 2000). This suggestive finding was not directly tested, however, until functional imaging studies revealed amygdala activation to subliminal presentations of fear faces (Whalen et al., 1998) and during the conditioning of responses to subliminal fear faces (Morris, Ohman, & Dolan, 1999). More recent studies show that the amygdala's response to subliminal fear faces is exaggerated in depression, but returns to normal after treatment (Sheline et al., 2001), and that amygdala lesions may block the attention-grabbing power of briefly presented affectively charged words (Anderson & Phelps, 2001).

In addition to its role in the acquisition of stimulus-response associations, the amygdala is important for consolidating explicit memory for emotionally arousing events.

Building on a large rodent literature linking the amygdala to arousal-mediated learning, Cahill and colleagues (1995) found with human subjects that drugs blocking norepinephrine release in the amygdala eliminated the memory enhancement typically found for the emotional components of a story. Subsequent imaging studies linked amygdala activation during encoding of both positively and negatively arousing photographs to subsequent memory for them (Hamann et al., 1999).

More broadly, amygdala activation has been observed in response to a wide variety of stimuli with intrinsic or learned affective significance, including pleasant and unpleasant odors (Zald & Pardo, 2000), foods (Zald et al., 1998), positive or negative photographs (Hamann et al., 1999; Hariri et al., 2003), film clips (Lane et al., 1997), and arousing musical passages (Blood & Zatorre, 2001). Amygdala lesions may skew perception of aversive or threatening stimuli in a positive direction, leading to impaired recognition of fear faces, a tendency to perceive as friendly people normatively appearing to be unfriendly, and a tendency to classify normatively unpleasant images as more pleasant (Adolphs, Tranel, & Damasio, 1998). As these data indicate, the amygdala may play a special role in perceiving social stimuli, a role further supported by the finding that amygdala lesions impair perception of subtle expressions of social emotions such as interest, boredom, and flirtation (Adolphs, Baron-Cohen, & Tranel, 2002), and that for Caucasian participants, amygdala activation to African-American faces is correlated with the amount of antiblack bias shown on an implicit measure of racial attitudes (Phelps et al., 2000). Heightened amygdala activation also may predict psychiatric outcomes, including the presence or severity of depression (Abercrombie et al., 1998), and accompanies evoked symptoms of anxiety, phobia, and obsessive-compulsive disorder (Phillips et al., 2003).

Beyond the amygdala, a number of other richly interconnected brain structures play an important role in affective evaluation. The first is the ventral striatum, which is intimately linked to the experience and prediction of rewards (Knutson et al., 2001). The second is the midportion of the cingulate cortex, which plays a special role in representing the unpleasant properties of physical pain, and may more generally serve to signal when current behavior is not meeting desired ends (Botvinick et al., 2004; Eisenberger & Lieberman, 2004; Ochsner et al., 2001; Peyron, Laurent, & Garcia-Larrea, 2000). The third is the insula, which like the cingulate is responsive to physical pain, and also signals the presence of disgust-related stimuli (Peyron et al., 2000). The fourth, and for present purposes, last, is the orbitofrontal/ventromedial prefrontal cortex (PFC; which includes ventral lateral PFC as well), which is sensitive to many of the same classes of stimuli processed by the amygdala, but seems to play a special role in altering existing affective associations. Thus, orbitofrontal lesions in primates (Dias,

Robbins, & Roberts, 1997) and in humans (Fellows & Farah, 2003) impair the ability to alter a stimulus-reinforcer association once it is learned, and single-unit recording studies in rats and primates suggest that orbitofrontal cortex neurons change their firing properties to previously rewarded (but now not rewarded) stimuli more rapidly than do amygdala neurons (Rolls et al., 1996). If the amygdala provides an initial appraisal of the affective relevance and potential threat value of stimulus, the ventral striatum, anterior cingulate, and insula may code additional affective properties relevant to specific types of stimuli, and the orbitofrontal cortex (OFC) constructs a higher-level representation that places the affective value of a stimulus in its current goal-relevant context (Lieberman, 2000; Ochsner & Feldman Barrett, 2001; Ochsner & Gross, 2004).

On balance, the available literature implicates the amygdala more than any other brain structure in affective evaluation, and especially in the generation of a negative evaluation that may bias perception of social targets and be a first step in the generation of a negative affective experience (Ochsner & Feldman Barrett, 2001).

Cognitive Regulation of Affective Evaluation

The use of thinking to control feeling, of cognition to regulate affect, for centuries has been a topic of central concern for philosophers, writers, and psychologists. Indeed, many classic stories involve the struggle to resist temptation, to control an impulse, to understand the nature of one's despair or desire, and thereby be free from its grasp. Whether it is biblical tales of the very first family and their inhibitory failures, a Roman emperor's meditations on the epistemology of pain, the legend of Siddhartha and his Buddhist renunciation of worldly wealth, stories of concentration camp survival, or gossip about the indiscretions of contemporary politicians, humans are always seeking information and telling stories about the way in which affect can and cannot be brought under the control of reason.

But which stories are correct? Which theories of the way in which cognitive control can impact emotion will survive empirical test, and what mechanisms are involved? One approach to answering these questions is to use several types of evidence to constrain theories. Until recently, information about the brain bases of affect and cognitive control were not brought to bear on these issues. The relative recency of neuroscience approaches to understanding affect regulation in humans, in comparison with long-standing interest in the issue, can be attributed to variety of reasons primarily having to do with availability of brain imaging technology and the establishment of a firm foundation of neuroscientific knowledge concerning the brain bases of simple behaviors (for discussion, see Ochsner, 2004; Ochsner & Lieberman, 2001).

A simple framework can be used to organize understanding of data from different domains of neuroscience research relevant to questions about affect regulation. Cognitive processes can be seen as impacting emotion generation processes at three different stages: the first involves initial attention to, and gating of, stimulus inputs for further processing; the second involves use of thinking to reinterpret or manipulate the meaning of an input once it has been gated into the system; and the third involves inhibition, selection, or modulation of expressive behavioral responses (Gross, 1998; Ochsner & Gross, 2004).

Building on findings from studies of "cold" forms of cognitive control, a working hypothesis has developed concerning the cognitive control of affect. The basic idea is that at whatever stage they might intervene, control processes used to regulate are supported by frontal and cingulate systems (Miller & Cohen, 2001) that may modulate affect generation systems such as the amygdala, insula, and pain-related cingulate cortex; which specific systems are involved may depend on the specific type of emotion and the specific kind of regulatory strategy employed (Ochsner & Gross, 2004, 2005; Ochsner et al., 2004).

Control processes used to regulate emotion depend on frontal and cingulate systems that implement cognitive control processes. Prefrontal cortex is thought to be involved in retrieving information from semantic memory, holding information in mind, and maintaining representations of strategic (e.g. regulatory) goals (Miller & Cohen, 2001). Anterior cingulate cortex is thought to monitor the extent to which continuing control is achieving its desired ends, signaling the extent to which desired and states have not yet been met (Lieberman et al., 2002; Miller & Cohen, 2001; Ochsner & Gross, 2004). As will be seen, available data generally support this hypothesis, with one important caveat: most studies investigated only the regulatory dynamics involved in the first two stages of affect processing—attentional control and cognitive change (which are the foci of the following review)—whereas none directly addressed regulation of affective behavioral responses per se.

Attentional Control

The use of attention to control the flow of sensory input long has been studied in the context of visual object recognition and vision more generally. Neuroimaging studies have examined the extent to which different stages of visual processing are modulated by attention. To the extent that a brain region's activation does not vary as a function of the allocation of attentional resources, computations carried out by that region may be characterized as taking place automatically (or at least with reduced need for conscious, on-line, attentional monitoring). Thus far, for simple nonaffective visual stimuli, findings indicate that attention may modulate systems that lie at numerous stages along the visual processing stream, including early systems (V1) that represent simple geometric spatial properties of the visual world (Culham, Cavanagh, & Kanwisher, 2001). In the context of effectively charged stimuli, the question becomes whether systems such as the amygdala or pain-related cingulate cortex, which presumably extract and encode the affective relevance as opposed to visual spatial properties of a stimulus, also are modulated by attention.

Selective Attention to Socioemotional Stimuli Five studies investigated amygdala activation in response to effectively charged compared with neutral faces during conditions of either full or diminished attention to the face stimuli. Two studies found that the amygdala's response to faces does not vary as a function of attentional resource deployment. In the first study (Vuilleumier et al., 2001), researchers asked participants make judgments about a display of four stimuli: two pictures of faces (neutral or fearful) and two pictures of houses. Face and house stimuli symmetrically flanked fixation, either above and below or to the left and right of fixation. On each trial participants were cued to judge whether the horizontal or vertical pair of stimuli was the same or different. This allows comparison of activation to attended (cued direction) and unattended (noncued direction) fearful faces. Fearful stimuli produced greater bilateral activation of the amygdala regardless of the level of attention. In contrast, when viewing faces, activation of the fusiform face area—a cortical region putatively specialized for processing face stimuli—did vary as a function of attention, as did activation of the parahippocampal place area—a cortical region specialized for processing spatial locations—when viewing houses.

Similar findings were obtained by Anderson and colleagues (2003), who also presented participants with faces and houses, semitransparently displayed on top of one another, under conditions of full and divided attention. Each type of grayscale stimulus was tinted slightly red or green enabling attentional selection of the face superimposed on the house (or vice versa). Participants were cued to judge the gender of the face or a feature of the house on each trial. As was the case for Vuilleumier et al., amygdala responses to fear faces were greater than to neutral faces, and the magnitude of this activation advantage was equivalent when participants attended to and judged face stimuli compared with attending to and judging house stimuli. Of interest, insula responses to disgust faces diminished as attentional allocation to them diminished, whereas the amygdala's response to disgust faces actually increased as attentional allocation to them diminished. This suggests that the tuning curve of the amygdala actually broadens when fewer attentional resources are available to identify and discriminate affectively relevant stimuli: when you're distracted, the amygdala responds both to fearful and disgust faces, but when you're paying full attention, the amygdala responds only to fear faces. It remains to be seen just how much the amygdala's response profile changes in different circumstances.

In direct contrast to these findings, two studies observed amygdala modulation in response to manipulations of attentional allocation. Pessoa et al. (2002) presented participants with fearful, happy, and neutral faces in the center of the screen and had them view the faces either with full attention, or when performing a concurrent line-orientation judgment task. For the orientation task, vertical or horizontal lines were presented above the face, one on the left and one on the right, and participants judged whether the orientation of the two lines was the same or different. When performing this judgment and paying less attention to the faces, amygdala activation to both happy and fearful faces dropped significantly and was indistinguishable from the response to neutral faces. Similar results were obtained in a study employing a very different paradigm (Phillips et al., 2004). These investigators presented disgusted, fearful, and neutral faces using either 30-msec subliminal presentations or 170-msec supraliminal presentations. All presentations were backward-masked with a consciously perceived neutral face of the same gender, which participants were instructed to view passively. During supraliminal presentations, fear faces activated the right amygdala and disgust faces activated the insula bilaterally, but both activations disappeared during subliminal presentations, suggesting that awareness of the presentation of affectively charged stimuli is necessary for the affective salience of the stimulus to be processed by the amygdala. It is striking that these findings stand in stark contrast to those of Whalen et al. (1998), who found amygdala activation to subliminal presentations of fear faces. Philips et al. suggest that the discrepancy may have to do with imperfect subliminal presentations by Whalen et al. that may have caused possible leakage of affectively charged stimuli into awareness. This was very unlikely in their paradigm, they argued, because they used rigorous psychophysical testing of detection thresholds for each participant to make sure that recognition of facial expression on subliminal trials was at chance. A problem for this account, however, is presented by research of Cunningham, Johnson et al. (2004), who found greater amygdala activation to black than to white faces in white participants when faces were presented subliminally, but that this difference disappeared when faces were presented supraliminally. Although Cunningham et al. did not verify unawareness of subliminal stimulus presentations as rigorously as did Philips et al., which according to Philips's account could explain amygdala activation to subliminally presented black faces, Philips et al.'s account cannot explain *failure* to find activation to supraliminal presentations of black faces. One possibility is that participants were motivated to actively regulate their responses to black faces when they could consciously perceive them (so as not to have a prejudicial response), and that there is little motivation to regulate responses to consciously perceived fearful faces, which in experimental contexts may only weakly signal the presence of actual threats. Consistent with this notion, Cunningham et al. found right prefrontal and anterior cingulate activation during supraliminal presentations.

Beyond the possible use of active regulation, another important factor may determine whether or not amygdala modulation as a function of attention to fear-relevant stimuli is observed. To the extent that low-level visual features simply are not being encoded as one fails to attend to and perceive those features, it is not likely that the information necessary to discriminate affectively relevant information would be available to the amygdala. Anderson et al. (2003) suggested just this sort of explanation for the findings of Pessoa et al. (2002), and it may be applicable to the findings of Philips et al. (2004) as well. On this account, performing a spatial orientation discrimination task that strongly depends on parietal systems, or simply failing to register the presentation of fear faces, would deprive temporal lobe systems that extract facial features of the information they need to pass along to the amygdala so that it may signal the presence of affectively charged fearful or disgusting face.

Attentional Distraction during Pain A second context in which attentional manipulations have been used to examine the regulation of affective responding is the experience of physical pain induced either by a heated thermode placed on the forearm or a painfully cold compress placed on a body part such as the foot. Rather than manipulating spatial attention, participants do two things at once: while experiencing painful stimulation, they perform an attention-demanding cognitive control task that limits the extent to which they can attend to the pain. Typically, reports of pain affect drop when participants were distracted, and the question is whether activation in pain-related processing regions drops as well. Frankenstein et al. (2001) were the first to observe this phenomenon. Participants experienced either a nonpainful cool or a painfully cold compress on their foot under conditions of full attention or when asked to generate silently proper names of objects from specific categories. When distracted by the verbal fluency task, reports of pain affect dropped, as did activation of regions of the midcingulate gyrus. By contrast, during distraction, activation of anterior cingulate regions associated with monitoring, as well as the left dorsolateral PFC, was observed, although it is not clear whether these regions were involved in active regulation of pain or simply in supporting performance of the verbal fluency task.

Others addressed this question by directly testing for an interaction between pain-processing and cognitive control regions. Participants were asked to experience either warm or painful heat while completing either blocks of neutral or interference Stroop trials (Bantick et al., 2002). Pain affect dropped during interference compared with neutral trials, and an activation interaction was observed between the presence of painful stimulation and the presence of cognitive distraction induced by the Stroop. Two regions associated with cognitive control and awareness of affect—right orbitofrontal cortex and rostral anterior cingulate cortex—were more active when resolving interference in the presence of pain than when resolving interference in the absence of pain, suggesting that they play a role in regulating pain responses. A number of regions associated with the experience of pain, including bilateral insula, thalamus, and midcingulate gyrus, showed a negative interaction and were more active during pain in the absence than in the presence of cognitive interference, suggesting that they were down-regulated by distraction.

An important question is the level or stage of pain processing (or affective processing more generally) that can be modulated by cognitive control. Bantick et al. (2002) demonstrated that distraction can modulated thalamic responses, and Tracey et al. (2002) extended these findings by showing that asking participants to "think of something else" during the experience of thermal pain could modulate activation in a brainstem region known as the periacqueductal gray (PAG), which is important for descending modulation of pain inputs from the spinal cord. Although Tracey et al. focused only on PAG activation, and so could not speak to the issue of which regions might have been involved in sending cortically based top-down regulatory instructions, others observed PAG-prefrontal interactions that speak to this issue (Valet et al., 2004). Valet et al. employed a task essentially similar to that employed by Bantick et al., and like that group, observed that rostral medial prefrontal regions were more active during cognitive distraction in the presence of pain than during cognitive distraction in the absence of pain. Although they did not report Bantick's negative interaction with pain-related processing systems, they did find that activation of pain-related thalamic nuclei as well as the PAG inversely correlated with activation of rostral medial regions differentially involved in resolving cognitive interference in the presence of pain.

Cognitive Change

In contrast to attentional control studies that examined the indirect effects on affective evaluation of performing a secondary task or diverting spatial attention, two other types of functional imaging studies had participants actively change their construal of the meaning of affectively charged stimuli. In the first type, participants engaged in cognitive reappraisal, or reinterpretation, of the meaning of photographs or films that elicited various negative emotions, so as to increase, maintain, or decrease their negative emotional response. In the second, participants selectively evaluated a specified semantic, perceptual or emotional stimulus dimension, such as judging the gender or the emotional expression of a face. Despite their differences, findings generated by these two types of studies have shown similar patterns of PFC–amygdala interactions, as discussed below.

Reappraisal The term reappraisal was first used by psychologist Richard Lazarus (1991) to describe our ability to alter the trajectory of a continuing emotional response by changing the way in which we reappraise its significance or relevance to current goals, wants, or needs. In an early study, he showed that providing a stress-reducing appraisal frame for an emotionally arousing and distressing film about a penile circumcision ritual (it is not that painful, those undergoing the ritual are honored to undergo it, etc.) led to diminished physiological activation compared with participants who did not receive this frame. Although subsequent studies examined the consequences of, and context for, reappraisal (see Gross, 1998, for review), until recently none has attempted to unpack the psychological and neural dynamics underlying it. One of the first studies to do so asked participants to view aversive photographs in one of two conditions: in a baseline condition participants were instructed simply to attend to and be aware of their emotional responses, but not to try and change them; in a reappraisal condition, they were instructed to reinterpret the meaning of each photo so as to lessen its emotional punch (Ochsner et al., 2002). For example, participants could reappraise an otherwise sad photograph of four women crying outside of a church as involving a wedding rather than a funeral, which would be an occasion for tears of joy rather than of sadness. During reappraisal, right amygdala activation to aversive compared with neutral photographs dropped significantly, as did subjective reports of negative affect, whereas regions of left dorsal and inferior PFC as well as anterior cingulate became active. Of importance, activation of left inferior PFC predicted amygdala deactivation, directly implicating these regions as the mechanistic substrate of reappraisal.

Compatible findings were obtained by Beauregard et al. who asked male participants to "inhibit" their emotional responses to sexually explicit film clips by becoming detached observers (Beauregard, Levesque, & Bourgouin, 2001). Right amygdala and hypothalamic activation for the arousing film was absent during attempted inhibition, and right dorsolateral prefrontal and anterior cingulate activation was found during inhibition but not during arousal. One problem with this study is that the condition in which participants inhibited their emotional responses while watching the film clips always occurred after the condition in which they simply watched the clips and let themselves respond in an unregulated fashion. This means that diminished activation of affect-related structures could be due to habituation to the film clips, to active regulation, or to both, and with the present design it is impossible to discriminate among these possibilities. In a subsequent study, these investigators used a similar distancing instruction to investigate the reappraisal of sad films that were presented in counterbalanced order with baseline unregulated viewing (Levesque et al., 2003). Consistent with earlier findings, they found that distancing activated regions of right dorsal lateral as well as lateral orbital frontal cortex, and compared with baseline, deactivated the left amygdala, insula, and right ventral PFC.

Thinking is used to regulate feeling not only in the service of decreasing feelings, but in the service of increasing them as well. Indeed, when we worry, make ourselves anxious, and ruminate about disappointments, and losses—real or imagined—we de facto up-regulate, or at least maintain, our negative emotion that might otherwise have diminished or dissipated. Ochsner and colleagues (2004) compared the use of reappraisal to increase and decrease negative emotion to determine whether the two depend on similar or different control systems, and whether they might modulate similar affect processing systems, albeit in different ways. They found (1) that cognitively increasing or decreasing negative emotion recruited left lateral prefrontal and anterior cingulate cortex, (2) that increasing selectively depended on left dorsal medial prefrontal systems involved in the metacognitive generation of negative information (Cato et al., 2004), (3) that decreasing selectively depended on right lateral and orbital frontal systems involved in response inhibition, and (4) that the right amygdala was modulated up or down in accordance with the goal of reappraisal. These findings substantially replicated those of Ochsner et al. (2002), although the earlier study did not find right lateral prefrontal activation when decreasing, a finding Ochsner et al. (2004) noted was present in the earlier study at a lower threshold and may have been detected in the later study because of greater power (a more sensitive pulse sequence combined with greater N). Schaefer et al. (2002) obtained consistent results by observing bilateral maintenance of amygdala activation after viewing aversive photographs when participants used reappraisal to maintain their negative feelings. This study did not constrain strategies, or report regions of prefrontal activation, so it is difficult to determine which regulatory mechanisms are responsible for these effects.

Selective Evaluation Some studies have manipulated the stimulus dimension specified for construal by contrasting a condition in which participants explicitly evaluated the affective valence of a stimulus or their response to it with a condition in which they judged some property that was not explicitly affective and/or could be inferred straightforwardly from perceptual features.

In one of the first of these studies, a matching task had participants judge which of two comparison stimuli presented at the bottom of the screen matched a target stimulus presented at the top of the screen (Hariri, Bookheimer, & Mazziotta, 2000). In the perceptual condition, all three stimuli were faces that expressed anger or fear. In the labeling condition only the target stimulus was a face, and comparison stimuli were expression labels (i.e. the words angry and afraid). Right amygdala activation was greater for matching perceptually, whereas right ventral lateral prefrontal activation was greater for matching to labels, and activation of right ventral PFC predicted activation of the amygdala. These findings were replicated and extended to other classes of negatively valenced stimuli, including the perception of African-American faces (Lieberman et al., unpublished observations) and aversive photographs (Hariri et al., 2003), for which inverse relationships between right ventral PFC and right amygdala were noted when participants matched to labels compared with percepts. As a group, these studies suggest that explicit attention to and semantic labeling of emotional properties of stimuli could down-regulate amygdala responses to them. A potential problem for this account, however, is that the perceptual condition in which more amygdala activation is observed includes three faces, whereas the labeling condition included a single face. This means that it is difficult to determine whether the number of face stimuli or the nature of the judgment is what drives the apparent amygdala modulation.

Be that as it may, explicit labeling of affective properties of stimuli down-regulates amygdala activation in three other paradigms, all of which equate stimulus properties in the affective labeling and non-affective labeling conditions. The first is a variant of the priming paradigm of Murphy and Zajonc (1993) in which supraliminally presented target faces with weak expressions of anger were preceded by subliminal primes that were angry faces, neutral faces, or blank control stimuli (Nomura et al., 2004). Participants judged whether the consciously perceived target face seemed to be angry, happy, or neutral. In general, greater right amygdala activation was observed for trials with angry primes, and right amygdala activation was correlated with the tendency to judge the target face as angry. Right ventral lateral PFC showed precisely the opposite pattern, having an inverse correlation with amygdala activation and attributions of angry expressions. In the second, decreased amygdala activation was observed by Taylor et al. (2003) when participants rated the valence of aversive and neutral photographs compared with viewing them passively. In the third, Critchley et al. (2000) had participants view happy, angry, or neutral faces, and judge either face gender or whether or not each face was emotionally expressive. In general, perception of emotional compared with neutral faces activated the left amygdala, and strikingly, right amygdala activation was diminished for explicit emotion compared with gender judgments.

Although all of these studies are consistent with the notion that explicit semantic labeling of emotional properties of stimuli can down-regulate amygdala responses—and presumably negative affect as well—this account has two salient problems. The first is that it cannot explain why retrieving semantic emotion knowledge and labeling affective properties of stimuli can sometimes boost negative affect as well as amygdala activation (Ochsner et al., 2004), as mentioned above. The second is that a number of studies failed to observe modulation of amygdala responses when participants explicitly attend to or semantically label affective stimulus properties.

For example, Gorno-Tempini et al. (2001) had participants judge either the gender or emotional expression of happy, disgust, or neutral faces. Although left amygdala activation generally was found for emotional faces, explicit expression judgments were not reported to diminish amygdala activation, and instead activated right dorsolateral PFC. It should be noted, however, that because the authors did not report the gender > expression judgment contrast, it is difficult to determine whether gender judgments might have produced greater amygdala activation, as was observed by Critchley et al. (2000). This ambiguity of analysis was not a problem for a study in which happy, sad, disgust, and fear faces were morphed with neutral faces to produce stimuli with high- and low-intensity expressions and were presented in pairs. Participants were asked to judge either which face of each pair was more male, or which was more emotionally expressive (Winston, O'Doherty, & Dolan, 2003). Although increasing intensity of expression activated the amygdala bilaterally, in contrast to results of Critchley et al. (2000), amygdala activation did not vary with judgment type.

In a similar study with different stimuli, this time using normatively trustworthy compared with untrustworthy-appearing neutral faces (that would be judged as nevertheless expressing a high degree of anger), Winston et al. (2002) had participants either judge whether faces were of high school or university students, or whether they were trustworthy or untrustworthy individuals. Bilateral amygdala activation to untrustworthy faces was observed that once again did not vary as a function of judgment type. Finally, Cunningham et al. (2003) had participants judge whether photos of famous people had been taken in the past or present or represented good (Martin Luther King) or bad (Osama bin Laden) people. "Bad" compared with "good" individuals activated the left amygdala and right ventral PFC, and amygdala activation did not vary as a function of judgment type.

If semantic selection-labeling sometimes, but not always, results in modulation of affective evaluation and amygdala activation, what determines when this modulation will take place? One possibility, suggested earlier in the context of attentional control, is that regulatory interactions between PFC and amygdala take place when one has the explicit motive or goal to regulate one's evaluative responses to stimulus dimensions that are the foci of attention. This is certainly the case when participants are instructed to regulate, as indicated by the fact that all of the reappraisal studies reported amygdala modulation, and all involved attention to, and the explicit goal-directed regulation of, affect. But it is also the case that even when not explicitly instructed to control one's evaluation, participants may do so spontaneously, and some cross-study variability could result from participants having the motivation to control in some experiments, but not in others. This possibility was strongly supported by a study in which participants judged whether words were either abstract or concrete, or whether they represented good- or bad-attitude objects. After the scanning session, participants rated each word for affective intensity, good-bad valence, and their motive to control their evaluations of each attitude object (Cunningham, Raye, & Johnson, 2004). When these ratings were used to predict brain activation to attitude objects during each judgment task, affectively intense words produced left amygdala activation. Critically, when making good-bad judgments, the motive to control predicted activation of the anterior cingulate, right orbitofrontal cortex, and lateral PFC, with activation of the last predicting amygdala deactivation.

Future Directions

The overarching question motivating this chapter has been, what are the neurocognitive mechanisms by which thinking controls negative feeling? Now we pause to take stock of what the preceding review suggests as an answer, or answers, to this question, and highlight specific questions that have yet to be addressed.

What Answers Have Emerged?

It does not take a neuroimaging study to tell us that distraction, inattention, reappraisal, and alterations of construal more generally, can change the way we affectively evaluate a stimulus. Careful behavioral observation and empirical experimentation, as well as good old-fashioned first-person experience, have long since confirmed this fact. What neuroimaging studies have begun to reveal are mechanisms that make these changes in affective construal possible. Three consistent findings have emerged that can be taken as answers to the question of what mechanisms mediate the cognitive regulation of affect.

1. Interactions between prefrontal and cingulate systems that implement control processes, and amygdala, cingulate, and other systems that implement affective appraisal processes, underlie the cognitive control of affect. This starting hypothesis, built by analogy to cold forms of cognitive control, was supported in studies of distraction during pain, reappraisal, and selective evaluation of the affective properties of stimuli.

2. The simplest alternative explanation for the effects of cognitive control on affective processing—namely, that engaging in cognitive processing, or thinking, diminishes affective evaluation in the same way that attending to a sound dampens vision—cannot account for the available data. If this account were correct, then amygdala activation should always decrease whenever participants perform some type of judgment, which clearly is not the case; amygdala activation can remain invariant with respect to variations in attention (Anderson et al., 2003) and judgment (Winston et al., 2002, 2003), and can even increase for some types of judgment (Ochsner et al., 2004).

3. Down-regulation of negative affect by cognitively changing the meaning of a stimulus may involve predominantly right lateralized systems. This inference is supported by a number of studies showing an inverse relationship between right ventrolateral PFC-OFC and right amygdala during conscious (and likely regulated) as opposed to nonconscious (and likely unregulated) perception of black faces (Cunningham et al., 2004), during reappraisal (Beauregard, Levesque, & Bourgouin, 2001; Levesque et al., 2003; Ochsner et al., 2004), and during the selective evaluation of the explicit affective properties of faces (Hariri et al., 1999; Lieberman et al., 2004; Nomura et al., 2004) or words representing attitude objects (Cunningham, Raye & Johnson, 2004).

4. Mechanisms mediating regulation of experience and regulation of perception-judgment may be highly overlapping. Virtually all studies investigating reappraisal instructed participants to change the way they felt about a stimulus by changing the way in which they made judgments about it. Virtually all studies involving the

selective evaluation of affective compared with non-affective properties of stimuli made no reference to experience in any way, instructing participants simply to attend to and discriminate stimulus features. Both of these forms of cognitive change rely on right PFC-OFC regions to mediate their regulatory impact on an affective outcome—experience in one case, judgment in the other—that is correlated with amygdala activation in both cases. It should be noted, however, that one reason regulation of experience and perception may recruit similar mechanisms is because most tasks involved both types of regulatory processing, and failed to discriminate them cleanly.

What Questions Remain Outstanding?

Research on the neurocognitive mechanisms of affect regulation has only been a primary focus of functional imaging research for the past five to ten years, so it is perhaps not surprising that the list of questions yet to be addressed is longer than the list of emerging answers.

1. Although some of the primary neural players in the cognitive regulation of affect have been identified, the precise functional nature of their regulatory interactions with the amygdala remains to be clarified, and may be context or task sensitive. Increased prefrontal-cingulate and decreased amygdala activation could result from at least two types of regulatory mechanisms: (1) processes involved in selecting appropriate, and/or limiting conflict between inappropriate, responses, processes that have been linked to right ventral PFC and anterior cingulate (Miller & Cohen, 2001); or (2) alterations in input to the amygdala that either (a) cut off or limit the flow of perceptual information indicating the presence of an aversive stimulus, as may be the case for distraction and selective attention, or (b) mentally generate an alternative set of semantic and/or perceptual inputs that feed forward to the amygdala, indicating the presence of a more neutral stimulus, as may be the case for both reappraisal and selective evaluation. It remains for future research that manipulates the psychological factors involved in a given task to discriminate these possibilities, which likely will include manipulations of both judgment and stimulus variables.

2. A second related question concerns the ways in which different types of strategies may involve different types of regulatory interactions. Distraction compared with reappraisal, for example, may recruit different systems but because to date studies of distraction involved pain, studies of reappraisal involved affectively charged photographs, and no published studies compared both strategies in a single task, it is difficult to address this question. Another strategic variable that may be important concerns the content that is in the focus of attention or is the focus of the reappraisal

strategy. Strategies that involve an inward focus on oneself and one's emotional experience may recruit medial prefrontal regions, as suggested by studies showing medial PFC activation for self-focused reappraisal strategies asking participants to distance themselves psychologically from pictured events (Ochsner et al., 2004) and for the regulation of pain experience (Bantick et al., 2002; Frankenstein et al., 2001; Valet et al., 2004), as opposed to studies showing lateral PFC activation for reappraisal strategies focused on situational features (Ochsner et al., 2004) and tasks involving selective evaluation of affective stimulus properties (Cunningham et al., 2003; Hariri et al., 1999, 2003; Lieberman et al., 2004).

3. To what extent must the conscious goal or motive be present in order for regulatory effects on the amygdala to be observed? Studies that explicitly gave participants the conscious goal to regulate (Ochsner et al., 2002, 2004) or have measured their spontaneous tendency to regulate (Cunningham et al., 2004) consistently observed amygdala activation, whereas other studies that neither manipulated nor measured explicit regulatory goals less consistently observed amygdala modulation. The simple fact that in everyday life thinking often influences feeling without our having the explicit intent for this to happen makes it seem unlikely that modulation always requires the conscious goal to change how one feels or perceives the world. What kind of processing goal is necessary is not clear, however.

4. Are the same control mechanisms and regulatory interactions involved in attentional control, cognitive change, and response modulation? To address this question, different types of strategies would have to be compared within the context of the same experiment, using the same participants and stimuli for each type of regulatory intervention. No published studies have examined the neurocognitive substrates of an explicitly behavior-focused affect-regulation strategy, such as suppression of emotion expressive behavior.

5. To what extent are regulatory dynamics such as those described here also involved in other types of cognition-affect interactions, including those not typically thought of as explicitly regulatory (for discussion, see Ochsner & Gross, in press)? Stimulus-reward reversal tasks and decision-making tasks, for example, may involve attentional control and cognitive change processes such as those involved in the experiments described here. Brain imaging studies could be used to determine whether similar or different mechanisms are involved. Expectations and beliefs also influence affect. Recent studies of the placebo effect suggest, for example, that prefrontal and anterior cingulate regions like those involved in reappraisal may modulate pain-related activations of midcingulate cortex. Future studies could directly compare reappraisal and placebo to determine whether or not they depend on similar mechanisms.

6. To what extent are mechanisms mediating affect regulation similar to, or different from, mechanisms mediating regulation of nonaffective inputs, including linguistic, visual-spatial, and auditory information? In other words, to what extent are systems classically thought to mediate shifts of spatial attention (Posner & Rothbart, 1998), working memory (D'Esposito, Postle, & Rypma, 2000), or response inhibition (Braver et al., 2001) like those involved in regulating an affective response? Their similarities and differences could illuminate the nature of the mechanisms involved in each.

7. How do individual differences in the tendency to generate affective responses of particular kinds, the ability to control, and other personality or mood-related variables, have an impact on the way in which affect regulation systems operate? This broad question may include specific questions about the development of regulatory capacity in children, its evolution across the lifespan as we age, its breakdown in individuals suffering from psychiatric disorders, as well as the range of variability found in normal healthy populations. Our models of the functional architecture of affect regulation will be stronger to the extent that they can accommodate, and make predictions about, the way in which specific neural mechanisms may differentially contribute to emotional responding and regulatory success in different individuals (Kosslyn et al., 2002).

Conclusion

The capacity Marcus Aurelius identified over two millennia ago, for our estimates of the value of a situation to determine our affective response to it, is central to our ability to adapt to stressful circumstances. Clearly, a complete understanding of mechanisms that give rise to this capacity is essential. A premise of this chapter is that a very useful kind of explanation speaks to several levels of analysis, linking socioemotional experience and behavior to theories of information-processing mechanisms, and linking those mechanisms to neural systems (Cacioppo et al., 2000; Ochsner & Lieberman, 2001). In doing so, brain data can constrain theories of psychological process, and vice versa. Although this chapter has focused on data from functional neuroimaging studies, the use of many other neuroscience techniques (including scalp electrode recording and analysis of patients with neuropsychological disorders) to converge on theories consistent with numerous methodologies is essential and necessary. With any luck, in the next decade our estimate of the progress we have made in characterizing the functional architecture of affect regulation will produce a greater measure of pleasure than of pain.

Acknowledgments

Supported by National Science Foundation grant BCS-93679 and NIH grant MH58147.

References

Abercrombie, H. C., Schaefer, S. M., Larson, C. L., Oakes, T. R., Lindgren, K. A., Holden, J. E., et al. (1998). Metabolic rate in the right amygdala predicts negative affect in depressed patients. *Neuroreport, 9*(14), 3301–3307.

Adolphs, R., Baron-Cohen, S., & Tranel, D. (2002). Impaired recognition of social emotions following amygdala damage. *Journal of Cognitive Neuroscience, 14*(8), 1264–1274.

Adolphs, R., Tranel, D., & Damasio, A. R. (1998). The human amygdala in social judgment. *Nature, 393*(6684), 470–474.

Anderson, A. K., Christoff, K., Panitz, D., De Rosa, E., & Gabrieli, J. D. (2003). Neural correlates of the automatic processing of threat facial signals. *Journal of Neuroscience, 23*(13), 5627–5633.

Anderson, A. K. & Phelps, E. A. (2001). Lesions of the human amygdala impair enhanced perception of emotionally salient events. *Nature, 411*(6835), 305–309.

Bantick, S. J., Wise, R. G., Ploghaus, A., Clare, S., Smith, S. M., & Tracey, I. (2002). Imaging how attention modulates pain in humans using functional MRI. *Brain, 125*(Pt 2), 310–319.

Beauregard, M., Levesque, J., & Bourgouin, P. (2001). Neural correlates of conscious self-regulation of emotion. *Journal of Neuroscience, 21*(18), RC165.

Blood, A. J. & Zatorre, R. J. (2001). Intensely pleasurable responses to music correlate with activity in brain regions implicated in reward and emotion. *Proceedings of the National Academy of Sciences of the United States of America, 98*(20), 11818–11823.

Botvinick, M. M., Cohen, J. D., & Carter, C. S. (2004). Conflict monitoring and anterior cingulate cortex: an update. *Trends Cognitive Science, 8*(12), 539–546.

Braver, T. S., Barch, D. M., Gray, J. R., Molfese, D. L., & Snyder, A. (2001). Anterior cingulate cortex and response conflict: Effects of frequency, inhibition and errors. *Cerebral Cortex, 11*(9), 825–836.

Buchel, C. & Dolan, R. J. (2000). Classical fear conditioning in functional neuroimaging. *Current Opinions in Neurobiology, 10*(2), 219–223.

Cacioppo, J. T. & Berntson, G. G. (1999). The affect system: Architecture and operating characteristics. *Current Directions in Psychological Science, 8*(5), 133–137.

Cacioppo, J. T., Berntson, G. G., Sheridan, J. F., & McClintock, M. K. (2000). Multilevel integrative analyses of human behavior: Social neuroscience and the complementing nature of social and biological approaches. *Psychological Bulletin, 126*(6), 829–843.

Cacioppo, J. T., Berntson, G. G., Taylor, S. E., & Schacter, D. L. (2002). *Foundations in Social Neuroscience*. Cambridge: MIT Press.

Cahill, L., Babinsky, R., Markowitsch, H. J., & McGaugh, J. L. (1995). The amygdala and emotional memory [letter]. *Nature, 377*(6547), 295–296.

Cato, M. A., Crosson, B., Gokcay, D., Soltysik, D., Wierenga, C., Gopinath, K., et al. (2004). Processing words with emotional connotation: An fMRI study of time course and laterality in rostral frontal and retrosplenial cortices. *Journal of Cognitive Neuroscience, 16*(2), 167–177.

Critchley, H., Daly, E., Phillips, M., Brammer, M., Bullmore, E., Williams, S., et al. (2000). Explicit and implicit neural mechanisms for processing of social information from facial expressions: A functional magnetic resonance imaging study. *Human Brain Mapping, 9*(2), 93–105.

Culham, J. C., Cavanagh, P., & Kanwisher, N. G. (2001). Attention response functions: Characterizing brain areas using fMRI activation during parametric variations of attentional load. *Neuron, 32*(4), 737–745.

Cunningham, W. A., Johnson, M. K., Gatenby, J. C., Gore, J. C., & Banaji, M. R. (2003). Neural components of social evaluation. *Journal of Personality and Social Psychology, 85*, 639–649.

Cunningham, W. A., Johnson, M. K., Raye, C. L., Gatenby, J. C., Gore, J. C., & Banaji, M. R. (2004). Separable neural components in the processing of black and white faces. *Psychological Science, 15*(12), 806–813.

Cunningham, W. A., Raye, C. L., & Johnson, M. K. (2004). Implicit and explicit evaluation: fMRI correlates of valence, emotional intensity, and control in the processing of attitudes. *Journal of Cognitive Neuroscience, 16*(10), 1717–1729.

Davidson, R. J., Putnam, K. M., & Larson, C. L. (2000). Dysfunction in the neural circuitry of emotion regulation—A possible prelude to violence. *Science, 289*(5479), 591–594.

D'Esposito, M., Postle, B. R., & Rypma, B. (2000). Prefrontal cortical contributions to working memory: Evidence from event-related fMRI studies. *Experimental Brain Research, 133*(1), 3–11.

Dias, R., Robbins, T. W., & Roberts, A. C. (1997). Dissociable forms of inhibitory control within prefrontal cortex with an analog of the Wisconsin card sort test: Restriction to novel situations and independence from "on-line" processing. *Journal of Neuroscience, 17*(23), 9285–9297.

Eisenberger, N. I., & Lieberman, M. D. (2004). Why rejection hurts: a common neural alarm system for physical and social pain. *Trends Cognitive Science, 8*(7), 294–300.

Feldman Barrett, L., Ochsner, K. N., & Gross, J. J. (in press). Automaticity and emotion. In J. A. Bargh (Ed.), *The Automaticity of Emotion*. New York: Oxford University Press.

Fellows, L. K. & Farah, M. J. (2003). Ventromedial frontal cortex mediates affective shifting in humans: Evidence from a reversal learning paradigm. *Brain, 126*(Pt 8), 1830–1837.

Frankenstein, U. N., Richter, W., McIntyre, M. C. & Remy, F. (2001). Distraction modulates anterior cingulate gyrus activations during the cold pressor test. *Neuroimage, 14*(4), 827–836.

Gorno-Tempini, M. L., Pradelli, S., Serafini, M., Pagnoni, G., Baraldi, P., Porro, C., et al. (2001). Explicit and incidental facial expression processing: an fMRI study. *Neuroimage, 14*(2), 465–473.

Gross, J. J. (1998). The emerging field of emotion regulation: An integrative review. *Review of General Psychologicy, 2,* 271–299.

Hamann, S. B., Ely, T. D., Grafton, S. T., & Kilts, C. D. (1999). Amygdala activity related to enhanced memory for pleasant and aversive stimuli. *Nature Neuroscience, 2*(3), 289–293.

Hariri, A. R., Bookheimer, S. Y., & Mazziotta, J. C. (2000). Modulating emotional responses: Effects of a neocortical network on the limbic system. *Neuroreport, 11*(1), 43–48.

Hariri, A. R., Mattay, V. S., Tessitore, A., Fera, F., & Weinberger, D. R. (2003). Neocortical modulation of the amygdala response to fearful stimuli. *Biological Psychiatry, 53*(6), 494–501.

Knutson, B., Fong, G. W., Adams, C. M., Varner, J. L., & Hommer, D. (2001). Dissociation of reward anticipation and outcome with event-related fMRI. *Neuroreport, 12*(17), 3683–3687.

Kosslyn, S. M., Cacioppo, J. T., Davidson, R. J., Hugdahl, K., Lovallo, W. R., Spiegel, D., et al. (2002). Bridging psychology and biology. The analysis of individuals in groups. *American Psychologist, 57*(5), 341–351.

LaBar, K. S., LeDoux, J. E., Spencer, D. D., & Phelps, E. A. (1995). Impaired fear conditioning following unilateral temporal lobectomy in humans. *Journal of Neuroscience, 15*(10), 6846–6855.

Lane, R. D., Reiman, E. M., Ahern, G. L., Schwartz, G. E., & Davidson, R. J. (1997). Neuroanatomical correlates of happiness, sadness, and disgust. *American Journal of Psychiatry, 154*(7), 926–933.

Lazarus, R. S. (1991). Progress on a cognitive-motivational-relational theory of emotion. *American Psychologist, 46*(8), 819–834.

LeDoux, J. E. (2000). Emotion circuits in the brain. *Annual Review of Neuroscience, 23,* 155–184.

Levesque, J., Eugene, F., Joanette, Y., Paquette, V., Mensour, B., Beaudoin, G., et al. (2003). Neural circuitry underlying voluntary suppression of sadness. *Biological Psychiatry, 53*(6), 502–510.

Lieberman, M. D. (2000). Intuition: A social cognitive neuroscience approach. *Psychological Bulletin, 126*(1), 109–137.

Lieberman, M. D., Gaunt, R., Gilbert, D. T., & Trope, Y. (2002). Reflexion and reflection: A social cognitive neuroscience approach to attributional inference. In M. P. Zanna (Ed.), *Advances in Experimental Social Psychology* (vol. 34, pp. 199–249). San Diego: Academic Press.

Miller, E. K. & Cohen, J. D. (2001). An integrative theory of prefrontal cortex function. *Annual Review of Neuroscience, 24,* 167–202.

Morris, J. S., Ohman, A., & Dolan, R. J. (1999). A subcortical pathway to the right amygdala mediating "unseen" fear. *Proceedings of the National Academy of Sciences of the United States of America, 96*(4), 1680–1685.

Murphy, S. T. & Zajonc, R. B. (1993). Affect, cognition, and awareness: Affective priming with optimal and suboptimal stimulus exposures. *Journal of Personality and Social Psychology, 64*(5), 723–739.

Nomura, M., Ohira, H., Haneda, K., Iidaka, T., Sadato, N., Okada, T., et al. (2004). Functional association of the amygdala and ventral prefrontal cortex during cognitive evaluation of facial expressions primed by masked angry faces: An event-related fMRI study. *Neuroimage, 21*(1), 352–363.

Ochsner, K. N. (2004). Current directions in social cognitive neuroscience. *Current Opinions in Neurobiology, 14*(2), 254–258.

Ochsner, K. N., Bunge, S. A., Gross, J. J., & Gabrieli, J. D. (2002). Rethinking feelings: An fMRI study of the cognitive regulation of emotion. *Journal of Cognitive Neuroscience, 14*(8), 1215–1229.

Ochsner, K. N. & Feldman Barrett, L. (2001). A multiprocess perspective on the neuroscience of emotion. In T. J. Mayne & G. A. Bonanno (Eds.), *Emotions: Currrent Issues and Future Directions* (pp. 38–81). New York: Guilford Press.

Ochsner, K. N. & Gross, J. J. (2004). Thinking makes it so: A social cognitive neuroscience approach to emotion regulation. In K. Vohs & R. Baumeister (Eds.), *The Handbook of Self-Regulation: Research, Theory, and Methods Hillsdale*, (pp. 229–255). NJ: Erlbaum.

Ochsner, K. N. & Gross, J. J. (2005). The cognitive control of emotion. *Trends in Cognitive Sciences, 9*(5), 242–249.

Ochsner, K. N. & Lieberman, M. D. (2001). The emergence of social cognitive neuroscience. *American Psychologist, 56*(9), 717–734.

Ochsner, K. N., Ray, R. D., Cooper, J. C., Robertson, E. R., Chopra, S., Gabrieli, J. D. E., et al. (2004). For better or for worse: neural systems supporting the cognitive down- and up-regulation of negative emotion. *Neuroimage, 23*, 483–499.

Pessoa, L., McKenna, M., Gutierrez, E., & Ungerleider, L. G. (2002). Neural processing of emotional faces requires attention. *Proceedings of the National Academy of Sciences of the United States of America, 99*(17), 11458–11463.

Peyron, R., Laurent, B., & Garcia-Larrea, L. (2000). Functional imaging of brain responses to pain. A review and meta-analysis (2000). *Neurophysiologic Clinique, 30*(5), 263–288.

Phelps, E. A., LaBar, K. S., Anderson, A. K., O'Connor, K. J., Fulbright, R. K., & Spencer, D. D. (1998). Specifying the contributions of the human amygdala to emotional memory: A case study. *Neurocase, 4*(6), 527–540.

Phelps, E. A., O'Connor, K. J., Cunningham, W. A., Funayama, E. S., Gatenby, J. C., Gore, J. C., et al. (2000). Performance on indirect measures of race evaluation predicts amygdala activation. *Journal of Cognitive Neuroscience, 12*(5), 729–738.

Phillips, M. L., Drevets, W. C., Rauch, S. L., & Lane, R. (2003). Neurobiology of emotion perception. II. Implications for major psychiatric disorders. *Biological Psychiatry, 54*(5), 515–528.

Phillips, M. L., Williams, L. M., Heining, M., Herba, C. M., Russell, T., Andrew, C., et al. (2004). Differential neural responses to overt and covert presentations of facial expressions of fear and disgust. *Neuroimage, 21*(4), 1484–1496.

Posner, M. I. & Rothbart, M. K. (1998). Attention, self-regulation and consciousness. *Philosophical Transactions of the Royal Society of London, Series B, Biological Sciences, 353*(1377), 1915–1927.

Rolls, E. T., Critchley, H. D., Mason, R., & Wakeman, E. A. (1996). Orbitofrontal cortex neurons: Role in olfactory and visual association learning. *Journal of Neurophysiology, 75*(5), 1970–1981.

Schaefer, S. M., Jackson, D. C., Davidson, R. J., Aguirre, G. K., Kimberg, D. Y., & Thompson-Schill, S. L. (2002). Modulation of amygdalar activity by the conscious regulation of negative emotion. *Journal of Cognitive Neuroscience, 14*(6), 913–921.

Scherer, K. R., Schorr, A., & Johnstone, T. (Eds.). (2001). *Appraisal Processes in Emotion: Theory, Methods, Research.* New York: Oxford University Press.

Sheline, Y. I., Barch, D. M., Donnelly, J. M., Ollinger, J. M., Snyder, A. Z., & Mintun, M. A. (2001). Increased amygdala response to masked emotional faces in depressed subjects resolves with antidepressant treatment: An fMRI study. *Biological Psychiatry, 50*(9), 651–658.

Taylor, S. F., Phan, K. L., Decker, L. R., & Liberzon, I. (2003). Subjective rating of emotionally salient stimuli modulates neural activity. *Neuroimage, 18*(3), 650–659.

Tracey, I., Ploghaus, A., Gati, J. S., Clare, S., Smith, S., Menon, R. S., et al. (2002). Imaging attentional modulation of pain in the periaqueductal gray in humans. *Journal of Neuroscience, 22*(7), 2748–2752.

Valet, M., Sprenger, T., Boecker, H., Willoch, F., Rummeny, E., Conrad, B., et al. (2004). Distraction modulates connectivity of the cingulo-frontal cortex and the midbrain during pain—An fMRI analysis. *Pain, 109*(3), 399–408.

Vuilleumier, P., Armony, J. L., Driver, J., & Dolan, R. J. (2001). Effects of attention and emotion on face processing in the human brain: An event-related fMRI study. *Neuron, 30*(3), 829–841.

Whalen, P. J., Rauch, S. L., Etcoff, N. L., McInerney, S. C., Lee, M. B., & Jenike, M. A. (1998). Masked presentations of emotional facial expressions modulate amygdala activity without explicit knowledge. *Journal of Neuroscience, 18*(1), 411–418.

Winston, J. S., O'Doherty, J., & Dolan, R. J. (2003). Common and distinct neural responses during direct and incidental processing of multiple facial emotions. *Neuroimage, 20*(1), 84–97.

Winston, J. S., Strange, B. A., O'Doherty, J., & Dolan, R. J. (2002). Automatic and intentional brain responses during evaluation of trustworthiness of faces. *Nature Neuroscience, 5*(3), 277–283.

Zald, D. H., Lee, J. T., Fluegel, K. W., & Pardo, J. V. (1998). Aversive gustatory stimulation activates limbic circuits in humans. *Brain, 121*(Pt 6), 1143–1154.

Zald, D. H. & Pardo, J. V. (2000). Functional neuroimaging of the olfactory system in humans. *International Journal of Psychophysiology, 36*(2), 165–181.

14 What Is Special about Social Cognition?

Ralph Adolphs

In the distant future I see open fields for far more important researches. Psychology will be based on a new foundation, that of the necessary acquirement of each mental power and capacity by gradation.
—Charles Darwin, *On the Origin of Species*

When Charles Darwin wrote those words in 1859, at the end of his famous book, he was presciently advocating the extension of evolutionary theory as applied to morphology—the topic of his book—to evolution of the mind. Sociobiology and evolutionary psychology have indeed taken up the baton, albeit with agendas that not even Darwin could have anticipated. The detailed claims of these disciplines remain debated, but their central question is as clear and urgent as ever. What distinguishes the human mind? What makes us essentially human?

It is easy to construe these questions so vaguely as to admit of little progress toward an answer; however, it is equally easy to point out some specific facts that provide purchase. Take the observation by Richard Passingham (1982) that, "Our species is unique because, in only 35,000 years or so, we have revolutionized the face of the earth." That observation is patently correct. *Homo sapiens* has changed the face of the earth and constructed societies on a scale never seen on this planet. In so doing, we have also become capable of destroying the majority of species, including ourselves. And all of that happened on a time scale far too rapid to have incorporated evolution by natural selection as Darwin had in mind. So how has this dramatic change been possible?

The answer, of course, lies with a different form of evolution than the one based on genes: the evolution of culture. Comparative psychologists and anthropologists have long pointed out the dramatically accelerated accumulation of information that is possible with cultural rather than genetic transmission. No surprise, the claim has been made that culture is indeed what distinguishes humans from all other species. Thus we find Michael Tomasello (1999) writing, "The basic fact is thus that human beings are able to pool their cognitive resources in ways that other species are not . . . made possible by a single very special form of social cognition, namely, the ability of individual organisms to understand conspecifics as beings like themselves who have intentional and mental lives like their own."

Although the claim that culture is unique to humans was challenged by studies in apes (Whiten et al., 1999), the idea that human social cognition is what makes the human mind special remains popular. One idea has been that intragroup or intergroup competition among early hominids fueled a need to anticipate and predict others' behavior. Both tactical deception (Whiten & Byrne, 1997) and social cooperativity are behavioral consequences of such a mechanism, and whereas precursors of both are present in other animals, they are not found remotely to the same extent as in humans. Could these mechanisms have fueled the expansion of the human brain and our distinctive cognitive abilities (Dunbar, 1998)?

We humans think of other people as having minds, having experiences, feelings, emotions, beliefs, and intentions, and as having a view of the world as we ourselves do. Broadly, this capacity has been dubbed "theory of mind," and there has been continuous debate regarding whether or not nonhuman primates might possess something similar (Povinelli & Vonk, 2003; Premack & Woodruff, 1978; Tomasello, Call, & Hare, 2003). This is an issue on which Tomasello himself recently changed his mind. Despite its prima facie plausibility, the idea still leaves unresolved the question of this chapter. Granted that our social cognitive abilities are special: are they the reason our minds are special, or are there more general capacities to which social cognition is derivative?

There is no shortage of examples of the latter possibility. Our ability to conceive of other minds could, for example, be part of a more general ability flexibly to adopt other points of view. Thus, we have the "mental time travel hypothesis" (Suddendorf & Corballis, 1997), the idea that only humans are capable of reexperiencing the past and imagining the future. Conceiving of ourselves from alternative points of view, conceiving of others as like ourselves—these are arguably merely specific instances of much more general meta-cognitive abilities. It is not clear that they are specifically social in any sense, or that they evolved solely as a consequence of social pressures.

Social Cognition and Modularity

Social cognition encompasses processing information relevant to guide social behavior, our interactions with other people. Of course, in some sense nearly all information is "relevant" for social behavior, making social cognition in that broad sense too general a term to be useful. Researchers generally restrict the term to that information directly perceived from, or inferred about, other people. A further restriction focuses on those aspects of cognition closely linked to emotion.

Social cognition, then, concerns perception of, attention to, memory for, and thinking about other people, and in a way that involves emotional and/or motivational processing. Examples include processing other people's faces and voices, judging their personality, predicting their likely behavior, and planning our own interactions with them. In one typical example, it begins with the sensory processing of social information (visual perception of someone's face or body posture); proceeds to formation of inferences and judgments about the social meaning and significance of this information, based on innate and acquired memories, current context, and future goals and plans; and culminates in the modulation of essentially all aspects of cognition and behavior. Although pervasive in the sense that it has the potential to influence essentially all aspects of cognition and behavior, social cognition draws on a circumscribed domain of information.

The degree to and manner in which different species engage in social interaction varies considerably. Even among primates, a highly social group of mammals, large differences range from the essentially solitary existence of orang-utans to close alliances and groups of chimpanzees, to large-scale human societies. No less variable is the occurrence of social cognition within an individual: we regularly engage in social interactions; but we also regularly drive our car, memorize phone numbers, and plan the decorating of our house, all of which are complex human activities, yet none of which is essentially social.

A natural question then arises regarding how social cognition fits into cognitive processing more generally construed. Is social cognition a particular type of cognition? If so, what distinguishes it, and how does it interact with nonsocial cognitive processes? In thinking about this question, people are often driven to one of two extremes: that social cognition is a way of processing information that is entirely distinct from nonsocial information processing; or that it is exactly like cognition in general, only applied to the domain of social stimuli.

This distinction was formalized in the concept of modularity. First detailed by the philosopher Jerry Fodor (1983), modularity has come to mean several things, and its current usage differs considerably from earlier ones (Coltheart, 1999; Cowie & Woodward, 2004). Without going into detail, we can itemize the following attributes of modular processes, as the term is used here. Modular processes

- Are to some extent specialized for processing certain kinds of information; that is, they are domain specific, rather than domain general.
- Evolved toward the above function.
- Are Best understood, from the view of cognitive psychology, as a distinct class of processes.
- Are best understood, from the view of neuroscience, as relying on a distinct set of brain structures.
- As a consequence of the prior two properties, can be disproportionately impaired after psychiatric or neurological disease.
- Show computational features indicating that they are specialized.

In conjunction with this incomplete list of attributes, modular processes also have certain properties that we think it is important not to ascribe to them. Modular processes

- Do not necessarily have a larger innate component than any other set of processes.
- Do not necessarily depend less on experience and development than any other set of processes.
- Are not limited in their application to the domain of information for which they are specialized or for which they evolved (teleologically speaking).
- Depending on the details, may be quite powerful and efficient in processing kinds of information for which they are not specialized, insofar as those make suitable computational demands (they may be preadapted for certain kinds of information processing).
- May interact extensively with other (nonsocial) cognitive processing; that is, they are not informationally encapsulated, to use Fodor's original term.

Whereas it seems clear that social cognition is not modular in a strong sense, if all of social cognition is considered together, certainly components appear strongly modular. Pheromonal signaling by way of the vomeronasal system (a pathway that prominently includes the amygdala), for instance, would appear paradigmatically modular. Language appears highly modular. All of these are components, but the entire collection of social cognitive abilities fails to be modular for the simple reason that it draws on processes that are too central—what Fodor called "horizontal." That is, pheromone reception and speech perception might both be modular insofar as they are inputs to social cognition; it is once we actually start judging and reasoning about such social information that the modularity breaks down.

A famous example purports to resist this negative conclusion, and claims to support evolutionary psychology's controversial idea that the mind is massively modular through and through. That is the Wason selection task (Barkow, Cosmides, & Tooby, 1992). In this task, content effects have been found such that people reason differently depending on the nature of the material they are reasoning about, even if the logical structure remains the same. In particular is the idea that we reason about social content in ways that reflect modules for detecting the violation of social contracts. The mechanism was even linked to specific regions of the brain (Stone et al., 2002). Debate continues about this particular example, but the Wason selection task aside, plenty of counterexamples exist in which certain quite circumscribed aspects of social cognition turn out, on further investigation, to lose the domain specificity that they initially appeared to support.

The Neuroscience of Social Cognition

Two good examples of current study in cognitive neuroscience are the social cognitive functions of the prefrontal cortex and of the amygdala. Both structures can be disproportionately activated in certain social tasks in functional imaging studies, and lesions in both can result in disproportionate impairments in social behavior. Nonetheless, good evidence also indicates that both structures participate prominently in processes that have nothing to do with social cognition as such. Are they examples of systems that evolved for social cognition that have been coopted for other functions? Or are they examples of domain-general systems on which social cognition also happens to draw? A similar question arises with regard to simulation, a mechanism much studied in relation to how we construe others, how we pick up their emotions, and how we generate empathy. Again, the basic mechanism appears to be derivative to more general aspects of motor control and imagery.

The Prefrontal Cortex

Let us begin with the prefrontal cortex. At the purely structural level, some evidence already shows that certain features of the human prefrontal cortex distinguish it from that of any other animal. For instance, particular sectors, such as frontal polar cortex (Broadman's area 10) are disproportionately larger in humans than in any other ape (Semendeferi et al., 2001). It is thought that this region of the brain subserves long-term planning, meta-cognition, self-relevant processing—in short, precisely those competences that Tomasello, Corballis, and others claimed to exist only in humans. The prefrontal cortex of humans, specifically anterior cingulate and frontoinsular cortex (lateral orbitofrontal cortex), contains morphologically specialized cells, so-called spindle cells, that are found only to a much lesser degree in other apes and not at all in other primates (Allman, 2002; Nimchinsky et al., 1999). One conjecture is that these cells serve rapid signaling of social information during error detection.

Lesions of the prefrontal cortex have been known to impair social functioning ever since the classic case of Phineas Gage (Damasio, 1994). Patients with damage centered in ventral and medial sectors of the prefrontal cortex have disproportionate difficulty in social behavior. Their social decision making is poor, their social relationships with others are severely compromised, they fail to detect social faux pas, and it is suggested that they may have a specific impairment on social or emotional "EQ" with sparing of the standard cognitive IQ (Bar-On et al., 2003; cf. Bechara this volume). In trying to get a handle on this issue, perhaps one of the most informative studies asked relatives of patients to fill out a detailed questionnaire of changes in personality and behavior after sustaining such a lesion (Barrash, Tranel, & Anderson, 2000). Several key attributes achieved statistical significance in being endorsed as different after damage to ventromedial prefrontal cortex, including lack of insight, lack of initiative, social inappropriateness, poor judgment, lack of persistence, indecisiveness, blunted emotional experience, apathy, inappropriate affect, and lack of planning.

Are all of these social? What might they all have in common? The claim has been made that this constellation of impairments is derivative to impaired emotional processing and, perhaps, especially impaired processing of social emotions. On the other hand, it is conceivable that more general defects, such as failure in response inhibition or global lack of motivation, could underlie the social impairments.

The evidence is more confusing yet when we consider functional imaging studies. Literally thousands of paradigms have been found to activate regions of prefrontal cortex. Some of these paradigms are specifically social, others are not. Two regions of considerable recent interest are the insula and the anterior cingulate cortex. The insula was proposed to be specialized in primates for the explicit representation of interoceptive information, the substrate of conscious feelings (Craig, 2002); and the anterior cingulate for error detection and response monitoring (Paus, 2001). In each case, social experiments resulted in activation in these structures, during empathy or social rejection, for instance. Yet reasons for activation in social experiments are presumed to be the use by social cognition of more general cognitive processes. The anterior cingulate cortex is activated by the pain of social rejection (Eisenberger, Lieberman, & Williams, 2003) or by observing someone else in pain (Singer et al., 2004), yet plays a general role in pain processing. In fact, it has a very general role in detecting any salient stimuli that might require interruption of continuing processing, of which pain may be an instance. The insula is activated by empathy for others (Singer et al., 2004), by observing others express disgust, as well as by experiencing pain, disgust, and other interoceptive information in oneself.

The Amygdala

The second structure, the amygdala, presents us with a parallel story. In the 1930s, Kluver and Bucy (1937, 1939) pointed to the impairments in social behavior in primates with complete amygdala lesions, emphasis that continued in several studies of the consequences of amygdala lesions on behavior in the real world. It was found that monkeys with bilateral amygdala lesions were so severely affected in their social behavior that they often died if left in the wild (Dicks, Myers, & Kling, 1969; Kling & Brothers, 1992). More recent studies made more focal amygdala lesions and found impairments in social behavior that are more subtle, as well as impairments in nonsocial behaviors such as approach tendencies toward novel objects (Emery et al., 2001). These results are broadly consistent with data from humans.

Several lesion studies (Adolphs et al., 1994; Anderson & Phelps, 2000; Anderson et al., 2000; Calder et al., 1996; Young et al., 1996), complemented by functional imaging studies (Breiter et al., 1996; Morris et al., 1996; Whalen et al., 2001), revealed that the human amygdala is critical for normal judgments about the internal states of others from viewing pictures of their facial expressions. Detailed analyses of some of these studies suggested that the impairment was disproportionately severe for recognition of expressions of fear, a conclusion supported also by some functional imaging studies.

The initial impression that amygdala damage results in a disproportionate impairment in recognition of fear must be tempered by more recent studies. A first flag was raised by many other studies of impairments in facial emotion recognition in a large number of pathological cases, including psychiatric illnesses and brain damage of various kinds: a common pattern across all studies is that fear is typically the emotion whose recognition is most severely impaired, and it is also the most difficult emotion to recognize on some tasks for normal individuals (Rapcsak et al., 2000). A second flag was raised by observing that some patients with complete bilateral amygdala damage appeared to perform normally on a task on which other patients with similar damage were impaired (Adolphs et al., 1999; Hamann et al., 1996; Schmolck & Squire, 2001). A third and related flag was the finding that, when data from several patients with bilateral amygdala damage were put together and analyzed in detail, the pattern that emerged was that they were indeed all impaired in their ability to make normal judgments regarding basic emotions shown in facial expressions, but they were not all impaired in the same way, on the same tasks, or on the same emotions (Adolphs, 1999; Adolphs et al., 1999; Schmolck & Squire, 2001).

The amygdala's role is not limited to making judgments about basic emotions, but includes making social judgments. This fact was suggested by earlier studies in nonhuman primates (Kling & Brothers, 1992; Kluver & Bucy, 1937; Rosvold, Mirsky, & Pribram, 1954), which revealed impaired social behavior after amygdala damage. Those results were corroborated by studies in monkeys with more selective amygdala lesions, and by using more sophisticated ways of assessing social behavior (Emery & Amaral, 1999; Emery et al., 2001), and they have been shown also in humans. Building on these findings, some authors suggest a general role for the amygdala in so-called theory of mind abilities: a collection of abilities whereby we attribute internal mental states, intentions, desires, and emotions to other people (Baron-Cohen et al., 2000; Fine, Lumsden, & Blair, 2001). Three sets of studies from our laboratory corroborate the view that the amygdala is important for generating social attributions: studies of judgments of trustworthiness, of social attributions to visual motion displays, and of the recognition of social emotions from faces and eyes.

In one study, we asked subjects to judge how much they would trust, or how much they would want to approach, an unfamiliar person (Adolphs, Tranel, & Damasio, 1998). Three subjects with bilateral damage to the amygdala were specifically impaired in their ability to judge the untrustworthiness and unapproachability of these stimuli; they performed normally when judging people who looked very trustworthy and approachable. Performances by these subjects became progressively more impaired the more untrustworthy or unapproachable the face was normally judged to be. These findings were corroborated by functional imaging in normal individuals (Winston et al., 2002). As expected, when normal individuals are shown faces of people judged to look untrustworthy, amygdala activation increases, compared with activation when they look at people judged to look trustworthy. Moreover, this pattern of activation was obtained regardless of whether subjects in the scanner were making explicit judgments about trustworthiness or making an unrelated judgment. Thus, the amygdala's role in evaluating the trustworthiness of unfamiliar people, as its role in making other social judgments, may be fairly automatic, rapid, and obligatory.

An even broader role for the amygdala in making social attributions comes from a study in which subjects were shown stimuli developed by social psychologist Fritz Heider, who designed short video clips depicting geometric shapes moving on a white background. Although the only cue available is visual motion (the stimuli do not otherwise look social), normal subjects immediately make social attributions to such stimuli (Heider & Simmel, 1944). The stimuli are interpreted as having intentions, emotions, and personalities, attributions assigned to them by the viewer in order to provide a compact, coherent, and relevant social description of them. It is in fact easier, and subjects normally cannot help but to see the stimuli in social terms. By contrast, a subject with selective bilateral amygdala damage (SM) gave a spontaneous description of the stimulus in purely geometric terms that lacked social attribution (Heberlein & Adolphs, 2004). This result was particularly striking because giving such a geometric description is normally more difficult than giving the social description. For our purposes, the finding is interesting because such a complex social judgment is at least not obviously based on simpler motivational processes.

Cited studies demonstrate that the human amygdala is important for processing information not only about basic emotions, but also about complex social states, intentions, and relationships. But what is the evidence that it might be disproportionately important for processing explicitly social information? It is difficult to know how to go about answering this question, but one study (Adolphs, Tranel, & Baron-Cohen, 2002) provides an approach. Certain classes of emotional states, so-called social emotions, require knowledge about complex social relationships and do not make sense independent of a social context. At least some basic emotions, such as fear and disgust, certainly apply to nonsocial situations; but social emotions, such as jealousy, pride, or embarrassment, necessarily require a social context and require a concept of a social self that is situated within a social group. Two subjects with complete bilateral amygdala damage, as well as thirty with unilateral amygdala damage, were impaired in their ability recognize social emotions from the face, as well as from just the eye region of the face. Further analysis revealed that the impairment was disproportionately severe for recognizing social emotions, compared with basic emotions, a dissociation that held up either for faces as a whole, or for just the eye region.

But does the amygdala play a disproportionately important role in processing social stimuli, as opposed to nonsocial stimuli? In many mammals, such as rodents, it seems likely that the answer to this question is no. No good evidence supports the idea that the rat amygdala, for instance, is more important to guide responses to other rats than it is to guide responses to food, water, or electric shock and the nonsocial stimuli with which these reinforcers can be associated. However, in primates, and especially in humans, the question is more intriguing. As noted, humans with bilateral amygdala damage do have pronounced impairments in making social judgments, and in some cases evidence shows that their impairments in making complex social judgments are worse and more pervasive than their impairments in making simpler emotional judgments. It is thus a reasonable hypothesis, albeit one that must be made more precise, that throughout phylogeny the role of the amygdala in processing specifically social stimuli has become progressively more elaborated, in tandem with the elaboration of social behavior. Perhaps the human amygdala does not play a role in social cognition merely and derivatively in virtue of its role in mediating general (nondomain specific) motivational processes; perhaps it is relatively specialized to process those stimuli (especially visual stimuli) that have an explicitly social significance. (Of course, insofar as all social stimuli are motivational, this is still consistent with the notion that the amygdala processes the motivational value of social stimuli, as opposed to some nonmotivational property.) Preliminary arguments can be made both in favor of and against this view.

An argument apparently in favor of the view comes from a study that found a three-way allometric correlation among the volume of the basolateral amygdala (relative to the size of the rest of the brain), the size of visual cortex, and the size and complexity of social groups (Barton & Aggleton, 2000). A different theory apparently against the idea comes from the observation that monkeys with selective bilateral amygdala damage, while manifesting abnormal fear and anxiety responses, still appear to have, contra Kluver and Bucy, a normal repertoire of social behaviors, if those can be triggered (Amaral et al., 2003). The same conclusion appears to hold in humans. Social judgments made by subject SM, for instance, were certainly compromised, but in no way abolished. But this evidence shows only that social cognition, as one would expect, relies not on a single type of mechanism but on a huge array of different strategies that depend on very many different neural structures, of which the amygdala is one. One strategy relies on recruiting basic motivational and emotional circuitry, but specifically for the domain of socially relevant stimuli, and it is this strategy that depends on the amygdala. Other strategies, perhaps those relying substantially on language and on declarative memory stores, may not rely on the amygdala.

Simulation

Rather than trying to establish that the above data argue for the domain specificity of social cognition, it may be more fruitful to acknowledge that social cognition draws on a host of cognitive resources, and to ask instead in which way social and nonsocial information processing support one another. A rich example adumbrated above is our ability to construct models of counterfactual situations, to imagine the impossible, to travel mentally in time. One instance of this ability is the ability to imagine what it is like to be another person whom we are observing. Interest in the mechanisms whereby we achieve this focuses on the notion of "simulation," the idea that we are able to run off-line some of the same processes that we would be engaging ourselves from the observation of another person's behavior. The ability may derive from basic motor control adaptations, and yet may have expanded vastly in the service of the need to predict social behavior accurately.

Considerable empirical evidence proposes that humans obtain knowledge about other people's emotional states, at least in part, through some kind of articulated emulation (see Grush, 2004, for a review). Premotor cortices are engaged when we observe others behaving emotionally, as are somatosensory cortices and insula. One interpretation of these findings is that we engage some of the same machinery during emulation as during actual emotions: the body outside the brain. Good evidence further supports this idea: observing other people express emotions results in some mirroring of the emotional state in the viewer. In this case it seems that the emulator is the same as the system in normal operation, although it may engage only a subset of a hierarchically structured system. The possibility of using the body itself as the emulator when we model another person's emotion would be not only economical, but suggests an interesting way in which actual, analog physical processes (state changes in various parameters of the body that normally make up an emotional response) can be used in information processing. The body might be thought of as a somatic scratchpad that we can probe with efferent signals. Given the complexity of interaction among multiple somatic parameters, it may not be feasible to emulate this entirely neurally.

Typically, of course, emulation of emotion should be less than the real thing. Thus it involves faint somatic changes that are a subset of having the emotion oneself and that involve active inhibition of expression of some of the components. Much the same happens when we dream. As in waking emulation, we construct models that include responses in our bodies, and whereas we have somatic responses during dreaming, these are actively inhibited from full expression. It would seem odd to have evolved such efferent processing and inhibition if the body itself did not play an important role in building the models that help us to predict the world.

The idea of using parts of the body as an emulator is in line with theories of situated or enactive cognition—the idea that the mind makes use of processes that include the external environment and an organism's interaction with it. Three interesting open issues remain. First, an emulation approach to understanding other minds is likely to be a very dynamic, iterative process. It seems unlikely that we obtain all the information in a single shot, and more plausible that we run various parts of the emulator, perhaps to different depths of detail, to approximate the answer we seek. Second, there is likely to be a collection of emulators rather than a single one; perhaps these are hierarchically structured in some way. One could imagine emotion emulation involving progressive layers of somatic involvement, depending on the detail of the modeling required, much as visual imagery involves different levels of the visual hierarchy, depending on the grain of the imagery. And third, we can well imagine extending the modeling outside the bounds of the body. To obtain social information, we may query not only our own bodies, but other people's. Clearly, this is the case in a general sense: we probe other people's reactions to initial and often subtle behaviors on own part, and use their response as feedback in constructing a more accurate model of the social world. All of these ways of creating knowledge about the world should probably be seen on a continuum ranging from simple lookup tables and systems of rules to models entirely internal to the brain or encompassing varying degrees of the body or external environment.

What Is Special about Social Cognition

In reviewing the above examples, I wish to draw two conclusions: first, that they cannot support the idea that social cognition is strongly modular; second, that they do support the idea that there is indeed something special about social cognition. The trick is to reconcile these two apparently disparate conclusions by steering a middle course that draws on both. Social cognition is too broad a capacity, and makes contact with the rest of cognition at too many places, to be considered anything like an encapsulated, impenetrable module. Yet it is reasonable to think that it comprises computational strategies that evolved specifically to guide social behavior; that some of those social computations are relatively involuntary, automatic, and below the level of our awareness; and that it draws on a restricted set of neural structures that are sufficiently well defined that we can speak of a neural system for social cognition.

In redefining the question, it may be necessary to abandon our predilection for reductive explanations, or at least the usual sort of reduction. It is usually assumed, without particular argument, that knowledge about the social meaning of a stimulus depends in part on knowledge about the basic motivational value of a stimulus, but not conversely. The assumption is that we could impair social knowledge while sparing knowledge about more basic emotional and motivational value; but that impairing knowledge about basic motivational value would necessarily entail impairments in social knowledge. What is puzzling about this picture is not merely that it is reductionist, but rather that we assume the reduction to proceed in one direction rather than another. Why not suppose that social cognition is the basic adaptive package, and that motivated behavior in general draws in part on that?

In general, there may be no single answer to the question of the relation between social and nonsocial cognition. Is one reducible to the other? Well, some components of social cognition may be reducible to some components of nonsocial cognition. On the other hand, some components of nonsocial cognition may be reducible to social cognition. It all depends on the particular component, since neither domain is monolithic. These considerations raise a key issue that may provide an answer to the question of this chapter: the level of analysis of a system. At a very molar level, social behavior is by definition special since it concerns interactions with other people rather than objects. At a very microscopic level, social cognition is clearly not special, since it depends on the same microphysical processes as do all other biological events. Is social cognition special? It depends on who is asking the question. Psychologically, behaviorally, aspects of it are indeed likely to be quite special and make our behaviors so different from those of any other animal. Neurobiologically, it will be more difficult to find similar evidence, since the special behaviors emerge out of vast collections of neural mechanisms that are themselves less clearly specialized.

Nonetheless, one might insist, if specialization exists at the behavioral level, that a trace of it must be found in all the subvenient levels—we should see the signature of such social specialization in the brain, in neurons, in genes—but only if we place those in a much broader context within which they contribute to behavior. That context, of course, is the social environment, culture, and social development. Given a long period of development and a complex social environment, it may indeed be the case that a particular social cognitive ability, or a particular aspect of social behavior, depends on the contribution made by specific genes (Paterson et al., 1999). But that does not mean that there are genes "for" those aspects of social cognition.

In much the same spirit, we should be reluctant to look for neural structures that are "for" social cognition. Rather, we should consider social cognition as emerging from a complex interplay among many structures, in the context of development, of a particular culture, and considering the brain as a system that generates behavior only through its interaction with the body and the social environment. What is special about social cognition may well be the extent to which it relies on relations among these different components and levels, and the extent to which it requires their integration.

References

Adolphs, R. (1999). The human amygdala and emotion. *Neuroscientist, 5*, 125–137.

Adolphs, R., Tranel, D., & Baron-Cohen, S. (2002). Amygdala damage impairs recognition of social emotions from facial expressions. *Journal of Cognitive Neuroscience, 14*, 1264–1274.

Adolphs, R., Tranel, D., & Damasio, A. R. (1998). The human amygdala in social judgment. *Nature, 393*, 470–474.

Adolphs, R., Tranel, D., Damasio, H., & Damasio, A. (1994). Impaired recognition of emotion in facial expressions following bilateral damage to the human amygdala. *Nature, 372*, 669–672.

Adolphs, R., Tranel, D., Hamann, S., Young, A., Calder, A., Anderson, A., et al. (1999). Recognition of facial emotion in nine subjects with bilateral amygdala damage. *Neuropsychologia, 37*, 1111–1117.

Allman, J. M. (2002). Two phylogenetic specializations of the human brain. *Neuroscientist, 8*, 335–346.

Amaral, D. G., Capitanio, J. P., Jourdain, M., Mason, W. A., Mendoza, S. P., & Prather, M. (2003). The amygdala: Is it an essential component of the neural network for social cognition? *Neuropsychologia, 41*, 235–240.

Anderson, A. K. & Phelps, E. A. (2000). Expression without recognition: Contributions of the human amygdala to emotional communication. *Psychological Science, 11*, 106–111.

Anderson, A. K., Spencer, D. D., Fulbright, R. K., & Phelps, E. A. (2000). Contribution of the anteromedial temporal lobes to the evaluation of facial emotion. *Neuropsychology, 14*, 526–536.

Barkow, J. H., Cosmides, L., & Tooby, J. (Eds.). (1992). *The Adapted Mind: Evolutionary Psychology and the Generation of Culture*. New York: Oxford University Press.

Baron-Cohen, S., Ring, H. A., Bullmore, E. T., Wheelwright, S., Ashwin, C., & Williams, S. C. R. (2000). The amygdala theory of autism. *Neuroscience and Biobehavioral Reviews, 24*, 355–364.

Bar-On, R., Tranel, D., Denburg, N., & Bechara, A. (2003). Exploring the neurological substrate of emotional and social intelligence. *Brain, 126*, 1790–1800.

Barrash, J., Tranel, D., & Anderson, S. W. (2000). Acquired personality disturbances associated with bilateral damage to the ventromedial prefrontal region. *Developmental Neuropsychology, 18*, 355–381.

Barton, R. A. & Aggleton, J. (2000). Primate evolution and the amygdala. In J. Aggleton (Ed.), *The Amygdala*, 2nd ed. (pp. 479–508). New York: Oxford University Press.

Breiter, H. C., Etcoff, N. L., Whalen, P. J., Kennedy, W. A., Rauch, S. L., Buckner, R. L., et al. (1996). Response and habituation of the human amygdala during visual processing of facial expression. *Neuron, 17*, 875–887.

Calder, A. J., Young, A. W., Rowland, D., Perrett, D. I., Hodges, J. R., & Etcoff, N. L. (1996). Facial emotion recognition after bilateral amygdala damage: Differentially severe impairment of fear. *Cognitive Neuropsychology, 13*, 699–745.

Coltheart, M. (1999). Modularity and cognition. *Trends in Cognitive Sciences, 3*, 115–119.

Cowie, F. & Woodward, J. F. (2004). Mental modules did not evolve by natural selection. In C. Hitchcock (Ed.), *Great Debates in Philosophy: Philosophy of Science.* New York: Blackwell.

Craig, A. D. (2002). How do you feel? Interoception: The sense of the physiological condition of the body. *Nature Reviews Neuroscience, 3*, 655–666.

Damasio, A. R. (1994). *Descartes' Error: Emotion, Reason, and the Human Brain.* New York: Grosset/Putnam.

Darwin, C. (1859). *On the Origin of Species by Means of Natural Selection.* London: John Murray.

Dicks, D., Myers, R. E., & Kling, A. (1969). Uncus and amygdala lesions: Effects on social behavior in the free ranging rhesus monkey. *Science, 165*, 69–71.

Dunbar, R. (1998). The social brain hypothesis. *Evolutionary Anthropology, 6*, 178–190.

Eisenberger, N. I., Lieberman, M. D., & Williams, K. D. (2003). Does rejection hurt? An fMRI study of social exclusion. *Science, 302*, 290–292.

Emery, N. J. & Amaral, D. G. (1999). The role of the amygdala in primate social cognition. In R. D. Lane & L. Nadel (Eds.), *Cognitive Neuroscience of Emotion* (pp. 156–191). Oxford: Oxford University Press.

Emery, N. J., Capitanio, J. P., Mason, W. A., Machado, C. J., Mendoza, S. P., & Amaral, D. G. (2001). The effects of bilateral lesions of the amygdala on dyadic social interactions in rhesus monkeys. *Behavioral Neuroscience, 115*, 515–544.

Fine, C., Lumsden, J., & Blair, R. J. R. (2001). Dissociation between "theory of mind" and executive functions in a patient with early left amygdala damage. *Brain, 124*, 287–298.

Fodor, J. A. (1983). *The Modularity of Mind.* Cambridge: MIT Press.

Grush, R. (2004). The emulation theory of representation: Motor control, imagery and perception. *Behavioral and Brain Sciences, 27*, 377–396.

Hamann, S. B., Stefanacci, L., Squire, L. R., Adolphs, R., Tranel, D., Damasio, H., et al. (1996). Recognizing facial emotion. *Nature, 379*, 497.

Heberlein, A. S. & Adolphs, R. (2004). Impaired spontaneous anthropomorphizing despite intact social knowledge and perception. *Proceedings of the National Academy of Sciences of the United States of America, 101*, 7487–7491.

Heider, F. & Simmel, M. (1944). An experimental study of apparent behavior. *American Journal of Psychology, 57*, 243–259.

Kling, A. S. & Brothers, L. A. (1992). The amygdala and social behavior. In J. P. Aggleton (Ed.), *The Amygdala: Neurobiological Aspects of Emotion, Memory, and Mental Dysfunction.* New York: Wiley–Liss.

Kluver, H. & Bucy, P. C. (1937). "Psychic blindness" and other symptoms following bilateral temporal lobectomy in rhesus monkeys. *American Journal of Physiology, 119*, 352–353.

Kluver, H. & Bucy, P. C. (1939). Preliminary analysis of functions of the temporal lobes in monkeys. *Archives of Neurology and Psychiatry, 42*, 979–997.

Morris, J. S., Frith, C. D., Perrett, D. I., Rowland, D., Young, A. W., Calder, A. J., et al. (1996). A differential neural response in the human amygdala to fearful and happy facial expressions. *Nature, 383*, 812–815.

Nimchinsky, E. A., Gilissen, E., Allman, J. M., Perl, D. P., Erwin, J. M., & Hof, P. R. (1999). A neuronal morphologic type unique to humans and great apes. *Proceedings of the National Academy of Sciences of the United States of America, 96*, 5268–5273.

Paterson, S. J., Brown, J. H., Gsoedl, M. K., Johnson, M. H., & Karmiloff-Smith, A. (1999). Cognitive modularity and genetic disorders. *Science, 286*, 2355–2357.

Paus, T. (2001). Primate anterior cingulate cortex: Where motor control, drive and cognition interface. *Nature Reviews Neuroscience, 2*, 417–424.

Povinelli, D. J. & Vonk, J. (2003). Chimpanzee minds: Suspiciously human? *Trends in Cognitive Science, 7*, 157–160.

Premack, D. & Woodruff, G. (1978). Does the chimpanzee have a theory of mind? *Behavioral and Brain Sciences, 1*, 515–526.

Rapcsak, S. Z., Galper, S. R., Comer, J. F., Reminger, S. L., Nielsen, L., Kaszniak, A. W., et al. (2000). Fear recognition deficits after focal brain damage. *Neurology, 54*, 575–581.

Rosvold, H. E., Mirsky, A. F., & Pribram, K. (1954). Influence of amygdalectomy on social behavior in monkeys. *Journal of Comparative and Physiological Psychology, 47*, 173–178.

Schmolck, H. & Squire, L. R. (2001). Impaired perception of facial emotions following bilateral damage to the anterior temporal lobe. *Neuropsychology, 15*, 30–38.

Semendeferi, K., Armstrong, E., Schleicher, A., Zilles, K., & Van Hoesen, G. W. (2001). Prefrontal cortex in humans and apes: A comparative study of area 10. *American Journal of Physical Anthropology, 114*, 224–241.

Singer, T., Seymour, B., O'Doherty, J., Kaube, H., Dolan, R. J., & Frith, C. D. (2004). Empathy for pain involves the affective but not sensory components of pain. *Science, 303*, 1157–1162.

Stone, V. E., Cosmides, L., Tooby, J., Kroll, N., & Knight, R. T. (2002). Selective impairment of reasoning about social exchance in a patient with bilateral limbic system damage. *Proceedings of the National Academy of Sciences of the United States of America, 99*, 11531–11536.

Suddendorf, T. & Corballis, M. (1997). Mental time travel and the evolution of the human mind. *Genetic, Social, and General Psychology Monographs, 123*, 133–167.

Tomasello, M. (1999). *The Cultural Origins of Human Cognition*. Cambridge: Harvard University Press.

Tomasello, M., Call, J., & Hare, B. (2003). Chimpanzees understand psychologial states—The question is which ones and to what extent. *Trends in Cognitive Science, 7*, 153–156.

Whalen, P. J., Shin, L. M., McInerney, S. C., Fischer, H., Wright, C. I., & Rauch, S. L. (2001). A functional MRI study of human amygdala responses to facial expressions of fear versus anger. *Emotion, 1,* 70–83.

Whiten, A., & Byrne, R. W. (Eds.). (1997). *Machiavellian Intelligence.* II. *Extensions and Evaluations.* Cambridge: Cambridge University Press.

Whiten, A., Goodall, J., McGrew, W. C., Nishida, T., Reynolds, V., Sugiyama, Y., et al. (1999). Cultures in chimpanzees. *Nature, 399,* 682–685.

Winston, J. S., Strange, B. A., O'Doherty, J., & Dolan, R. J. (2002). Automatic and intentional brain responses during evaluation of trustworthiness of faces. *Nature Neuroscience, 5,* 277–283.

Young, A. W., Hellawell, D. J., Van de Wal, C., & Johnson, M. (1996). Facial expression processing after amygdalotomy. *Neuropsychologia, 34,* 31–39.

15 Social Neuroscience: A Perspective

Marcus E. Raichle

Bridging the gap between descriptions of human behaviors and underlying neural events in a central theme in this book on social neuroscience. As well, it has been a dream of both psychologists and neuroscientists for quite some time.

William James, in his monumental two-volume work *The Principles of Psychology*, devotes two insightful chapters to the brain and its relationship to behavior. The prescience of his remarks is particularly remarkable given the fact that they were written 1890. In a fashion typical of much of his writing, James clearly identified the challenge: "A science of the mind must reduce . . . complexities (of behavior) to their elements. A science of the brain must point out the functions of its elements. A science of the relations of mind and brain must show how the elementary ingredients of the former correspond to the elementary functions of the latter" (page 28, vol. 1).

The importance of meeting this challenge was appreciated at about the same time by no less a neuroscientist than Sir Charles Sherrington (1906) who wrote, ". . . physiology and psychology, instead of prosecuting their studies, as some now recommend, more strictly apart one from another than at present, will find it serviceable for each to give to the results achieved by the other even closer heed than has been customary hitherto" (page 385). Although it is clear that progress was made from the time of James and Sherrington until the present through studies in experimental animals and in patients with various diseases afflicting the brain, the opportunity to relate normal behavior to normal brain function in humans was largely nonexistent until the latter part of the twentieth century.

None other than the great experimental physiologist (some might say biological psychologist) Ivan Pavlov envisioned what was needed: "If we could look through the skull into the brain of a consciously thinking person, and if the place of optimal excitability were luminous, then we should see playing over the cerebral surface, a bright spot with fantastic, waving borders constantly fluctuating in size and form, surrounded by a darkness more or less deep, covering the rest of the hemispheres

(1928; for translation, see Brugger, 1997). For many years such a vision seemed mere fantasy.

It was the introduction of brain-imaging techniques in the 1970s that permitted us to realize Pavlov's vision of monitoring human brain function in a safe yet increasingly detailed and quantitative way.[1] This began a revolution in the relationship between the brain and cognitive science that resulted in the birth of cognitive neuroscience (for a review, see Posner & Raichle, 1994).

Cognitive neuroscience with surprising success has combined the experimental strategies of cognitive psychology with brain-imaging techniques to examine how brain function supports mental activities. The number of papers reporting results of such studies is increasing at an exponential rate (Illes, Krischen, & Gabrieli, 2003), as is the investment in brain-imaging centers worldwide (many now in psychology departments) devoted to the study of brain–behavior relationships (Raichle, 2003a). Unheard of in the late 1980s, when the James S. McDonnell Foundation and the Pew Charitable Trusts initiated their program in cognitive neuroscience, were faculty positions in psychology departments for cognitive neuroscientists. They are now common, and young scientists with combined qualifications in imaging and psychology are eagerly sought.

Social neuroscience greatly extends the potential range of behaviors well beyond those heretofore explored by most cognitive neuroscientists. This exciting opportunity, however, comes with a clear requirement. Social scientists and neuroscientists must understand each other even more fully than at present in order to determine the questions and approaches that might work best. This is clearly a challenging and a very multidisciplinary *work in progress* for all concerned.

Brain Imaging

It is apparent from this book that, among the tools of neuroscience that has made social neuroscience a possibility and will fuel its growth, imaging of the human brain, especially its functional activity using functional magnetic resonance imaging (fMRI), has the potential to play a prominent role. Therefore, it seems appropriate to consider the present status of functional brain imaging as well as some of the important trends in its development. It must be kept in mind that this field is itself in a very active state of development that is likely to continue many years into the future (for a recent survey, see Raichle, 2003b).

To begin, it is useful to consider the intended goal of functional brain imaging. This may seem self-evident to most. Yet, interpretations frequently stated or implied about relevant data (Nichols & Newsome, 1999; Uttal, 2001) suggest that, if not careful, the procedure could be viewed as no more than a modern and extraordinarily expensive version of nineteenth-century phrenology (for alternative views, Posner, 2003; Raichle, 2003b). In this regard, it is worth noting that functional imaging researchers themselves have occasionally unwittingly perpetuated this notion. Data presented in journal articles, at meetings, and in lay publications have focused on specific areas of the brain, frequently without including the entire data set. These areas are then discussed in terms of complex mental functions—just what the phrenologists did!

It was Korbinian Brodmann (1909), one of the pioneers in studying the organization of the human brain from both microscopic and macroscopic points of view, whose perspective I find appealing even though it was written well in advance of the discovery of modern imaging technology (1909 to be exact). He said, "Indeed, recently theories have abounded which, like phrenology, attempt to localize complex mental activity such as memory, will, fantasy, intelligence or spatial qualities such as appreciation of shape and position to circumscribed cortical zones." He went on, "These mental faculties are notions used to designate extraordinarily involved complexes of mental functions.... One cannot think of their taking place in any other way than through an infinitely complex and involved interaction and cooperation of numerous elementary activities.... in each particular case [these] supposed elementary functional loci are active in differing numbers, in differing degrees and in differing combinations.... Such activities are ... always the result ... of the function of a large number of suborgans distributed more or less widely over the cortical surface ..." (Garey, 1994, pp. 254–255).

With this prescient admonition in mind, the task of functional brain imaging would seem to be clear. First, identify the network of regions of the brain and its relationship to the performance of a well-designed task. This is a process that is well under way in the functional imaging community and is complemented by a long history of lesion behavior work from the neuropsychology community and, in some instances, neurophysiological and neuroanatomical studies in laboratory animals. Second, and definitely more challenging, identify the elementary operations performed within such a network and relate these operations to the task of interest. This agenda has progressed in interesting ways.

Functional brain-imaging studies have clearly and, no surprise, revealed that networks of brain areas rather than single areas are associated with the performance of tasks. Probably of even greater interest is the fact that elements of these networks can be seen in different tasks, indicating the presence of elementary operations that are common across tasks. Whereas some might view this as evidence of lack of specificity in functional imaging data, others, including me, see this as an important clue to the manner in which the brain allocates its finite processing resources to the accomplishment of an infinite range of behaviors. A model of how one might think about this observation comes from studies in invertebrate neurobiology, in which ensembles of neurons were observed to reconfigure their relationships depending on the task at hand, thus multiplying the potential of finite resources to serve the needs of the organism (Marder & Weimann, 1991). It would not be too surprising if this rather sensible strategy had been conserved in an expanded version in the vertebrate brain. Obviously, success in pursuing information of this type will require very careful task analysis, with particular attention to the presence of common elements in tasks that might superficially seem quite different.

In pursuit of these ideas, debate has arisen about the level of representation that might be expected in particular areas (i.e., how elementary is elementary when seeking the operations performed in a particular area?). This is exemplified in interesting results and discussions about the specificity of face perception in the human brain (for contrasting views, see Duchaine et al., 2004; Grill-Spector, Knouf, & Kanwisher, 2004; Hanson, Matsuka, & Haxby, 2004; Haxby et al., 2001). We must be careful in adjudicating such a debate pending deeper understanding of the detailed anatomy of the area under consideration. In the putative face area of the medial temporal lobe of humans, we lack such detailed information.

A contrasting example exists in the ventral medial prefrontal cortex, where processes related to emotion are suspected (Bush, Luu, & Posner, 2000; Drevets et al., 1997; Mayberg et al., 1999; Simpson et al., 2001). Extensions of detailed cytoarchitectonics maps from monkey to human (Gusnard et al., 2003; Ongur, Ferry, & Price, 2003) illustrate the type of information that will lay the foundation for more detailed understanding of exactly what is going on. This across levels of analysis, as others so eloquently pointed out (Cacioppo & Berntson, 1992), will clearly play an important role in future research.

Integration Across Levels of Analysis

Nowhere has the integration of levels of analysis been more important to the functional imaging agenda than in a pioneering group of studies that sought to provide an understanding of relationships between the neurophysiological activity of cells within the brain and circulatory and metabolic signals that are observed with functional imaging techniques (Lauritzen, 2001; Logothetis et al., 2001). To put this work in perspective it is important to note that for neurophysiologists, task-induced changes in brain activity are traditionally recorded as changes in the firing rate or spiking activity of individual neurons, which reflects their output to cells elsewhere in the brain. For them spiking has been the gold standard. They have most often ignored the signals associated with the input to cells that are reflected in very complex electrical signals (often called local field potentials) generated by millions of tiny but very important cell processes (dendrites, axon terminals together occasionally referred to as the neuropil). New data directly correlating neurophysiology with imaging signals (Lauritzen, 2001; Logothetis et al., 2001) reveal that imaging signals reflect changes in the neuropil and not spiking. As is the case in so much of science, these findings were clearly anticipated by others (Creutzfeldt, 1975; Schwartz et al., 1979).

Thus, for neurophysiologists and social scientists using functional brain imaging to communicate, they must first realize that the aspects of brain function they routinely measure are different. Whereas they may at times correlate nicely, they are not causally related in any simple manner (see Raichle, 2003b, for recent discussions). As a result, both neurophysiologists and social neuroscientists have necessary but not sufficient information to obtain a complete picture of events critical to an ultimate understanding the function of the brain at a systems level. A balanced perspective without claims to a privileged access to the complete truth will be most helpful as we seek to understand these issues. Social neuroscientists who remain informed on this evolving dialogue will be able to draw much more from the work they do.

Individual Differences

Another issue likely to be of increasing importance to our understanding of human brain function is the matter of individual differences. From the perspective of cognitive neuroscience and functional brain imaging, an early worry was that individual differences would be sufficiently great that attempts to average imaging data across individuals to improve signals and diminish noise would be doomed to failure. These fears were quickly put to rest by the first attempts to average functional imaging data across individuals. The results were stunning (Fox et al., 1988; for a historical review, see Raichle, 2000, p. 57). The approach has since dominated the field of cognitive neuroscience, and with great success. However, for all who have examined such data in detail (particularly high-quality fMRI data), the existence of individual differences begins to emerge with exciting prospects for even deeper understanding of human behavior (Gusnard et al., 2003). Coupled with a long-standing interest in and techniques for characterization of personality differences from psychology and psychiatry, we are poised to make major advances in this area that should certainly be of interest to social scientists.

Development

Cognitive neuroscience has focused largely on the adult human brain, examining how it functions normally and how it changes when focally damaged. It is important to be reminded of the importance of considering brain function from a developmental perspective. The developmental psychology literature is rich with details about milestones associated with the maturation of the human brain. Missing, however, is satisfactory understanding of the maturation process within systems of the human brain. Whether it is the development of attention, language, memory, management of distress, or personality development more generally, we need information about brain maturation at a systems level. Such information would enrich not only our understanding of development itself but also the end result of that development; that is, the organization of the adult brain. The paucity of extant information reflects not only the challenge of safely and accurately accessing the necessary information in humans, but also a general focus on the cellular and molecular levels within developmental neurobiology to the relative exclusion of a more integrated systems approach. Tools available to cognitive neuroscientists are becoming available that allow safe access to the information in humans. Various imaging techniques applicable to children can access not only functional information but also anatomical information, including the development of fiber pathways. Data from a small group of pioneering investigators (Dehaene-Lambertz, Dehaene, & Hertz-Pannier, 2002; McKinstry et al., 2002; Miller et al., 2003; Munakata, Casey, & Diamond, 2004; Schlagger et al., 2002) provide exciting glimpses of the future. Parallel studies in nonhuman primates, where changes down to a cellular level in the brain can be correlated with detailed analysis of complex social behaviors, should be encouraged.

The "Resting" Brain

Finally, it will be critically important to maintain a sense of proportion when it comes to viewing functional imaging signals. As discussed by Gusnard earlier in this volume, the human brain is an extremely costly organ in energy terms, whereas the activity changes observed in functional imaging studies are very small. Although this high cost of continuing brain function has been known for a long time (Clark & Sokoloff,

1999), it has only recently received the attention that it deserves (for a recent brief review, see Raichle & Gusnard, 2002), primarily because of the realization that most of this cost is related to functional aspects of information processing and trafficking, and not such things as protein synthesis and axonal transport, which appear to account for a relatively small fraction of the overall cost of running the brain. These facts present a challenge to those wishing to understand the function of the brain. How do you study functionality of a state in which there is neither a controlled input nor an observed output? Chapter 3 presents experimental observations that provide interesting avenues for future research in this important area.

On a more general level, the "cost analysis" of human brain function should orient one to the view, long espoused by Llinás (1974, 2001; Llinás & Pare, 1991), that the brain operates intrinsically with sensory information, often in an impoverished form, modulating rather than informing the system. As William James (1890) observed, "whilst part of what we perceive comes through our senses from the object before us, another part (and it may be the larger part) always comes . . . out of our own head" (p. 103, vol. 2). This view led many to posit that the brain operates as a Bayesian inference machine (Olshausen, 2003), maintaining a state of "priors" related both to experience and to our genetic endowment. A number of beautiful reminders of the latter have appeared (Kuhl, 2003; Meltzoff & Decety, 2003). As well, an experiment in the cat visual cortex is a direct demonstration of the phenomenon I am describing. In the absence of sensory input, the cat visual cortex appeared to be creating, in its spontaneous activity, representations of anticipated visual stimuli (Kenet et al., 2003). These and other observations led us to endorse the general view that the brain develops and maintains a probabilistic model of anticipated events, and most continuing neuronal activity is an internal representation of that model with which sensory information is compared. As social neuroscience contemplates its agenda, appreciating this perspective will be important.

At the moment social neuroscience is clearly a work in progress but one that is immensely interesting to many of us because it offers the potential of extending the relationship of brain science to such a broad range of important and exciting issues. Taking a lesson from the immense success of cognitive neuroscience, those interested in social neuroscience must focus on training a new generation of social scientists who understand and can use effectively the new tools of neuroscience such as functional brain imaging. These new researchers must not only understand the major questions to be asked but also how relevant information on brain function can be obtained to help in their quest for answers. The challenge will be to understand how best to integrate the potential of tools such as brain imaging with the fascinating yet complex issues of interest to social scientists. We will all be the beneficiaries of the success of this venture.

Note

1. Like most things in science, the introduction of brain imaging had important antecedents dating back to the 1800s and continuing up to the introduction of imaging as we now know it. Interested readers may find a review of this literature of interest (Raichle, 2000).

References

Brodmann, K. (1909). *Vergleichende lokalisationlehre der grosshirnrinde.* Leipzig: J. A. Barth.

Brugger, P. (1997). Pavlov on neuroimaging. *Journal of Neurology, Neurosurgery and Psychiatry, 62,* 636.

Bush, G., Luu, P., & Posner, M. I. (2000). Cognitive and emotional influences in anterior cingulate cortex. *Trends in Cognitive Sciences, 4*(6), 215–222.

Cacioppo, J. T. & Berntson, G. G. (1992). Social psychological contributions to the decade of the brain. Doctrine of multilevel analysis. *American Psychologist, 47,* 1019–1028.

Clark, D. D. & Sokoloff, L. (1999). Circulation and energy metabolism of the brain. In G. J. Siegel, B. W. Agranoff, R. W. Albers, S. K. Fisher, & M. D. Uhler (Eds.), *Basic Neurochemistry. Molecular, Cellular and Medical Aspects,* 6th ed. (pp. 637–670). Philadelphia: Lippincott–Raven.

Creutzfeldt, O. D. (1975). Neurophysiological correlates of different functional states of the brain. In D. H. Ingvar & N. A. Lassen (Eds.), *Brain Work: The Coupling of Function, Metabolism and Blood Flow in the Brain* (pp. 22–47). Copenhagen: Munksgaard.

Dehaene-Lambertz, G., Dehaene, S., & Hertz-Pannier, L. (2002). Functional imaging of speech perception in children. *Science, 298,* 2013–2015.

Drevets, W. C., Price, J. L., Simpson, J. R., Jr., Todd, R. D., Reich, T., Vannier, M. W., et al. (1997). Subgenual prefrontal cortex abnormalities in mood disorders. *Nature, 386,* 824–827.

Duchaine, B. C., Dingle, K., Butterworth, E., & Nakayama, K. (2004). Normal greeble learning in a severe case of developmental prosopagnosia. *Neuron, 43,* 469–473.

Fox, P. T., Mintun, M. A., Rieman, E. M., & Raichle, M. E. (1988). Enhanced detection of focal brain responses using intersubject averaging and change-distribution analysis of subtracted PET images. *Journal of Cerebral Blood Flow and Metabolism, 8,* 642–653.

Garey, L. J. (1994). *Brodmann's "Localization in the Cerebral Cortex"* (L. J. Garey, Trans.). London: Smith–Gordon.

Grill-Spector, K., Knouf, N., & Kanwisher, N. (2004). The fusiform face area subserves face perception, not generic within-category indentification. *Nature Neuroscience, 7,* 555–562.

Gusnard, D. A., Ollinger, J. M., Shulman, G. L., Cloninger, C. R., Price, J. L., Van Essen, D. C., et al. (2003). Persistence and brain circuitry. *Proceedings of the National Academy of Sciences of the United States of America, 100*, 3479–3484.

Hanson, S. J., Matsuka, T., & Haxby, J. V. (2004). Combinatorial codes in ventral temporal lobe for object recognition: Haxby (2001) revisited: Is there a "face" area? *Neuroimage, 23*, 156–166.

Haxby, J. V., Gobbini, M. I., Furey, M. L., Ishai, A., Schouten, J. L., & Pietrini, P. (2001). Distributed and overlapping representation of faces and objects in ventral temporal cortex. *Science, 293*, 2425–2430.

Illes, J., Krischen, M. P., & Gabrieli, J. D. E. (2003). From neuroimaging to neuroethics. *Nature Neuroscience, 6*, 205.

James, W. (1890). *Principles of Psychology* (pp. 97–99). New York: Henry Holt.

Kenet, T., Bibitchkov, D., Tsodyks, M., Grinvald, A., & Ariell, A. (2003). Spontaneously emerging cortical representations of visual attributes. *Nature, 425*, 954–956.

Kuhl, P. K. (2003). Human speech and birdsong: Communication and the social brain. *Proceedings of the National Academy of Sciences of the United States of America, 100*, 9645–9646.

Lauritzen, M. (2001). Relationship of spikes, synaptic activity and local changes of cerebral blood flow. *Journal of Cerebral Blood Flow and Metabolism, 21*, 1367–1383.

Llinás, R. (1974). La forme et la fonction des cellules nerveuses. *Recherche, 5*, 232–240.

Llinás, R. (2001). *I of the Vortex: From Neurons to Self.* Cambridge: MIT Press.

Llinás, R. & Pare, D. (1991). Of dreaming and wakefulness. *Neuroscience, 44*, 521–535.

Logothetis, N. K., Pauls, J., Augath, M., Trinath, T., & Oeltermann, A. (2001). Neurophysiological investigation of the basis of the fMRI signal. *Nature, 412*, 150–157.

Marder, E. & Weimann, J. M. (1991). Modulatory control of multiple task processing in the stomatogastric nervous system. In J. Kien, C. McCrohan, & B. Winlow (Eds.), *Neurobiology of Motor Program Selection: New Approaches to Mechanisms of Behavioral Choice* (pp. 3–19). Manchester, U.K.: Manchester University Press.

Mayberg, H. S., Liotti, M., Brannan, S. K., McGinnis, S., Mahurin, R. K., Jerabek, P. A., et al. (1999). Reciprocal limbic-cortical function and negative mood: Converging PET findings in depression and normal sadness. *American Journal of Psychiatry, 156*, 675–682.

McKinstry, R. C., Mathur, A., Miller, J. H., Ozcan, A., Snyder, A. Z., Schefft, G. L., et al. (2002). Radial organization of developing preterm human cerebral cortex revealed by non-invasive water diffusion anisotropy MRI. *Cerebral Cortex, 12*, 1237–1243.

Meltzoff, A. N. & Decety, J. (2003). What imitation tells us about social cognition: A rapprochement between developmental psychology and cognitive neuroscience. *Philosophical Transactions of the Royal Society of London, Series B, Biological Sciences, 358*, 491–500.

Miller, J. H., McKinstry, R. C., Philip, J. V., Mukherjee, P., & Neil, J. J. (2003). Diffusion-tensor imaging of normal brain maturation: A guide to structural development and myelination. *American Journal of Roentgenology, 180*, 851–859.

Munakata, Y., Casey, B. J., & Diamond, A. (2004). Developmental cognitive neuroscience: Progress and potential. *Trends in Cognitive Sciences, 8*, 122–128.

Nichols, M. J. & Newsome, W. T. (1999). The neurobiology of cognition. *Nature, 402*, C35–C38.

Olshausen, B. A. (2003). Principles of image representation in visual cortex. In L. M. Chalupa & J. S. Werner (Eds.), *The Visual Neurosciences* (pp. 1603–1615). Cambridge: MIT Press.

Ongur, D., Ferry, A. T., & Price, J. L. (2003). Architectonic subdivisions of the human orbital and medial prefontal cortex. *Journal of Comparative Neurology, 460*, 425–449.

Posner, M. I. (2003). Imaging a science of mind. *Trends in Cognitive Sciences, 7*, 450–453.

Posner, M. I. & Raichle, M. E. (1994). *Images of Mind.* New York: Freeman.

Raichle, M. E. (2000). A brief history of human functional brain mapping. In A. W. Toga & J. C. Mazziotta (Eds.), *Brain Mapping. The Systems* (pp. 33–75). San Diego: Academic Press.

Raichle, M. E. (2003a). Functional brain imaging and human brain function. *Journal of Neuroscience, 23*, 3959–3962.

Raichle, M. E. (2003b). Functional brain imaging and human brain function (miniseries). *Journal of Neuroscience, 23*, 3959–4011.

Raichle, M. E. & Gusnard, D. A. (2002). Appraising the brain's energy budget. *Proceedings of the National Academy of Sciences of the United States of America, 99*, 10237–10239.

Schlagger, B. L., Browm, T. T., Lugar, H. M., Visscher, K. M., Miezen, F. M., & Petersen, S. E. (2002). Functional neuroanatomical differences between adults and school-age children in the processing of single words. *Science, 296*, 1476–1479.

Schwartz, W. J., Smith, C. B., Davidsen, L., Savaki, H., Sokoloff, L., Mata, M., et al. (1979). Metabolic mapping of functional activity in the hypothalamo-neurohypophysial system of the rat. *Science, 205*, 723–725.

Sherrington, C. (1906). *The Integrative Action of the Nervous System.* New Haven: Yale University Paperback (1961) p. 385.

Simpson, J. R. J., Drevets, W. C., Snyder, A. Z., Gusnard, D. A., & Raichle, M. E. (2001). Emotion-induced changes in human medial prefrontal cortex. II. During anticipatory anxiety. *Proceedings of the National Academy of Sciences of the United States of America, 98*, 688–691.

Uttal, W. R. (2001). *The New Phrenology*. Cambridge: Bradford Books, MIT Press.

Contibutors

Ralph Adolphs
California Institute of Technology

Nalini Ambady
Tufts University

Mahzarin R. Banaji
Harvard University

Reuven Bar-On
University of Texas

Antoine Bechara
University of Iowa

Jennifer S. Beer
University of California, Davis

Gary G. Berntson
Ohio State University

Joan Y. Chiao
Harvard University

Pearl Chiu
Harvard University

Joshua Correll
University of Colorado

Patricia Deldin
University of Michigan

Naomi I. Eisenberger
University of California, Los Angeles

Debra A. Gusnard
Washington University

Tiffany A. Ito
University of California, Los Angeles

Matthew D. Lieberman
University of California, Los Anyeles

C. Neil Macrae
Dartmouth College

Malia F. Mason
Dartmouth College

Jason P. Mitchell
Dartmouth College and Harvard University

Howard C. Nusbaum
University of Chicago

Kevin N. Ochsner
Columbia University

Elizabeth A. Phelps
New York University

Marcus E. Raichle
Washington University School of Medicine

Rebecca Saxe
Harvard University

Steven L. Small
University of Chicago

Valerie E. Stone
University of Queensland

Geoffrey R. Urland
University of Colorado

Eve Willadsen-Jensen
University of Colorado

Index